OF THE ELEMENTS

Weights, International Union of Pure and Applied Chemistry.

CARBON - 12

			IIIA	IVA	VA	VIA	VIIA		He
									Helium
									0.179 g/l
									−268.9 −272
									$1s^2$
			5 10.81	6 12.011	7 14.007	8 15.9994	9 19.00	10 20.183	
			B	**C**	**N**	**O**	**F**	**Ne**	
			Boron	Carbon	Nitrogen	Oxygen	Fluorine	Neon	
			2.45	2.22	1.25 g/l	1.43 g/l	1.69 g/l	0.90 g/l	
			2530 2300		−195.8 −210.0	−182.9 −218.7	−187.9 −223	−245.9 −248.5	
			$(He)2s^22p$	$(He)2s^22p^2$	$(He)2s^22p^3$	$(He)2s^22p^4$	$(He)2s^22p^5$	$(He)2s^22p^6$	
			13 26.98	14 28.09	15 30.974	16 32.064	17 35.453	18 39.948	
			Al	**Si**	**P**	**S**	**Cl**	**Ar**	
			Aluminum	Silicon	Phosphorus	Sulfur	Chlorine	Argon	
			2.71	2.33	1.82	1.96	3.2 g/l	1.78 g/l	
	IB	IIB	2270 658.6	2355 1414	280 44.2	444.6 119.0	−34.05 −100.9	−185.8 −189.3	
			$(Ne)3s^23p$	$(Ne)3s^23p^2$	$(Ne)3s^23p^3$	$(Ne)3s^23p^4$	$(Ne)3s^23p^5$	$(Ne)3s^23p^6$	
58.71 29	63.54 30	65.37 31	69.72 32	72.59 33	74.92 34	78.96 35	79.909 36	83.80	
Ni **Cu**	**Zn**	**Ga**	**Ge**	**As**	**Se**	**Br**	**Kr**		
Nickel Copper	Zinc	Gallium	Germanium	Arsenic	Selenium	Bromine	Krypton		
8.9 8.9	7.14	5.91	5.36	5.7	4.6	3.12	3.7 g/l		
1455 2310 1083	907 419.5	2070 29.78	2700 960	684.8 217.4	58.78 −7.3	−151.7 −169			
Ar)$3d^84s^2$ $(Ar)3d^{10}4s$	$(Ar)3d^{10}4s^2$	$(Ar)3d^{10}4s^24p$	$(Ar)3d^{10}4s^24p^2$	$(Ar)3d^{10}4s^24p^3$	$(Ar)3d^{10}4s^24p^4$	$(Ar)3d^{10}4s^24p^5$	$(Ar)3d^{10}4s^24p^6$		
106.4 47 107.870	48 112.40	49 114.82	50 118.69	51 121.75	52 127.60	53 126.90	54 131.30		
Pd **Ag**	**Cd**	**In**	**Sn**	**Sb**	**Te**	**I**	**Xe**		
alladium Silver	Cadmium	Indium	Tin	Antimony	Tellurium	Iodine	Xenon		
12.0 10.5	8.6	7.3	7.31	6.58	6.24	4.93	5.85 g/l		
1555 1950 961	767 321	1450 156.4	2337 232	1440 630	1380 450	183 113.6	−109.1 −140		
(Kr)$4d^{10}$ $(Kr)4d^{10}5s$	$(Kr)4d^{10}5s^2$	$(Kr)4d^{10}5s^25p$	$(Kr)4d^{10}5s^25p^2$	$(Kr)4d^{10}5s^25p^3$	$(Kr)4d^{10}5s^25p^4$	$(Kr)4d^{10}5s^25p^5$	$(Kr)4d^{10}5s^25p^6$		
195.09 79 196.97	80 200.59	81 204.37	82 207.19	83 208.98	84 (210)	85 (210)	86 (222)		
Pt **Au**	**Hg**	**Tl**	**Pb**	**Bi**	**Po**	**At**	**Rn**		
latinum Gold	Mercury	Thallium	Lead	Bismuth	Polonium	Astatine	Radon		
21.45 19.3	13.546	11.85	11.34	9.80	9.4		9.73 g/l		
1770 2660 1063	356.6 −38.87	1457 303.5	1750 327.4	1420 271	970 252	−62 −71			
)$4f^{14}5d^96s$ $(Xe)4f^{14}5d^{10}6s$	$(Xe)4f^{14}5d^{10}6s^2$	$(Xe)4f^{14}5d^{10}6s^26p$	$(Xe)4f^{14}5d^{10}6s^26p^2$	$(Xe)4f^{14}5d^{10}6s^26p^3$	$(Xe)4f^{14}5d^{10}6s^26p^4$	$(Xe)4f^{14}5d^{10}6s^26p^5$	$(Xe)4f^{14}5d^{10}6s^26p^6$		

152.0 64 157.25	65 158.92	66 162.50	67 164.93	68 167.26	69 168.93	70 173.04	71 174.97
Eu **Gd**	**Tb**	**Dy**	**Ho**	**Er**	**Tm**	**Yb**	**Lu**
uropium Gadolinium	Terbium	Dysprosium	Holmium	Erbium	Thulium	Ytterbium	Lutetium
5.24 7.95	8.33	8.56	8.76	9.16	9.35	7.01	9.74
1150 ca. 1100	ca. 1100	ca. 1100	ca. 1200	1250	3500 ca. 1600	ca. 1800	ca. 1800
Xe)$4f^76s^2$ $(Xe)4f^75d6s^2$	$(Xe)4f^85d6s^2$	$(Xe)4f^{10}6s^2$	$(Xe)4f^{11}6s^2$	$(Xe)4f^{12}6s^2$	$(Xe)4f^{13}6s^2$	$(Xe)4f^{14}6s^2$	$(Xe)4f^{14}5d6s^2$
(243) 96 (247)	97 (247)	98 (251)	99 (254)	100 (253)	101 (256)	102 (254)	103 (257)
Am **Cm**	**Bk**	**Cf**	**Es**	**Fm**	**Md**	**No**	**Lw**
nericium Curium	Berkelium	Californium	Einsteinium	Fermium	Mendelevium	Nobelium	Lawrencium
11.7 7 ?							
)$5f^76d^17s^2$ $(Rn)5f^76d^17s^2$	$(Rn)5f^86d^17s^2$	$(Rn)5f^96d^17s^2$					

Courtesy of Central Scientific Company, Chicago, Ill.

STRUCTURE AND PROPERTIES
OF ENGINEERING ALLOYS

STRUCTURE AND PROPERTIES OF ENGINEERING ALLOYS

William F. Smith

Professor of Engineering
University of Central Florida

McGraw-Hill Book Company

New York   St. Louis   San Francisco   Auckland   Bogotá   Hamburg
London   Madrid   Mexico   Montreal   New Delhi
Panama   Paris   São Paulo   Singapore   Sydney   Tokyo   Toronto

This book was set in Times Roman by Science Typographers, Inc.
The editors were Julienne V. Brown and J. W. Maisel;
the production supervisor was John Mancia.
R. R. Donnelley & Sons Company was printer and binder.

STRUCTURE AND PROPERTIES OF ENGINEERING ALLOYS

7890 DODO 8987

Library of Congress Cataloging in Publication Data

Smith, William Fortune, date
    Structure and properties of engineering alloys.

    (McGraw-Hill series in materials science and engineering)
    Includes bibliographies and index.
    1. Alloys. I. Title.
TA483.S64   620.1′6   80-12949
ISBN 0-07-058560-1

CONTENTS

PREFACE

Engineering alloys constitute an important segment of our modern industrial economy, and therefore a knowledge of these materials is especially essential for many practicing engineers. This book has been written primarily for junior or senior student metallurgical, materials, and mechanical engineers who will be entering industry after obtaining a bachelor's degree. The book is also intended as a refresher text and reference for practicing engineers who have become specialized or who have left the mainstream of technical work.

The principal *objectives* of this book are the following:

1. To familiarize the reader with the chemical compositions of the various types of engineering alloys and their applications. Wherever possible, explanations of why different alloying elements are used in the alloys are given.
2. To provide descriptions of the structures of different alloys in various fabricated and heat-treated conditions using photomicrographs. Explanations of the origin of the structures are given whenever possible.
3. To relate the microstructures of the alloys to their main engineering properties such as ultimate tensile and yield strengths, elongation, toughness, and corrosion resistance.
4. To provide a summary of the main production methods for the major metal systems and to give some information concerning their primary fabrication.
5. To describe the various heat treatments given engineering alloys and how the structures are modified by these treatments. Explanations for the behavior of the alloys are given whenever possible within the scope of the book. The effects of heat treatment on major engineering properties are also provided.
6. To be a reference source in regard to the alloy systems for further in-depth study.

It is recommended that the reader of this book have some previous basic knowledge of materials science. This can be obtained either by taking an introductory materials science course at a university or college or by self-study of a basic materials science textbook.

This book is organized according to individual engineering alloy systems. The first four chapters are related to carbon and alloy steels since economically they are the most important engineering alloys. Following chapters are on aluminum alloys, copper alloys, stainless steels, cast irons, tool steels, titanium alloys, and nickel and cobalt alloys.

The material in the book represents a compilation of the research and developmental results of the work of countless engineers and scientists who have contributed to our present knowledge of engineering alloys. The author would like to acknowledge those distinguished engineers and scientists who have directly contributed personal papers and micrographs. Special mention should be given to Professor Morris Cohen of the Massachusetts Institute of Technology; Professor R. W. K. Honeycombe of Cambridge University; Professor K. T. Aust of the University of Toronto; Professor J. F. Wallace of Case-Western Reserve University; Professor G. Krauss of Colorado School of Mines; R. M. Fisher, J. H. Gross, and G. R. Speich of U.S. Steel Research Laboratories; A. R. Marder, R. W. Hinton, and G. J. Roe of Bethlehem Steel Co.; A. T. Davenport and D. R. DiMicco of Republic Steel Co.; L. Mair and M. V. Balakrishnan of Inland Steel Co.; A. J. Heckler of Armco Steel Co.; J. Stepanic of Latrobe Steel Co.; G. F. Ruff of General Motors Co.; H. R. Smartt of Ford Motor Co.; T. I. Trezick of Central Foundry; C. T. Sims, E. Koch, and G. Welsch of General Electric Co.; A. P. Bond and P. Grobner of Climax Molybdenum Co.; H. D. Kessler and S. R. Seagle of Reactive Metals, Inc.; J. C. Chesnutt of Rockwell International; F. Krill of Kaiser Aluminum Co.; R. H. Stevens, H. Y. Hunsicker, and D. Robinson of Aluminum Co. of America; G. C. Tilburg of Amax Copper, Inc.; D. Bain of Anaconda American Brass Co.; T. S. Howald of Chase Brass Co.; and D. J. Tillack of Huntington Alloys.

Special thanks are due Professor Walter Owen of the Massachusetts Institute of Technology for his encouragement in regard to finishing the book. Grateful acknowledgment is due Advisory Editors Professor Charles A. Wert of the University of Illinois and Professor Michael B. Bever of the Massachusetts Institute of Technology for their helpful suggestions and encouragement. Finally, grateful appreciation is due Dean R. D. Kersten of the University of Central Florida and Editor in-Chief B. J. Clark for their encouraging support during the writing of the book.

William F. Smith

IRON-CARBON ALLOYS I

Iron is by far the least expensive of all the metals and, next to aluminum, the most plentiful. Iron and its many alloys constitute about 90 percent of the world's production of metals. Pure iron itself is used only for a relatively few special applications. Most iron is used in the form of *plain-carbon steels*, which are alloys of iron and carbon with small amounts of other elements.

In 1978, the United States produced 137 million tons of steel.[1] Plain-carbon steels accounted for 85.3 percent of this production. The reasons for the importance of plain-carbon steels is that they are strong, tough, ductile, and *inexpensive* materials that can be cast, worked, machined, and heat-treated to a wide range of properties. Unfortunately, plain-carbon steel has poor atmospheric corrosion resistance. But it can easily be protected by painting, enameling, or galvanizing. No other engineering material offers such a desirable combination of properties at such a low cost as does plain-carbon steel. Indeed, basically speaking, the highly industrialized countries of the world are still living in an "iron age," and will continue to do so into the foreseeable future.

In this book on engineering alloys, iron and steel topics constitute over half of the subject matter, which is justified in view of the importance of ferrous alloys. In Chaps. 1 and 2, iron-carbon (Fe-C) alloys are treated from a fundamental standpoint of their structure and heat treatment. Chapter 3 deals with the structure and properties of plain-carbon steels, and Chap. 4 with alloy steels. Other groups of ferrous alloys such as stainless steels, cast irons, and tool steels are dealt with in Chaps. 7, 8, and 9, respectively.

---

[1] "1978 Annual Statistical Report," American Iron and Steel Institute, Washington, D.C., 1979.

1-1 ELEMENTAL IRON

Very pure iron is produced only in small quantities and is used principally for research purposes. By zone refining, it can be made more than 99.99 percent pure. The yield strength of this pure iron is very low, being about 7500 psi (Table 1-1). Slightly less pure iron (99.9 percent) is produced commercially and has a higher yield strength (10,000 to 20,000 psi). Small quantities of elements such as carbon, manganese, phosphorus, and sulfur produce this great increase in the strength of elemental iron. The mechanical properties of some fully annealed irons are listed in Table 1-1, while their chemical compositions are given in Table 1-2.

Pure iron exists in three allotropic forms: alpha (α), gamma (γ), and delta (δ). Figure 1-1 shows an idealized cooling curve for pure iron, indicating the temperature ranges over which each of these crystallographic forms are stable at atmospheric pressure. From room temperature to 910°C, pure iron has a body-centered cubic (BCC) crystal structure, and is called α iron. α iron is ferromagnetic, but on heating to 768°C (Curie point), the ferromagnetism disappears but the crystal structure remains BCC. Nonferromagnetic α iron is stable up to 910°C and then is transformed into face-centered cubic (FCC) γ iron. Upon heating to 1403°C, the γ iron is transformed back again into the BCC structure as δ iron, which is stable up to the melting point of pure iron,

Table 1-1 Mechanical properties of some fully annealed irons at 21°C†

Type of iron	Tensile strength, psi	Yield strength, psi	Elongation, %
Zone-refined iron	28,000	7,000	
Electrolytic iron (vacuum melted)	35,000–40,000	10,000–20,000	40–60
Ingot iron (Armco)	41,000	18,000	47

† Data from *Metals Handbook*, vol. 1: "Properties," p. 211, American Society for Metals, Metals Park, OH, 1961.

Table 1-2 Chemical compositions of some relatively pure irons†

Type of iron	Chemical composition, %								
	C	Mn	P	S	Si	Cu	Ni	O_2	N_2
Armco ingot iron	0.012	0.017	0.005	0.025	trace	. . .	. . .	. . .	. . .
Electrolytic	0.006	. . .	0.005	0.004	0.005	. . .	. . .	. . .	. . .
H_2 purified	0.005	0.028	0.004	0.003	0.0012	. . .	. . .	0.003	0.0001

† Data from Ref. 2.

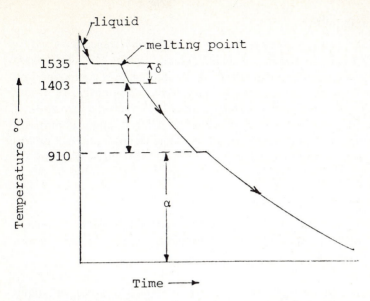

Figure 1-1 Idealized cooling curve for pure iron at atmospheric pressure.

Table 1-3 Crystallographic properties of pure iron†

Allotropic forms	Crystallographic form	Unit cube edge, Å	Temperature range
Alpha	BCC	2.86(70°F)	Up to 910°C(1670°F)
Gamma	FCC	3.65(1800°F)	910–1403°C(1670–2557°F)
Delta	BCC	2.93(2650°F)	1403–1535°C(2557–2795°F)
Density, 7.868 g/cm³	Melting point, 1535°C(2795°F)		Boiling point, 3000°C(5432°F)

† Data from Ref. 2.

1535°C. The high-temperature BCC iron has a longer cube edge than BCC α iron. Table 1-3 summarizes the crystallographic properties of pure iron.

1-2 THE Fe-Fe₃C ALLOY SYSTEM

Fe-C alloys containing from a trace to about 1.2% carbon (abbreviated "1.2% C") and with only minor amounts of other elements are termed *plain-carbon steels*. However, for purposes of this first chapter the plain-carbon steels will be treated as essentially binary Fe-C alloys. The effects of other alloying elements and impurities will be discussed in subsequent chapters.

Fe-Fe₃C Phase Diagram

The phases present at various temperatures for very slowly cooled Fe-C alloys with up to 6.67% C are shown in the phase diagram of Fig. 1-2. This phase diagram is not a true equilibrium diagram since the intermetallic compound iron carbide (Fe₃C), or *cementite* as it is called, is not a true equilibrium phase. Under certain conditions cementite will decompose into the more stable phases of graphite and iron. However, once Fe₃C is formed, it is for all practical purposes very stable and therefore can be treated as an "equilibrium" phase. For this reason, the phase diagram shown in Fig. 1-2 is a metastable phase diagram.

Solid Phases in the Fe-Fe₃C Phase Diagram

The Fe-Fe₃C phase diagram (Fig. 1-2) contains four solid phases: α ferrite, austenite, cementite (Fe₃C), and δ ferrite. A description of each of these phases follows.

α **Ferrite** The solid solution of carbon in α iron is termed α *ferrite*, or simply ferrite. This phase has a BCC crystal structure, and at 0% C it corresponds to α iron. The phase diagram indicates that carbon is only slightly soluble in ferrite since the maximum solid solubility of carbon in α ferrite is 0.02 percent at

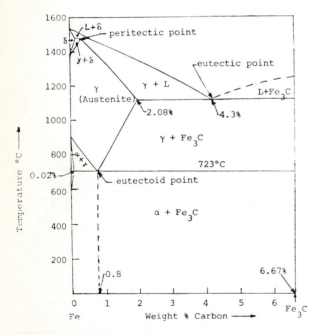

Figure 1-2 The Fe-Fe₃C metastable system. In this binary system, there are three important invariant reactions: peritectic at 1495°C, eutectic at 1148°C, and eutectoid at 723°C.

723°C. The solubility of carbon in α ferrite decreases with decreasing temperature until it is about 0.008 percent at 0°C. The carbon atoms, because of their small size, are located in the interstitial spaces in the iron crystal lattice.

Austenite The solid solution of carbon in γ iron is designated *austenite*. It has a FCC crystal structure and a much greater solid solubility for carbon than α ferrite. The solubility of carbon in austenite reaches a maximum of 2.08 percent at 1148°C and then decreases to 0.8 percent at 723°C (Fig. 1-2). As in the case of α ferrite, the carbon atoms are dissolved interstitially, but to a much greater extent in the FCC lattice. This difference in the solid solubility of carbon in austenite and α ferrite is the basis for the hardening of most steels.

Cementite The intermetallic Fe-C compound Fe_3C is called *cementite*. Iron carbide (Fe_3C) has negligible solubility limits and contains 6.67% C and 93.3% Fe. Cementite, which is a hard and brittle compound, has an orthorhombic crystal structure with 12 iron atoms and four carbon atoms per unit cell (Fig. 1-3).

δ Ferrite The solid solution of carbon in δ iron is called *δ ferrite*. It has a BCC crystal structure, but with a different lattice parameter than α ferrite. The maximum solid solubility of carbon in δ ferrite is 0.09 percent at 1495°C.

Invariant Reactions in the Fe-Fe_3C Phase Diagram

The Fe_3C phase diagram of Fig. 1-2 has three invariant reactions, each of which occurs at constant temperature and involves three phases. These reactions are *peritectic*, *eutectic*, and *eutectoid*.

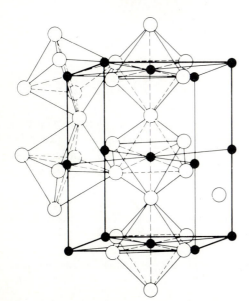

Figure 1-3 The atomic structure of cementite (Fe_3C). Iron carbide (Fe_3C) has an orthorhombic unit cell consisting of 12 iron atoms and 4 carbon atoms. Positions of the carbon atoms are indicated by the solid circles and those of the iron atoms by open circles. *(After S. B. Hendricks, Zeit. Kristal. 74 (1930) 534, as shown in "The Making, Shaping and Treating of Steel," 9th ed., United States Steel Co., 1971, p. 1077.)*

Peritectic reaction At the peritectic reaction point, liquid of 0.53% C combines with δ ferrite of 0.09% C to produce γ austenite of 0.17% C. This reaction can be written as

$$\text{Liquid (0.53\% C)} + \delta(0.09\% \text{ C}) \xrightarrow{1495°C} \gamma(0.17\% \text{ C})$$

Since this reaction occurs at such high temperatures, no δ ferrite will normally be present in plain-carbon steels at room temperature.

Eutectic reaction At the eutectic reaction point, liquid of 4.3% C decomposes to produce γ austenite with 2.08% C and the intermetallic compound Fe₃C (cementite), which contains 6.67% C. This reaction can be written as

$$\text{Liquid (4.3\% C)} \xrightarrow{1148°C} \gamma \text{ austenite (2.08\% C)} + \text{Fe}_3\text{C (6.67\% C)}$$

Since plain-carbon steels do not contain more than about 1.2% C, the eutectic reaction will not be treated in the Fe-C alloy and steel chapters. This reaction will, however, be important in the study of cast irons, which contain above 2% C and which will be the subject of a later chapter.

Eutectoid reaction At the eutectoid reaction point, solid austenite of 0.8% C decomposes into α ferrite with 0.02% C and cementite with 6.67% C. This reaction can be written as

$$\gamma \text{ austenite (0.8\% C)} \xrightarrow{723°C} \alpha \text{ ferrite (0.02\% C)} + \text{Fe}_3\text{C (6.67\% C)}$$

Critical Temperatures

The temperature of 723°C is the critical temperature above which austenite becomes unstable when slowly heated under conditions approaching equilibrium. It is designated the A_1 line[1] and is shown in Fig. 1-4. The symbol A is derived from the thermal *arrests* which are observed upon heating and cooling pure iron (Fig. 1-1). If high-purity plain-carbon steels with less than 0.8% C are heated above the A_3 line,[1] all the ferrite in the steel is transformed into homogeneous austenite. Similarly, if high-purity plain-carbon steels are heated above the A_{cm} line, all the cementite is transformed into homogeneous austenite.

When plain-carbon steels are heated or cooled through the transformation temperatures at faster than equilibrium rates, the transformation temperatures are displaced as indicated in Fig. 1-4. The thermal hysteresis (lag) which occurs upon rapid heating is indicated by the subscript c from the French word "chauffage" for heating. The thermal hysteresis which occurs on cooling is

---

[1] The designations A_{e_1} and A_{e_3} are sometimes used; the e subscript indicates *equilibrium* heating or cooling.

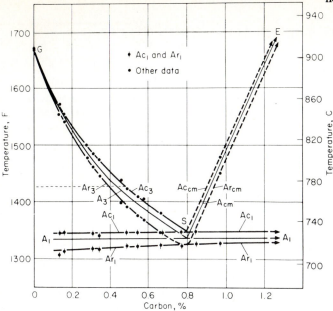

Figure 1-4 Transformation temperatures in high-purity iron-carbon alloys. *(After E. C. Bain and H. W. Paxton, "Alloying Elements in Steel," 2nd ed., American Society for Metals, 1966, p. 20.)*

indicated by the subscript r from the French word "refroidissement" for cooling.

For example, the designation A_{r_3} indicates the transformation temperature on rapid cooling a plain-carbon steel through the A_3 transformation temperature. Thermal hysteresis is common in the industrial heat treatment of steel since the rapid heating and cooling of steels is frequently practiced.

Eutectoid, Hypoeutectoid, and Hypereutectoid Plain-Carbon Steels

A plain-carbon steel containing 0.8% C is termed a *eutectoid* steel since the eutectoid transformation of austenite to cementite and ferrite occurs at this composition. If the carbon content of the steel is less than 0.8% C, it is designated a *hypoeutectoid* steel. Most steels produced commercially arc hypoeutectoid steels.

Steels containing more than 0.8% C are called *hypereutectoid* steels. Hypereutectoid steels with up to about 1.2% C are produced commercially. When the carbon content of the steel goes beyond 1.2 percent, the steel becomes very brittle, and thus few steels are made with more than 1.2% C. In order to increase the strength of steels, other alloying elements are added which increase the strength as well as maintaining ductility and toughness.

1-3 SLOW COOLING OF PLAIN-CARBON STEELS

Eutectoid Plain-Carbon Steels

If a sample of a 0.8% plain-carbon steel is heated to about 750°C and held for a sufficient time, its structure will become homogeneous austenite. That is, the whole structure will become FCC austenite with the exception of some insoluble high-melting carbides or other impurity compounds. This process is called *austenitizing*.

If this eutectoid steel is slowly cooled under conditions approaching equilibrium, the structure will remain austenitic until just above the eutectoid temperature, as is indicated in Fig. 1-5 at point *a*. At the eutectoid temperature or slightly below it, if there is any undercooling, the entire structure will be transformed from austenite to a lamellar structure of alternate plates of α ferrite and cementite (Fe₃C). Just below the eutectoid temperature the lamellar structure will appear, as indicated at point *b* in Fig. 1-5.

Figure 1-6 shows a micrograph of an eutectoid steel that was slowly cooled in a furnace. Since this eutectoid structure as seen in the optical microscope resembles mother of pearl, it has been named *pearlite*. It should be noted that pearlite is not a single phase but a mixture of two phases, α ferrite and cementite. The details of the nucleation and growth of this structure will be discussed in Sec. 1-5.

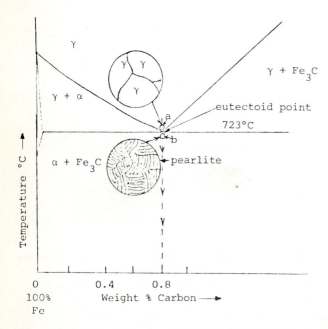

Figure 1-5 Transformation of an eutectoid steel (0.80% C) with slow cooling.

Figure 1-6 Microstructure of a slowly cooled eutectoid steel. The microstructure consists of lamellar eutectoid pearlite. The dark etched phase is cementite and the white phase is ferrite. *(United States Steel Co., as presented in Metals Handbook, 8th ed., vol. 8, American Society for Metals, 1973, p. 188.)*

If the lever rule is applied to a slowly cooled 0.80% eutectoid steel at a temperature just under the eutectoid temperature of 723°C, the alloy should be composed of the following weight percentages of ferrite and cementite:

$$\text{Wt\% ferrite} = \frac{6.67 - 0.80\%}{6.67 - 0.02} \times 100\% = 88\%$$

$$\text{Wt\% cementite} = \frac{0.80 - 0.02}{6.67 - 0.02} \times 100\% = 12\%$$

Thus the pearlitic structure should consist of approximately 88% ferrite and 12% cementite at room temperature since the solubilities change very little from 723°C to room temperature. Also, since the densities of ferrite and cementite are approximately the same, the area ratio of the ferrite lamellae to those of the cementite should be about 7 : 1.

Hypoeutectoid Plain-Carbon Steels

If a sample of a 0.4% plain-carbon steel (hypoeutectoid steel) is heated to about 900°C (point *a* in Fig. 1-7) for a sufficient time, its structure will become homogeneous austenite, as in the case of the eutectoid plain-carbon steel previously discussed. If this 0.4% C steel is then slowly cooled to the temperature

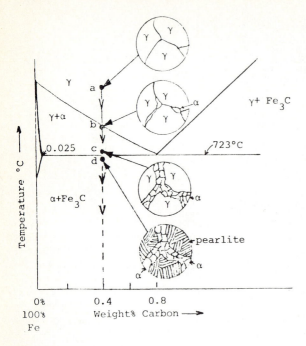

Figure 1-7 Transformation of a 0.4% carbon hypoeutectoid plain-carbon steel with slow cooling.

shown at point *b* in Fig. 1-7 (about 775°C), *proeutectoid ferrite*[1] will begin to nucleate heterogeneously at the austenitic grain boundaries. As the alloy is continuously cooled from the temperature at point *b* to that at *c* in Fig. 1-7, the proeutectoid ferrite will continue to grow into the austenite until about 50 percent of the sample is transformed. The excess carbon from the ferrite which is formed will be rejected at the austenite-ferrite interface into the remaining austenite, which becomes richer in carbon. While the alloy is cooled from the temperature at point *b* to that at *c*, the carbon content of the remaining austenite will be increased from 0.4 to 0.8 percent. At 723°C, if conditions approaching equilibrium prevail, the remaining austenite will be converted to pearlite by the eutectoid reaction: austenite → ferrite + cementite. The ferrite in the pearlite is called *eutectoid ferrite*, as contrasted to the proeutectoid ferrite which formed first. Both types of ferrite have the same composition under conditions approaching equilibrium.

Using the lever rule just slightly above 723°C at point *c* in Fig. 1-7, the weight percent proeutectoid ferrite and weight percent austenite can be calculated as

$$\text{Wt\% proeutectoid ferrite} = \frac{0.80 - 0.40}{0.80 - 0.02} \times 100\% = 50\%$$

$$\text{Wt\% austenite} = \frac{0.40 - 0.02}{0.80 - 0.02} \times 100\% = 50\%$$

---

[1] The prefix "pro-" means "before," and thus the term *proeutectoid ferrite* is used to distinguish this constituent, which forms earlier, from eutectoid ferrite, which forms by the eutectoid reaction later in the cooling.

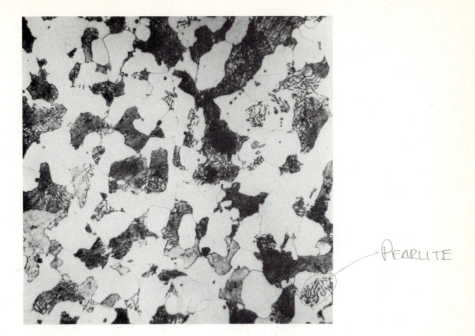

PEARLITE

Figure 1-8 Microstructure of a 0.35% C hypoeutectoid steel. The white constituent in this microstructure is proeutectoid ferrite; the dark constituent is pearlite. (Etchant: 2% nital; X500.)

Since all the remaining austenite will react to form pearlite at the eutectoid temperature of 723°C, the weight percent pearlite just slightly below 723°C at point c in Fig. 1-7 will be equal to the weight percent austenite just slightly above 723°C at point b. Thus there will be about 50% pearlite present in the 0.4% C steel at just under 723°C, if conditions approaching equilibrium exist. Since the decrease in solid solubility of the carbon from the eutectoid temperature to room temperature is very slight (i.e., 0.02 percent to near zero), there will be essentially no difference in the relative amounts of proeutectoid ferrite and pearlite at room temperature. Figure 1-8 shows the microstructure of a 0.35% C hypoeutectoid steel which was austenitized and slowly cooled to room temperature.

Hypereutectoid Plain-Carbon Steels

If a hypereutectoid plain-carbon steel is slowly cooled, the proeutectoid phase in this case is *cementite*, as contrasted to the proeutectoid ferrite phase that was formed in hypoeutectoid steels. Consider the cooling of a 1.2% plain-carbon steel which has been austenitized at 950°C (point a in Fig. 1-9). If this steel is slowly cooled to the temperature at point b in Fig. 1-9, proeutectoid cementite will begin to nucleate and grow at the austenitic grain boundaries. As the alloy is continuously cooled from the temperature at point b to that at c in Fig. 1-9, proeutectoid cementite will continue to form and deplete the carbon from the

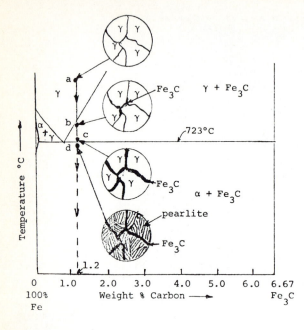

Figure 1-9 Transformation of a 1.2% carbon hypereutectoid plain-carbon steel with slow cooling.

remaining austenite at the austenite-cementite interfaces. If conditions approaching equilibrium are present during cooling from the temperature at point *b* to that at *c*, the overall carbon content of the austenite will be decreased from 1.2 to 0.8 percent. At 723°C, the remaining austenite will be transformed to pearlite by the eutectoid reaction. The cementite in the pearlite is referred to as "eutectoid cementite" to differentiate it from the proeutectoid cementite.

Using the lever rule just slightly above 723°C at point *c* in Fig. 1-9, the weight percent proeutectoid cementite and austenite can be calculated as

$$\text{Wt\% proeutectoid cementite} = \frac{1.2 - 0.80}{6.67 - 0.80} \times 100\% = 6.8\%$$

$$\text{Wt\% austenite} = \frac{6.67 - 1.2}{6.67 - 0.80} \times 100\% = 93.2\%$$

Since all the remaining austenite will be transformed into pearlite at the eutectoid temperature of 723°C, the weight percent pearlite just below 723°C at point *d* will be equal to the weight percent austenite just slightly above 723°C at point *c*. Thus there will be about 93.2% pearlite present in the 1.2% C steel at just under 723°C, if conditions approaching equilibrium exist. Since the decrease in solid solubility of carbon in ferrite from 723°C to room temperature is very slight, the same relative amounts of cementite and pearlite will be present at room temperature.

It is interesting to note that at 0.4% C in the hypoeutectoid steel, there is 50% proeutectoid ferrite, while in the 1.2% hypereutectoid steel, there is only 6.8% proeutectoid cementite. This difference is quite apparent in comparing the microstructure of the 0.4% C steel of Fig. 1-8 with that of the 1.2% C steel in Fig.

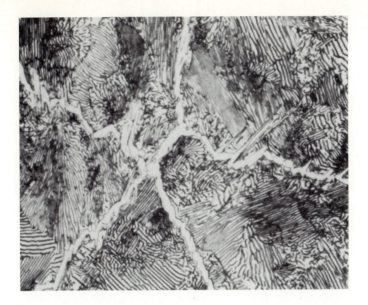

Figure 1-10 Microstructure of a 1.2% carbon hypereutectoid steel. In this structure the cementite appears as the white platelike constituent which has formed at the former austenite grain boundaries. The remaining structure consists of lamellar pearlite. (Etchant: picral; X1000.) *(Courtesy of United States Steel Research Laboratory.)*

1-10. The reason for this difference in proeutectoid constituent is that in the case of the 0.4% C steel, the ($\gamma + \alpha$) phase field extends *only from* 0.025 to 0.8% C. However, in the case of the 1.2% C steel, the ($\gamma + Fe_3C$) phase field extends from 0.8 to 6.67% C.

1-4 ISOTHERMAL TRANSFORMATION OF AN EUTECTOID PLAIN-CARBON STEEL

In the previous section, an eutectoid steel sample was allowed to cool slowly to room temperature under conditions approaching equilibrium, and it thereby produced a coarse pearlitic structure. Let us now consider what happens to the microstructure of an austenitized eutectoid steel when it is rapidly cooled to temperatures below the eutectoid temperature and isothermally transformed. In this way the austenitic decomposition can be followed as the transformation progresses by examining the microstructures of the samples after various intervals.

In such isothermal transformation experiments of an eutectoid steel, which were first made by Davenport and Bain,[1] small thin-steel samples about the size

---

[1]   E. S. Davenport and E. C. Bain, *Trans. AIME* 90(1930):117.

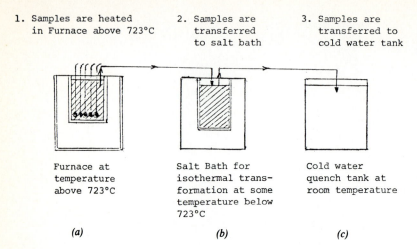

1. Samples are heated in Furnace above 723°C

2. Samples are transferred to salt bath

3. Samples are transferred to cold water tank

Furnace at temperature above 723°C

Salt Bath for isothermal transformation at some temperature below 723°C

Cold water quench tank at room temperature

(a)   *(b)*   *(c)*

Figure 1-11 Experimental arrangement for determining the microscopic changes that occur during the isothermal transformation of austenite in an eutectoid plain-carbon steel.

of a dime are first austenitized in a furnace at a temperature above the eutectoid temperature (Fig. 1-11*a*). The samples are then rapidly quenched (cooled) in a salt bath at the desired temperature below the eutectoid temperature (Fig. 1-11*b*). After various time intervals, the samples are removed from the salt bath, one at a time, and quenched into water at room temperature (Fig. 1-11*c*). The microstructure after each transformation time can then be examined at room temperature.

Consider the isothermal transformation of a 0.80% eutectoid plain-carbon steel, as is illustrated in Fig. 1-12. The steel samples are first austenitized at

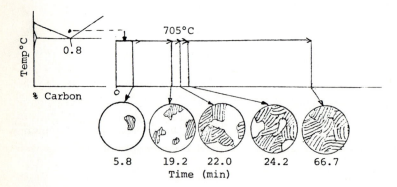

Figure 1-12 Experiment for following the microstructural changes which occur during the isothermal transformation of an eutectoid plain-carbon steel at 705°C. After austenitizing, samples are quenched in salt bath at 705°C, held for times indicated, and then quenched in water at room temperature.

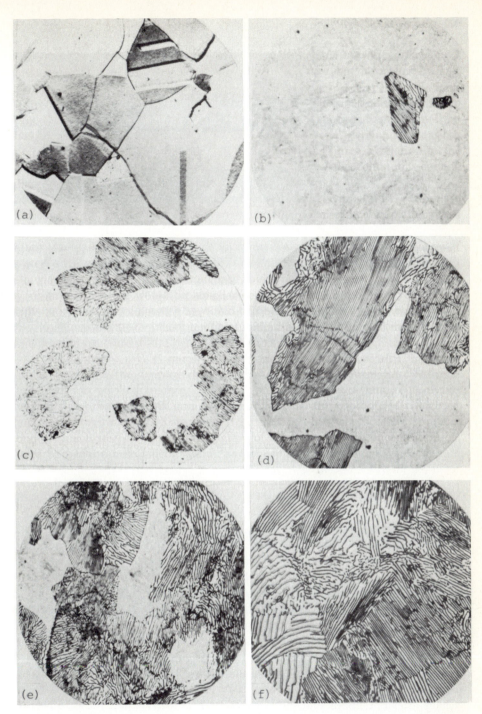

Figure 1-13 Microstructures showing the changes in the isothermal transformation of austenite to pearlite in an eutectoid plain-carbon steel at 705°C. (*a*) Austenite, (*b*) after 5.8 min, (*c*) after 19.2 min, (*d*) after 22 min, (*e*) after 24.2 min, (*f*) after 66.7 min. (Etchant: picral; X1000.) *(After J. Vilella, E. C. Bain and H. W. Paxton, in "Alloying Elements in Steel," 2nd ed., American Society for Metals, 1966, pp. 21–26.)*

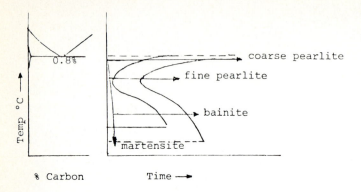

Figure 1-14 Isothermal transformation diagram for an eutectoid plain-carbon steel showing its relationship to the Fe-Fe₃C phase diagram.

760°C, which results in a polycrystalline austenitic structure as shown in Fig. 1-13a. Five small (dime-shaped) samples are then quenched into a salt bath at 705°C. After 5.8 min at 705°C, pearlite begins to nucleate and grow at the austenitic grain boundaries (Fig. 1-13b). After 19.2 min about 25 percent of the austenite is transformed into pearlite (Fig. 1-13c). The transformation now accelerates and after 22 min, 50 percent of the austenite transforms into pearlite (Fig. 1-13d). In just over 2 min more, making a total of 24.2 min, 75 percent of the austenite is transformed to 75% pearlite (Fig. 1-13e). The reaction now slows down due to the impingement of the pearlite nodules, and after 66.7 min it is complete (Fig. 1-13f).

By repeating the same procedure for progressively lower temperatures, an isothermal transformation diagram such as is shown schematically in Fig. 1-14 can be constructed. Since this diagram involves time, temperature, and transformation, it is sometimes called a TTT diagram. However, this type of diagram is best referred to as an IT (isothermal transformation) diagram to distinguish it from a CCT or "continuous-cooling transformation" diagram.

When eutectoid plain-carbon steels are isothermally transformed at temperatures in the upper section of the IT diagram, from about 550 to 723°C, austenite transforms to *pearlite*. If the eutectoid steel is quenched from the austenitizing temperature to room temperature, a new metastable phase called *martensite* is formed. Martensite is essentially a supersaturated solid solution of carbon in α ferrite. If the eutectoid steel is quenched to some temperature between about 250 and 550°C and is isothermally transformed, an intermediate structure between pearlite and martensite is formed. This structure is called *bainite* after E. C. Bain, and shows characteristics intermediate between pearlite and martensite. A summary of pearlitic, martensitic, and bainitic microstructures associated with the isothermal transformation diagram are shown in Fig. 1-15.

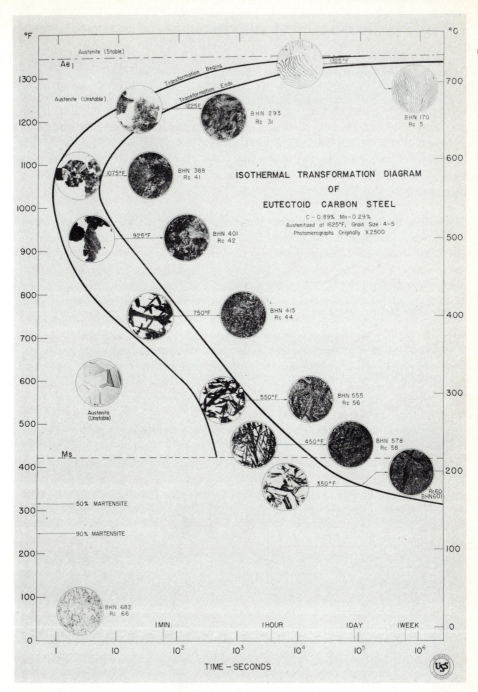

Figure 1-15 Isothermal transformation diagram of an eutectoid steel. *(Courtesy of United States Steel Co., Research Laboratory.)*

1-5 TRANSFORMATION OF AUSTENITE TO PEARLITE

Mechanism and Morphology

If a homogeneous sample of an eutectoid steel is quenched from the austenite region to some particular temperature between 723 and 550°C, the nucleation and growth of pearlite will occur as indicated in the isothermal transformation diagram of Fig. 1-16. The microstructures of the progressive stages of the formation of coarse pearlite by isothermal transformation at 705°C have already been shown in Fig. 1-13. In this section some of the details of the austenite-to-pearlite transformation will be discussed.

Nucleation and growth of pearlite lamellae If the austenite is homogeneous, nucleation of the pearlite will occur first at the grain boundaries, since they are energetically favorable nucleation sites and have faster paths for diffusion than areas within the grains. Since the nucleation of pearlite occurs so rapidly in a plain-carbon steel, it is extremely difficult to determine the nucleation mechanism. However, by slowing down the nucleation process by the addition of 12% Mn to an eutectoid steel, it has been shown by thin-foil microscopy that either ferrite or cementite can nucleate pearlitic nodules (see Ref. 10).

Let us assume in the nucleation of pearlite that a cementite lamella is formed first (Fig. 1-17a). The austenite in the region adjacent to the cementite will be depleted of carbon and, as a consequence, when the carbon content of the austenite decreases to a low enough level, adjacent layers of ferrite will form (Fig. 1-17b). The ferrite plate will grow straight ahead into the austenite as well as sideways until a sufficient amount of carbon is rejected so that new carbon lamellae can be produced (Fig. 1-17c). Eventually, a new side nucleus of cementite can form and start a new colony advancing in another direction (Fig.

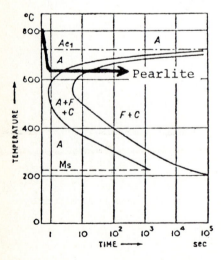

Figure 1-16 Isothermal transformation diagram for an eutectoid steel indicating the cooling path for the formation of pearlite.

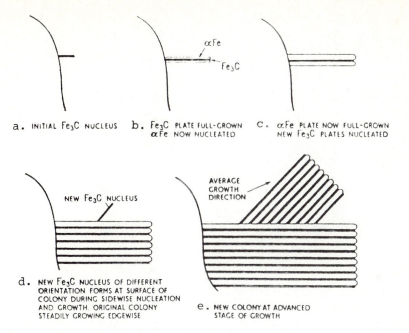

a. INITIAL Fe₃C NUCLEUS   b. Fe₃C PLATE FULL-GROWN   C. αFe PLATE NOW FULL-GROWN
                          αFe NOW NUCLEATED   NEW Fe₃C PLATES NUCLEATED

d. NEW Fe₃C NUCLEUS OF DIFFERENT
ORIENTATION FORMS AT SURFACE OF
COLONY DURING SIDEWISE NUCLEATION
AND GROWTH. ORIGINAL COLONY   e. NEW COLONY AT ADVANCED
STEADILY GROWING EDGEWISE   STAGE OF GROWTH

Figure 1-17 The nucleation and growth of pearlite. *(After R. F. Mehl, Trans. ASM, 29 (1941): 813. as presented in "Progress in Metal Physics," vol. 6, 1956, p. 92, Pergamon. Reprinted with permission.)*

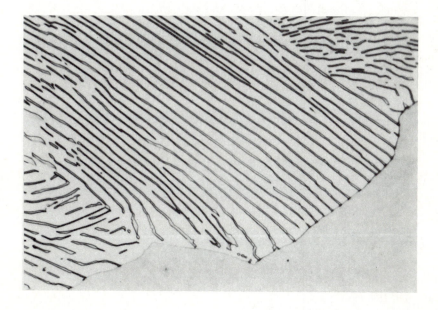

Figure 1-18 Eutectoid steel that has partially transformed to pearlite at 700°C. *(After J. R. Vilella, United States Steel Co., Research Laboratory.)*

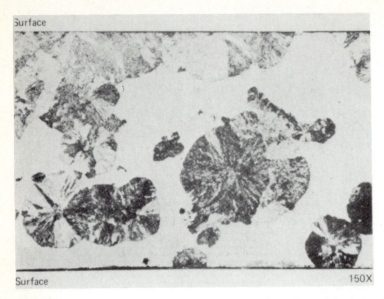

Figure 1-19 Cross-section view of the microstructure of pearlite nodules in partially transformed hot-stage specimen showing nodules forming at both specimen surface and interior. [*After B. L. Bramfitt and A. R. Marder, Met. Trans. 4(1973):2291.*]

1-17*d* and *e*). The structure of an eutectoid steel that has been partially transformed into pearlite at 700°C is shown in Fig. 1-18.

The growth of pearlite from austenite during isothermal transformation is always nodular, as is shown in the hot-stage micrograph in Fig. 1-19. The pearlitic nodules grow from the nuclei along the austenitic boundaries and continue to grow radially until they impinge on one another. Within each pearlitic nodule there are colonies of alternating ferrite and cementite lamellae, all with the same orientation.

A plot of the percent pearlite transformed versus time produces a sigmoidal curve (Fig. 1-20*a*). In the first stage, the transformation rate of austenite to

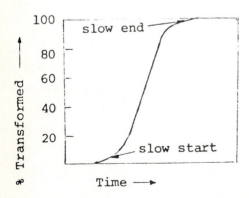

Figure 1-20a Idealized isothermal reaction curve.

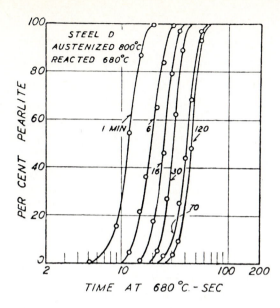

Figure 1-20b Isothermal reaction curves at 680°C for eutectoid steel austenitized at 800°C for 1, 6, 16, 30, 70, and 120 min. (Note how the transformation of austenite to pearlite begins slowly and later ends slowly.) [*After G. A. Roberts, Trans. AIME, 154(1943):318.*]

pearlite is slow since only a few pearlitic nodules are nucleated and grow. This stage may be considered an incubation period. In the second stage, the transformation rate is greatly accelerated since many new nuclei are formed and grow, while the growth of the existing nodules continues. The growth rate of the pearlite at any instant is proportional to the area of austenite-pearlite interface at that time. Finally, a third stage is reached when the transformation rate decreases since the nucleation rate is decreased and continued growth of the existing pearlitic nodules is hampered by their impinging on each other (Fig. 1-20b).

Effects of Temperature

The decomposition of austenite to pearlite involves two important variables, both of which are ultimately temperature-dependent. These are the nucleation rate N of the pearlite, and the growth rate G. The nucleation rate, which is the number of nuclei formed in a unit volume in a unit time, increases as the temperature of the transformation is lowered. Thus, as the ΔT of undercooling below the A_{e1} is increased, more nuclei are available to form pearlite. The growth rate is diffusion-dependent, and hence, as the temperature of the transformation decreases, the growth rate decreases also.

The transformation rate of austenite to pearlite at a specific temperature will therefore depend on the rate of nucleation of the pearlite, N, and the growth rate G of the pearlitic nodules. At relatively high temperatures, i.e., slightly below 723°C, the A_{e1} temperature, the nucleation rate will be relatively low due to the small ΔT and the growth rate high due to a high diffusion rate (Fig. 1-21). Thus the ratio N/G will be small. At this temperature, the nuclei will grow rapidly

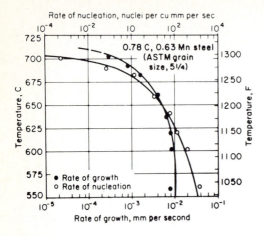

Figure 1-21 Rates of nucleation and growth of pearlite colonies in an eutectoid steel as a function of temperature. *(After R. F. Mehl and W. C. Hagel from "Progress in Metal Physics," vol. 6, Pergamon, 1956, p. 102, as presented in the Metals Handbook, 8th ed., vol. 8, American Society for Metals, 1973, p. 189.)*

into large pearlitic nodules which can cross grain boundaries and consume large numbers of austenitic grains before impinging on one another.

As the transformation temperature is lowered, the nucleation rate increases at a faster rate than the rate of growth, as is also indicated in Fig. 1-21. Thus the ratio N/G will become larger. As a result, at the early stages of the transformation, the austenitic grain boundaries will be outlined by pearlitic nodules that form by nucleation and growth. This is clearly shown in the eutectoid steel of Fig. 1-22, which is in the early stages of transformation at 550°C.

The interlamellar spacing of the pearlite decreases as the transformation temperature decreases since, with a high nucleation rate, the carbon atoms do not have to migrate as far in forming the ferrite-cementite lamellae. The increased free energy from the transformation due to the large degree of

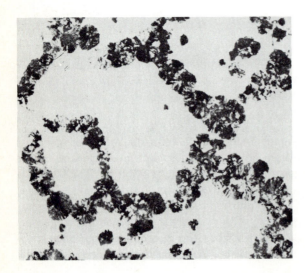

Figure 1-22 Eutectoid steel partially transformed to pearlite at 550°C. (Note how the pearlite outlines the austenite grain boundaries.) *(After H. Aaronson, from P. G. Shewmon, "Transformations in Metals" McGraw-Hill, 1969, p. 228.)*

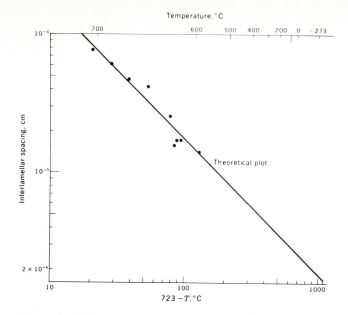

Figure 1-23 Relationship between interlamellar spacing and degree of undercooling in an eutectoid steel. *(After G. E. Pellisier, M. F. Hawkes, W. A. Johnson, and R. F. Mehl, Trans. ASM, 29(1942):1049, as presented in P. G. Shewmon, "Transformations in Metals," McGraw-Hill, 1969, p. 232.)*

undercooling (ΔT) supplies sufficient energy to provide the large interfacial energy of the fine ferrite-cementite lamellae formed at the lower temperatures. Figure 1-23 shows how the interlamellar spacing of the pearlite in an eutectoid steel decreases with decreasing temperature.

Effects of Grain Size

The grain size of the austenite will affect the austenite-to-pearlite transformation since the nucleation rate is structure-sensitive. That is, nucleation occurs in regions of high energy. In homogeneous austenite, pearlitic nucleation occurs almost exclusively at grain boundaries. Thus a finer austenitic grain size will provide more nuclei for pearlitic nodules and will form a finer pearlitic structure. The prior austenitic grain size has been found to have no effect on the interlamellar spacing of the pearlite.[1] The interlamellar spacing is determined by the temperature of the transformation.[2] For a particular temperature, the nucleation and diffusion rates will determine this spacing (Fig. 1-23).

[1] F. C. Hull, R. A. Colten, and R. F. Mehl, *Trans. AIME* 150(1942):185.
[2] G. E. Pellisier, M. F. Hawkes, W. A. Johnson, and R. F. Mehl, *Trans. ASM* 29(1942):1049.

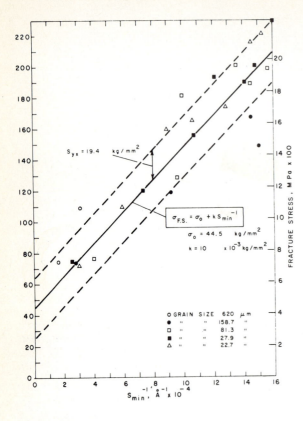

Figure 1-24 Effect of minimum interlamellar spacing of pearlite in an eutectoid steel on yield strength. [*After A. R. Marder and B. L. Bramfitt, Met. Trans. 7A(1976):365.*]

The Strength of Pearlite[1]

In general, the pearlitic structure is softer than the martensitic or bainitic types. As the transformation temperature of austenite to pearlite is decreased within the pearlitic range (723 to 550°C), the interlamellar spacing of the pearlite decreases (Fig. 1-23). The strength of the fine pearlite is greater than that of the coarse pearlite since dislocations have more difficulty passing through the fine lamellar structure of cementite and ferrite.

For eutectoid steels, the increase in yield strength varies inversely as the interlamellar spacing of the pearlite. The effect of minimum interlamellar spacing on the yield strength of a high-purity eutectoid steel is shown in Fig. 1-24. For this eutectoid steel, the yield strength can be related to the interlamellar spacing by the following expression:

$$\sigma_y \text{ (MPa)} = 139 + 46.4 S^{-1}$$

[1] See Ref. 12.

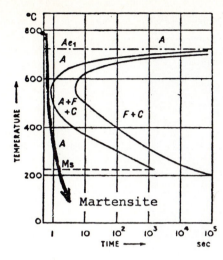

Figure 1-25 Isothermal transformation diagram for an eutectoid steel indicating the cooling path for the formation of martensite.

1-6 TRANSFORMATION OF AUSTENITE TO MARTENSITE

If a plain-carbon eutectoid steel (Fe–0.8% C) is cooled rapidly from the austenitic region so that it misses the nose of the IT curve (Fig. 1-25), a new phase called *martensite* is formed at temperatures below about 220°C. Martensite in steels is a metastable structure consisting of a supersaturated solid solution of carbon in α ferrite. The study of the martensitic transformation in steels is of great engineering importance because of the ability of martensite to harden and strengthen many steels.

This section on martensite will begin with a description of its characteristics, which will be followed by an examination of its morphological changes with variation in carbon content. Then, a brief discussion of its mechanisms and kinetics of formation will be given. Finally, some information on the strength of martensite will be presented.

Characteristics of the Martensitic Transformation in Plain-Carbon Steels

1. An important characteristic of the martensitic transformation in plain-carbon steels is that various microscopically observable structures are produced by it and that *the type of martensitic structure obtained depends on the carbon content of the steel* (see Refs. 13 and 14). If the carbon content of the steel is low (i.e., about 0.2 wt%), then well-defined laths of martensite are observed in the optical microscope (Fig. 1-26a). As the carbon content is increased (i.e., to about 0.6 wt%), plates of martensite begin to form, as is pointed out in Fig. 1-26b. If the carbon content is increased still more to about 1.2 wt%, the martensite appears as an array of well-defined plates (Fig. 1-26c). This

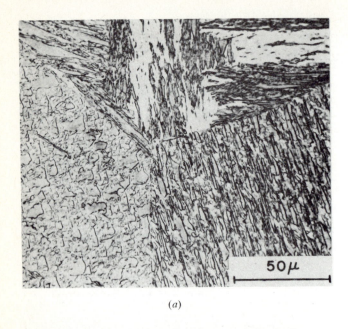

(a)

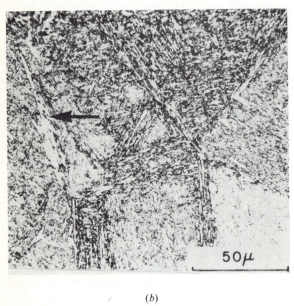

(b)

Figure 1-26 Effect of carbon content on the structure of martensite in plain-carbon steels: (a) lath type; (b) mixed lath and plate types, arrow points to a plate; (c) plate type. (Etchant: sodium bisulfite; optical micrographs.) [*After A. R. Marder and G. Krauss, Trans. ASM, 60(1967):651.*]

(c)

Figure 1-26 Continued

sequence of optical micrographs shows how the type of martensite formed depends on the carbon content of the steel.

2. Another important characteristic of the martensitic transformation is that it is *diffusionless*. That is, the reaction takes place so rapidly that atoms do not have time to intermix. There appears to be no thermal activation energy barrier to prevent its formation.

3. There appears to be *no compositional change* in the parent phase after the martensitic reaction, and each atom tends to preserve its own original neighbors. The relative positions of the carbon atoms with respect to the iron atoms are the same in the martensite as they were in the austenite.

4. *The crystal structure produced by the martensitic transformation in plain-carbon steels changes from BCC to body-centered tetragonal (BCT) as the carbon content of the steel is increased.* For low carbon contents less than about 0.2 weight percent, the austenite transforms to a BCC α ferrite crystal structure. When the carbon content of the steel is increased, the BCC structure is distorted into a BCT crystal structure. The BCT structure is produced primarily because of the greater solid solubility difference of carbon in FCC austenite and BCC ferrite iron. Figures 1-27a and b show that the interstitial spaces for the carbon atoms are much larger in the FCC unit cell than in the BCC unit cell. This change in interstitial spacing leads to the distortion of the BCC unit cell along the c axis to accommodate the excess carbon atoms (Fig. 1-27c).

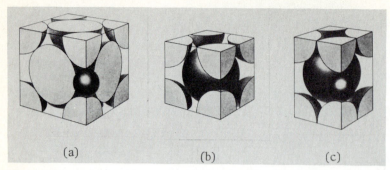

Figure 1-27 Interstitial positions of carbon atoms in FCC, BCC, and BCT iron-crystal structure unit cells. *(After E. R. Parker and V. F. Zackay, "Strong and Ductile Steels," Scientific American, November, 1968, p. 36. Used by permission.)*

5. *The martensitic transformation in steel starts at a definite temperature called the* M_s (Fig. 1-28). When austenitized Fe-C alloys are quenched, martensite starts to form as the temperature of the alloys reaches the M_s. As the temperature continues to be lowered during cooling, more and more of the austenite is transformed into martensite until the M_f (martensitic finish) temperature is reached. However, not all plain-carbon steels can be transformed into 100 percent martensite, and as the carbon content of the steel is increased, more and more *retained austenite* is formed upon quenching.

6. *In the higher-carbon plain-carbon steels, martensitic plates are formed by a displacive or shearlike transformation process* which causes a shape deformation on a flat surface.

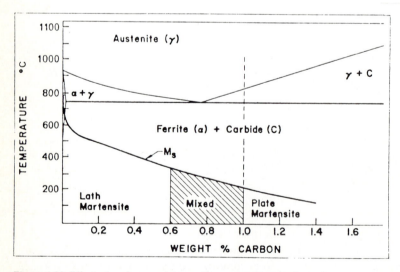

Figure 1-28 Effect of carbon content on the martensite transformation start temperature, M_s, for iron-carbon alloys. *(After A. R. Marder and G. Krauss, as presented in "Hardenability Concepts with Applications to Steel," AIME, 1978, p. 238.)*

The Morphology of Martensite in Fe-C Alloys

Two major types of martensite form in alloys, depending on the carbon content of the plain-carbon steels. These are Type I, or _lath_ martensite, and Type II, or _plate_ martensite. The lath martensite is predominant in Fe-C alloys with up to about 0.6% C (Fig. 1-28). Above about 1.0% C, the plate-type martensite predominates, while between 0.6 and 1.0% C, a transition from lath to plate types takes place. In this region, therefore, mixed structures of both types appear.

Type I—lath martensite In low-carbon Fe-C alloys, the martensite consists of domains which have groups of laths separated by low-angle or high-angle grain boundaries (Fig. 1-26a). The structure within the martensitic laths is highly distorted, consisting of regions with high densities of dislocation tangles (Fig. 1-29). The structure of low-carbon lath martensites consists of a regular repetition of laths of different but limited orientation through a whole domain of martensite. The formation of a lath domain is believed to be the result of a phase front that has propagated through a whole region of austenite matrix, resulting in the almost complete transformation of the parent austenite to martensite. As a result, only a small amount of retained austenite is present at room temperature in the low-carbon lath-type martensite (Fig. 1-30).

Figure 1-29 Structure of lath martensite in an Fe-0.2% C alloy. (Note the parallel alignment of the laths.) [*After A. R. Marder and G. Krauss, Trans. ASM. 60(1967):651.*]

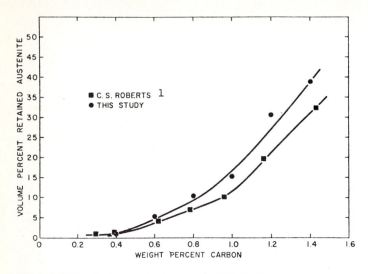

Figure 1-30 Retained austenite in quenched Fe-C alloys at room temperature as a function of carbon content. Note that the amount of retained austenite does not become significant until about 0.4% C. [*After A. R. Marder and G. Krauss, Trans. ASM, 60(1967):651.*]

[1]C. S. Roberts, *Trans. AIME* 197 (1953)203.

Figure 1-31 Plate martensite showing fine transformation twins. [*After M. Oka and C. M. Wayman, Trans. ASM, 62(1969):370*]

Type II—plate-type martensite The structure of martensite in high-carbon Fe-C alloys consists of needlelike plates of martensite often surrounded by large amounts of retained austenite. The plates are found to have irrational habit planes ranging from {225} to {259} as the carbon content increases. Above 1% C, the structure of Fe-C martensites is found to be exclusively plate martensite and retained austenite. In contrast to the low-carbon martensite, the plates found in the high-carbon martensites are formed independently on specific habit planes within the austenite and terminate or originate on one another. The plates in high-carbon martensite vary in size and have a fine structure of parallel *twins* which are of the {112} type (Fig. 1-31).

Mixed Lath and Plate Martensite There is a transition from lath- to plate-type martensite between 0.6 and 1.0% C in Fe-C alloys (Fig. 1-28). As the carbon content is increased, the size and frequency of the martensite plates increases and the amount of the lath martensite decreases. Kelly and Nutting (Ref. 13) found that the factor that determined whether martensite would form as laths or plates was the transformation temperature at which a particular martensitic grain was formed. They believed that if the M_s temperature for an Fe-C alloy was below the critical transformation temperature, mostly plate-type martensite would be formed. The carbon content of an Fe-C alloy, since it controls the M_s temperature, would thus determine whether lath- or plate-type martensite would be formed. Therefore, there exists a range of temperatures for the formation of mixed lath and plate martensites, corresponding to a range of carbon contents from about 0.6 to 1.0 percent. This temperature band for Fe-C alloys is approximately 200 to 320°C.

Mechanism of Formation of Martensite in Plain-Carbon Steels

Features At present the mechanism of formation of martensite in plain-carbon steels is not completely understood. The martensitic transformation in Fe-C alloys is very complicated. So while much research has been done on it, much more will be required in the future.

The transformation of austenite to martensite occurs because below a critical temperature, designated the M_s temperature, martensite is the structurally stable state of the alloy and has a lower free energy (Ref. 17). As stated before, the martensitic reaction in Fe-C alloys is diffusionless and therefore takes place without atom mixing. The martensitic reaction occurs with a cooperative rearrangement of the atoms, so that the relative displacement of the atoms is not more than the interatomic distance.

Certain features of the martensitic reaction are well established. Several of these are:

1. *The degree of tetragonality of the martensitic lattice increases as the carbon content increases.* Figure 1-32 shows how the c axis of a martensitic unit cell increases from 2.86 Å for the BCC structure to 3.08 Å for martensite with

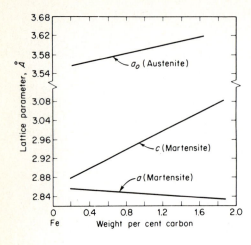

Figure 1-32 Variation of the lattice parameters of austenite and martensite as a function of carbon content. [*After C. S. Roberts, Trans. AIME 197(1953):203, as presented in "Physical Metallurgy Principles", 2nd ed., by R. E. Reed-Hill, ©1973 by Litton Educational Publishing, Inc., Reprinted by permission of D. Van Nostrand Co.*]

1.8% C. Correspondingly, the *a* axis decreases from 2.86 to 2.83 Å in the same carbon range. This distortion of the BCC unit cell to a slightly BCT unit cell is a direct consequence of the carbon atoms straining the martensitic lattice into tetragonality.

2. *The change in morphology of Fe-C alloys with increasing carbon content is accompanied by a change in deformation mode from slip to twinning.* What causes the change is not completely understood. It is observed that, the higher the carbon concentration in the martensite, the greater is the tendency to form twinned plates. Lower M_s temperatures lead to increased twinning in martensitic structures. Increasing the strength of martensite and lowering the M_s temperature with more carbon solute favors twinning as the preferred shear mode.

Deformation modes Explanations to account for slip and twinning have been proposed:

Slip as a deformation mode in the formation of martensite To account for the experimentally observed high dislocation densities found in lath-type martensite, Wasilewski (Ref. 19) has proposed that *accommodation dislocations* are created, which are effectively sessile. These dislocations would form to relieve the very high levels of local strain energy formed by the rapid rate of formation of martensite. They would form more or less uniformly throughout single domains of martensite to accommodate the increased strain energy. As a result, domains with a high density of tangled accommodation dislocations would be formed, as is observed in lath martensite.

Twinning as a deformation mode in the formation of martensite With higher carbon contents, the martensitic reaction becomes increasingly more difficult, as is indicated by the increase in amount of retained austenite (Fig. 1-30) under those conditions and by the effects of lowering the M_s temperature. As has been described, twinning becomes more and more the predominant mode of deformation as the carbon content increases.

As the transformation temperature is lowered, slip becomes increasingly more difficult. The important factor appears to be the relative magnitudes of the critical resolved shear stresses for twinning and slip at a given temperature and alloy composition. Wasilewski (Ref. 19) suggests that twinning occurs as a means of preventing the failure of the lattice when sufficiently high strain energies accumulate locally. Thus a significant part of the elastic strain energy generated by the martensitic reaction at lower temperatures can be accommodated in the twin/matrix coherent boundary, thereby lowering the average energy in the martensite. Even so, many times the elastic energy in the martensitic plates of the higher-carbon-containing Fe-C alloys cannot be accommodated by twinning, and cracking in the plates is observed, as shown by Marder and Krauss (Fig. 1-33).

The shape change which occurs during the formation of a martensitic plate in a high-carbon Fe-C alloy is shown schematically in Fig. 1-34. Deformation twinning has to occur in order for this type of a distortion to take place during the transformation of austenite to plate martensite. In this process, the surface of the crystal is tilted as shown in Fig. 1-35 for an Fe–1.86% C alloy. The relationship between the planes of the austenitic lattice and those of the martensite is not clear since many different relations have been proposed (Ref. 4, p. 445). However, the relationship usually cited for Fe-C alloys with 0.5 to 1.4% C is

$$(111)_\gamma \| (101)_\alpha \quad \text{with} \quad [1\bar{1}0]_\gamma \| [11\bar{1}]_\alpha$$

More research on this subject is necessary to clarify the situation.

Figure 1-33 Microstructure of a 0.93% C iron-carbon martensite specimen showing cracks in a martensite plate. Note the crack indicated by the arrow. (Etchant: sodium bisulfite.) [*After A. R. Marder and G. Krauss, Trans. ASM 60(1967):651.*]

50μ

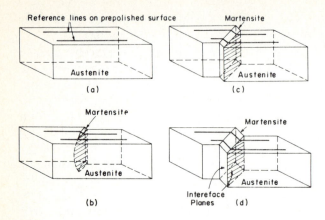

(a)

(b)

(c)

(d)

Figure 1-34 Schematic representation of the martensite transformation in high-carbon iron-carbon alloys. [*After M. Cohen, Trans. AIME 224(1962):638.*]

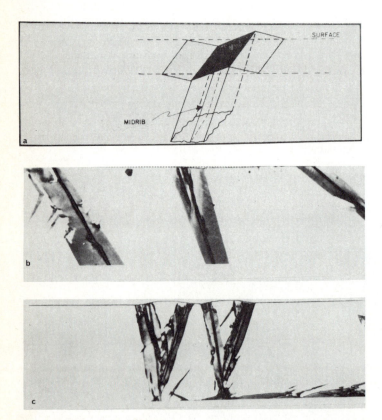

Figure 1-35 Shape deformation of plate martensite. (*a*) Schematic of shape change. (*b*) Shape of deformation in an Fe-1.86% C alloy. (*c*) Bulge in the austenite caused by martensite plate formation. (Etchant: nital; X1000.) [*After G. Krauss and A. R. Marder, Met. Trans. 2(1971):2343.*]

Nucleation and growth of martensite It is generally accepted that martensitic embryos must form in the austenitic matrix at some high temperature and that the embryos must be activated to grow immediately when the M_s is reached. Favorable nucleation sites, according to Olson and Cohen (Ref. 20), are grain boundaries, incoherent twin boundaries, and inclusion particle interfaces. Also groups of dislocations can interact by plastic deformation to provide a suitable nucleation site. Martensite formed in one region of a sample can provide a combination of strain-induced nucleation and also provide elastic stresses to assist an existing nucleation site. This is called the *autocatalytic effect.*

The driving force for the growth of the martensite is the energy released in forming a lower-energy structure. The strain energy necessary to produce martensitic laths or plates is more than compensated for by the volume free energy released by the formation of the martensitic phase. Observation shows that the austenite-martensite interface moves rapidly. After martensitic plates are nucleated, they grow rapidly until they strike another plate or grain boundary, and sometimes thicken slightly before stopping.

Kinetics of Formation of Martensite in Plain-Carbon Steels

Rapidity of the transformation Since the formation of martensite in Fe-C alloys is diffusionless, it only forms during the cooling process and not isothermally. The fraction of martensite that is formed depends only on the temperature to which it is cooled. Thus in the I-T diagram, horizontal lines are drawn to indicate percent martensite formed at a particular temperature. Such a reaction, i.e., one that is only dependent on the temperature to which the martensite is quenched, is termed an *athermal transformation.* The percent martensite formed at a particular temperature does not depend on the cooling rate once a critical cooling rate is attained, but depends on the alloy composition and thermal and mechanical history.

The fact that the martensitic plates grow so rapidly must mean that there is little or no activation energy needed for their growth. The velocity of the growth of the plates in some cases approaches the speed of sound. The formation of the plates sometimes occurs by bursts such that the stresses produced by the formation of the initial plates catalyzes the nucleation of others.

Stabilization Consider a sample of an austenitic plain-carbon steel which is being cooled rapidly to form martensite. If cooling is stopped for some time interval, say 1 s, at some temperature below the M_s, then the martensitic reaction will also stop, and the martensite is said to be *stabilized.* No further transformation of the austenite to martensite will occur even though there is sufficient free-energy difference for the reaction to continue. If the cooling of a stabilized steel is resumed, a certain amount of undercooling is necessary to start the austenite-to-martensite reaction again, and when it occurs it frequently begins with a burst. One theory to account for stabilization is that the carbon atoms diffuse to the dislocations in the martensite, thus preventing their further movement so that the reaction is stopped.

Strength and Hardness of Martensite in Fe-C Alloys

The hardening produced in Fe-C martensites is directly related to their carbon contents. This relationship is clearly demonstrated in Fig. 1-36, which shows the hardness of fully hardened martensitic plain-carbon steels as a function of carbon content.

Low-carbon martensites Four different strengthening effects of carbon in low-carbon martensites have been identified by Speich and Warlimont (Ref. 22). These are

1. Refinement of the martensitic cell size with increasing carbon content
2. Segregation of carbon to the martensitic cell walls during the quench
3. Solid-solution hardening
4. Dispersion hardening due to precipitation of carbide during the quench

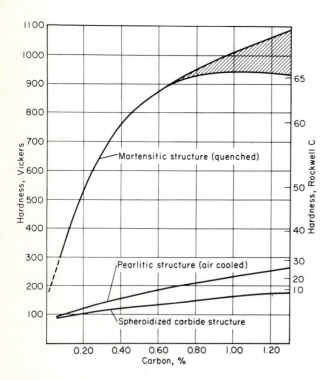

Figure 1-36 Approximate hardness of fully hardened martensitic plain-carbon steel as a function of carbon content. The cross-hatched region indicates some possible loss of hardness due to the formation of retained austenite, which is softer than martensite. [*After E. C. Bain and H. W. Paxton, "Alloying Elements in Steel," 2nd ed., American Society for Metals, 1966, p. 37.*]

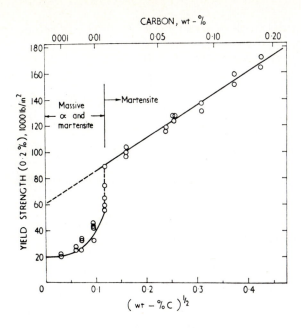

Figure 1-37 Variation of the yield strength of low-carbon martensites with the square root of the carbon content. [*After G. R. Speich and H. Warlimont, J. Iron & Steel Inst. 206(1968):385.*]

At low-carbon contents, lath martensite is formed which contains a high density of dislocations arranged in platelike cells. This cell structure is associated with an increase in strength in the low-carbon martensites. The stress required to move dislocations through dense dislocation networks and the finely spaced cell walls of the lath martensite would certainly be an important strengthening mechanism in low-carbon martensites. Items 2 to 4 above would also contribute in varying degrees to the strength of low-carbon martensites.

Figure 1-37 shows the net effect of all these mechanisms on the strength of low-carbon martensites in the 0.01 to 0.18% C range.

High-carbon martensites With higher carbon contents in Fe-C alloys, solid-solution hardening becomes the dominant hardening mechanism, as is evidenced by the distortion of the BCC iron lattice into tetragonality. The increase in hardness can be correlated with the increased distortion of the iron lattice. However, the introduction of numerous twinned interfaces in plate martensite would also be another hardening mechanism.

PROBLEMS

1. Describe the three allotropic forms of pure iron. Indicate how these are related to a cooling curve for pure iron.

2. Draw the phase diagram for the Fe-Fe$_3$C (metastable) alloy system and indicate the equilibrium critical temperatures.

3. Describe the following solid phases which occur in the Fe-Fe$_3$C diagram: (*a*) α ferrite, (*b*) austenite, (*c*) cementite, and (*d*) δ ferrite

4. Write the invariant reactions which occur in the Fe-Fe$_3$C diagram.

5 Explain the meaning of the following designations on the Fe-Fe$_3$C diagram when fast-cooling and -heating rates are involved: A_{c_1}, A_{r_1}, A_{c_3}, A_{r_3}, $A_{c_{cm}}$, $A_{r_{cm}}$.

6. Define eutectoid, hypoeutectoid, and hypereutectoid plain-carbon steels.

7. Define the term "austenitizing."

8. Distinguish between eutectoid and proeutectoid ferrite in the Fe-Fe$_3$C diagram.

9. Describe the structural changes which occur when a 0.6% plain-carbon steel is slowly cooled from the austenitic region.

10. Repeat question 9 for a 1.1% hypereutectoid steel.

11. Describe the procedure for making an isothermal transformation experiment for an eutectoid steel.

12. Draw an isothermal transformation diagram for an eutectoid plain-carbon steel and indicate the cooling conditions necessary to form (a) pearlite, (b) bainite, and (c) martensite.

13. Using a diagram describe the nucleation and growth of pearlite.

14. Explain how the transformation of austenite to pearlite plotted as the percent pearlite transformed versus time leads to a sigmoidal curve.

15. How does the rate of nucleation and growth in an eutectoid steel vary as a function of temperature? How can this relationship be explained qualitatively?

16. How does the interlamellar spacing in pearlite vary with decreasing temperature in the isothermal transformation of an eutectoid steel?

17. How does grain size affect the austenite-to-pearlite transformation of an eutectoid steel?

18. How is the strength of pearlite affected by the interlamellar spacing?

19. Describe the principal characteristics of the martensitic transformation in plain-carbon steels.

20. Describe the morphological changes that occur in Fe-C martensites as the carbon content is increased from 0.2 to 1.2 percent.

21. Describe the microstructure of lath Fe-C martensites.

22. Describe the microstructure of plate Fe-C martensites.

23. What is believed to be the cause of the change from lath to plate martensite?

24. What explanation can be given for the change from slip to twinning deformation modes in martensite as the carbon content is increased?

25. What is the main cause of the cracking in martensitic plates in high-carbon Fe-C martensites?

26. What shape changes take place during the formation of martensitic plates in high-carbon Fe-C martensites?

27. Describe a possible mechanism for the nucleation and growth of a martensitic plate.

28. Describe what is meant by the term "athermal transformation." How can this term be applied to a martensitic transformation?

29. What is meant by the stabilization of martensite?

30. What are some of the strengthening mechanisms which are believed to be related to the strength of (a) low-carbon martensite and (b) high-carbon martensite?

31. Using the Fe-Fe$_3$C metastable phase diagram, calculate the following for a slowly cooled 0.60% C hypoeutectoid steel:

   (a) The weight percent austenite and proeutectoid ferrite just above 723°C
   (b) The weight percent proeutectoid ferrite just below 723°C
   (c) The weight percent cementite just below 723°C
   (d) The weight percent eutectoid ferrite just below 723°C

32. (a) Using the Fe-Fe$_3$C metastable phase diagram, calculate the following:

   (i) The weight percent proeutectoid ferrite just below 723°C for a 0.7% hypoeutectoid steel
   (ii) The weight percent proeutectoid cementite just below 723°C for a 0.9% hypereutectoid steel

   (b) Why is the weight percent proeutectoid ferrite considerably larger than the weight percent proeutectoid cementite?

33. The stimulating autocatalytic effect of Fe-C plate martensitic formation on the further progression of the transformation has been attributed to a combination of strain-induced nucleation and an elastic stress assist of existing nucleation sites. Suggest a possible third contribution to this autocatalysis. (See Ref. 20.)

34. During the formation of plate martensite in a Fe-1.39% C alloy, the increase in amount of plate martensite has been shown to be predominantly by the nucleation of new plates. By what other mode could the amount of plate martensite be increased as the temperature is decreased? (See Ref. 21.)

REFERENCES

1. E. C. Bain and H. W. Paxton: "Alloying Elements in Steel," 2d ed., American Society for Metals, Metals Park, OH, 1966.
2. H. E. McGannon (ed.): "The Making, Shaping, and Treating of Steel," 9th ed., United States Steel Corporation, Pittsburgh, 1971.
3. R. E. Reed-Hill: "Physical Metallurgy Principles," 2d ed., Van Nostrand, New York, 1973.
4. R. E. Smallman: "Modern Physical Metallurgy," 3d ed., Butterworths, London, 1970.
5. R. F. Mehl and W. C. Hagel: "The Austenite-Pearlite Reaction." In Bruce Chalmers and R. King (eds.): *Progress in Metal Physics*, vol. 6, Pergamon Press, Elmsford, NY, 1956.
6. L. Kaufman and M. Cohen: "Thermodynamics and Kinetics of Martensite Transformations." In Bruce Chalmers and R. King (eds.): *Progress in Metal Physics*, vol. 7, Pergamon Press, Elmsford, NY, 1958.
7. P. G. Shewmon: "Transformations in Metals," McGraw-Hill, New York, 1969.
8. V. F. Zackay and H. L. Aaronson (eds.): "Decomposition of Austenite by Diffusional Processes," John Wiley, New York, 1962.
9. R. M. Brick, A. W. Pense, and R. B. Gordon, "Structure and Properties of Engineering Materials," McGraw-Hill, New York, 1977.
10. R. J. Dippenaar and R. W. K. Honeycombe: "The Crystallography and Nucleation of Pearlite," *Proc. Roy. Soc. Lond.* A333(1973):455.
11. B. L. Bramfitt and A. R. Marder: "Effect of Cooling Rate and Alloying on the Transformation of Austenite," *Met. Trans.* 4(1973):2291.
12. A. R. Marder and B. L. Bramfitt: "The Effect of Morphology on the Strength of Pearlite," *Met. Trans.* 7A(1976):365.
13. P. M. Kelly and J. Nutting: "The Martensite Transformation in Carbon Steels," *Proc. Roy. Soc.*, A259(1960):45.
14. A. R. Marder and G. Krauss: "The Morphology of Martensite in Iron-Carbon Alloys," *Trans. ASM* 60(1967):651.
15. M. Oka and C. M. Wayman: "Electron Metallography of the Substructure of Martensite in High-Carbon Steels," *Trans. ASM* 62(1969):370.
16. Morris Cohen: "The Strengthening of Steel," *Trans. AIME* 224(1962):638.
17. G. V. Kurdjumov: "Phenomena Occurring in the Quenching and Tempering of Steels," *J. Iron Steel Inst.* 195(1960):26.
18. O. Johari and G. Thomas: "Factors Determining Twinning in Martensite," *Acta Met.* 13(1965):1211.
19. R. J. Wasilewski: "On the Nature of the Martensitic Transformation," *Met. Trans.* 6A(1975):1405.
20. G. B. Olson and Morris Cohen: "A General Mechanism of Martensitic Nucleation: Part II FCC BCC and other Martensitic Transformations," *Met. Trans.* 7A(1976):1905.
21. G. Krauss and A. R. Marder: "Morphology of Martensite in Iron Alloys," *Met. Trans.* 2(1971):2343.
22. G. R. Speich and H. Warlimont: "Yield Strength and Transformation Substructure of Low-Carbon Martensite," *J. Iron Steel Inst.* 206(1968):385.

CHAPTER TWO

IRON-CARBON ALLOYS II

2-1 TRANSFORMATION OF AUSTENITE TO BAINITE

If a plain-carbon eutectoid steel is quenched from the austenitic region to some intermediate temperature between 250 and 550°C and is isothermally trans-formed, a structure called *bainite* is formed (Fig. 2-1). Bainite is named after E. C. Bain,[1] who was one of the early investigators to explore this type of transformation. The bainitic reaction in Fe-C alloys is not completely under-stood and is a subject of controversy (Ref. 10). Moreover, it is difficult to study in plain carbon-steels since it is so complex and occurs so rapidly.

Bainite can be defined in terms of its microstructure as the product of a *nonlamellar* eutectoid reaction, in contrast to pearlite which is the product of a *lamellar* eutectoid reaction. In a eutectoid plain-carbon steel, bainite, like pearlite, is a mixture of two phases, ferrite and cementite (Fe_3C).

The austenite-to-bainite reaction has a dual nature. It has some characteris-tics of a nucleation and growth transformation similar to the austenite-to-pearlite transformation, while in other ways it shows characteristics of the austenite-to-martensite transformation. The bainite formed about 350 to 550°C is termed *upper* bainite and that which is formed between about 250 to 350°C is called *lower* bainite.

---

[1] E. S. Davenport and E. C. Bain, *Trans. AIME* 90(1930):117.

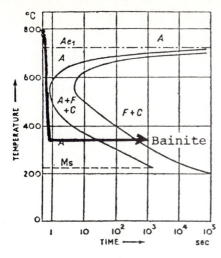

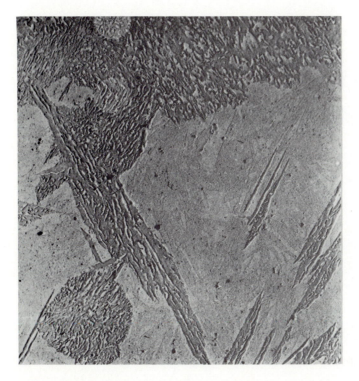

Figure 2-1 Isothermal transformation diagram for an eutectoid steel indicating the cooling path for the formation of bainite.

Figure 2-2 Microstructure of upper bainite formed in a 0.8% C eutectoid plain-carbon steel isothermally transformed at 445°C. (Electron replica micrograph; X10,000.) *(Courtesy R. M. Fisher, United States Steel Co., Research Laboratories.)*

Upper Bainite

Upper bainite is formed at the intermediate temperature range of the isothermal transformation diagram (350 to 550°C) of an eutectoid plain-carbon steel. In plain-carbon steels near eutectoid composition, bainite consists of a ferrite-cementite two-phase structure, as shown in Fig. 2-2. However, the cementite is in the form of rods rather than, as in the case of pearlite, having the form of lamellae.

Experimental work on upper bainite by Shackleton and Kelly (Ref. 11) led them to conclude that the cementite and the ferrite nucleate independently in the austenite. They found the rate-controlling step in the formation of upper bainite to be the diffusion of carbon in the austenite. Thus the cementite nucleated in the austenite will grow and deplete the surrounding region of carbon so that the austenite then can transform to ferrite. If ferrite is nucleated first, it will reject carbon ahead of the advancing ferrite-austenite interface, and allow cementite to be formed in the austenite in front of the growing bainite-ferrite interface. In this way, the nucleation of the ferrite will take place immediately adjacent to the cementite (or vice versa).

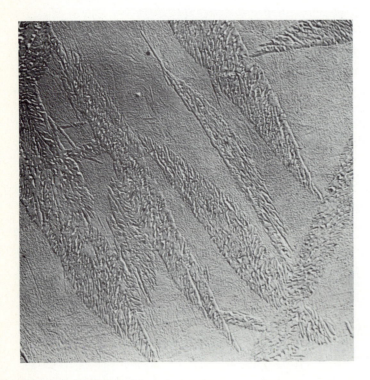

Figure 2-3 Microstructure of lower bainite formed in a 0.8% C eutectoid plain-carbon steel isothermally transformed at 315°C. (Electron replica micrograph; X10,000.) *(Courtesy of R. M. Fisher, United States Steel Co., Research Laboratories.)*

Lower Bainite

Lower bainite in a plain-carbon steel forms at temperatures below about 350°C, and its appearance is quite different from upper bainite (Fig. 2-3). Since diffusion rates are low at temperatures between 250 to 350°C, the iron carbide in lower bainite is precipitated internally in the ferrite plates. In contrast to martensite, which precipitates carbides in two or more orientations, lower bainite precipitates the carbides predominantly in one orientation, which is about 55° to the longitudinal axis of the ferrite. Also, in contrast to high-carbon martensites, lower bainite does not show the twinning characteristics.

The mechanism operating in the formation of lower bainite is believed to be identical to that produced by the formation and tempering of martensite (Ref. 11). That is, supersaturated ferrite is formed from the austenite by a shear process and cementite subsequently precipitates inside the ferrite.

Surface Relief Effects

The nucleation and growth of bainite plates in a 0.66% C–3.3% Cr alloy steel using a hot stage microscope at 350°C has been studied by Speich (see Ref. 12). The formation of bainite in this alloy after various time intervals is shown in the micrographs of Fig. 2-4. Light and electron micrographs at higher magnifications show the morphology of the bainite formed (Fig. 2-5). The progressive development of the bainitic plates is indicative of a nucleation and growth-type mechanism similar to the austenite-to-pearlite reaction. However in this case, bainitic plates are formed. The bainitic plates are random and acicular (needle-like), and in some ways resemble the martensitic plates produced in high-carbon plain-carbon steels. Surface relief effects caused by the surface tilting of the bainitic plates are produced during this bainitic reaction. This surface relief is similar to that produced during the Fe-C martensitic transformation and suggests a shear mechanism (see Fig. 1-35). However, the exact mechanism of the bainitic reaction is still not clear and will require further research.

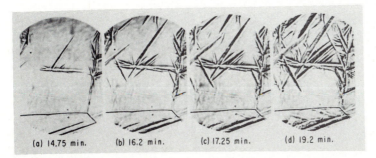

(a) 14.75 min.   (b) 16.2 min.   (c) 17.25 min.   (d) 19.2 min.

Figure 2-4 Hot-stage micrographs of the formation of bainite in a 0.66% C–3.3% Cr steel at 350°C after (*a*) 14.8 min, (*b*) 16.2 min, (*c*) 17.2 min, and (*d*) 19.2 min. The surface contrast is due to surface-relief effects. (X350.) *(Courtesy of G. Speich, United States Steel Co., Research Laboratories.)*

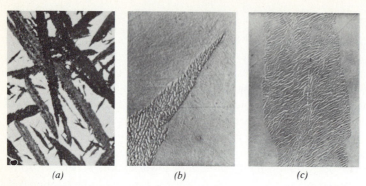

(a)   *(b)*   *(c)*

Figure 2-5 Light and electron micrographs of bainite formed in a 0.66% C–3.3% Cr steel at 350°C. (*a*) Light micrograph at X700 and (*b*) and (*c*) electron micrographs at X16,000. *(Courtesy of G. Speich, United States Steel Co, Research Laboratories.)*

2-2 ISOTHERMAL TRANSFORMATION OF NONEUTECTOID PLAIN-CARBON STEELS

Isothermal transformation diagrams have been determined for noneutectoid plain-carbon steels. The IT diagram for a 0.47% C hypoeutectoid plain-carbon steel is shown in Fig. 2-6. Several differences between this IT diagram and the diagram for the eutectoid plain-carbon steel (Fig. 1-15) are apparent. One major difference is that the diagram has been shifted to the left for the hypoeutectoid steel, so that it is not possible to quench a steel from the austenitic region and produce an entirely martensitic structure.

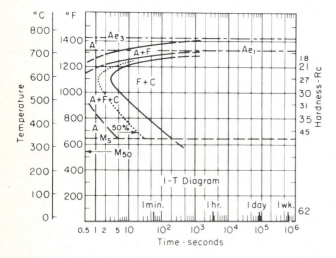

Figure 2-6 Isothermal transformation diagram for a hypoeutectoid steel containing 0.47% C and 0.57% Mn (austenitizing temperature: 843°C). [*After R. A. Grange, V. E. Lambert, and J. J. Harrington, Trans. ASM 51(1959):377.*]

A second major difference is that another transformation line is added to the upper part of the eutectoid steel IT diagram for the start of the formation of proeutectoid ferrite. At temperatures between 723 and about 765°C, only proeutectoid ferrite can be produced by isothermal transformation.

If a sample of the 0.47% C steel is quenched to 690°C from an austenitic temperature of 843°C and isothermally transformed, an almost equilibrium structure consisting of proeutectoid ferrite and coarse pearlite (some of it spheroidized) is produced (Fig. 2-7a). If another sample of this steel is quenched to the lower temperature of 650°C and isothermally transformed, the amount of proeutectoid ferrite is suppressed and the amount of pearlite is increased (Fig. 2-7b). In order for this change to be possible, the amount of ferrite in the pearlite is correspondingly increased as the amount of proeutectoid ferrite is decreased. This occurs since the reaction is a nonequilibrium irreversible transformation. Quenching to a lower temperature of 538°C and partially isothermally transforming the steel produces nodules of pearlite with some upper bainite (Fig. 2-7c). Quenching to 425°C and partially isothermally transforming at that temperature produces essentially a lower-bainitic structure (Fig. 2-7d). Rapid

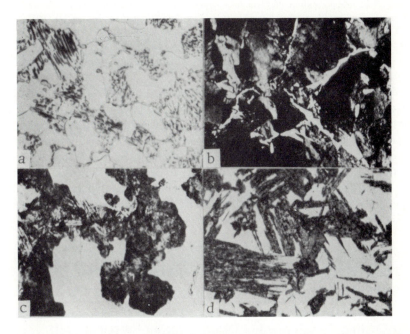

Figure 2-7 Microstructure of 0.47% C hypoeutectoid steel after isothermal transformations as follows: (a) after complete transformation at 690°C [structure shows proeutectoid ferrite (white) and coarse pearlite]; (b) after complete transformation at 650°C [structure shows proeutectoid ferrite (white) and pearlite (black)], (c) after partial transformation at 538°C [structure shows pearlite nodules (black) with some upper bainite needles and martensite (white)], (d) after partial transformation at 425°C [structure shows lower bainite (black) and martensite (white)]. [*After R. A. Grange, V. E. Lambert, and J. J. Harrington, Trans. ASM 51(1959):377.*]

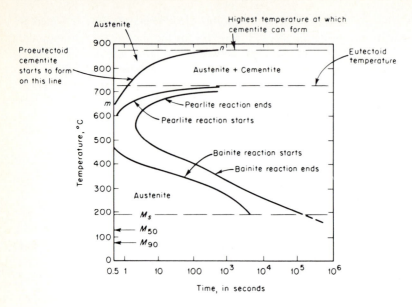

Figure 2-8 Isothermal transformation diagram of a hypereutectoid steel containing 1.13% carbon and 0.30% manganese. *(From "Isothermal Transformation Diagrams," United States Steel Co., 1951, as presented in R. E. Reed-Hill, "Physical Metallurgy Principles," 2nd ed., ©1973 by Litton Educational Publishing, Inc. Reprinted by permission of D. Van Nostrand Co.)*

quenching directly to martensite will produce a martensitic structure with small amounts of ferrite and pearlite.

Isothermal transformation diagrams have also been made for hypereutectoid plain-carbon steels. They are similar to the hypereutectoid steel diagrams except that the proeutectoid phase is cementite rather than proeutectoid ferrite. An isothermal transformation diagram for a 1.13% C hypereutectoid steel is shown in Fig. 2-8. As in the case of the hypoeutectoid steels, quenching hypereutectoid plain-carbon steels and isothermally transforming them will produce structures with mixed constituents.

2-3 CONTINUOUS-COOLING TRANSFORMATIONS IN PLAIN-CARBON STEELS

Continuous-Cooling Transformation Diagram for an Eutectoid Steel

In industrial heat-treating operations in most cases, a steel is not isothermally transformed at some temperature above the martensitic start temperature but is continuously cooled from the austenitic temperature to room temperature. Thus in continuously cooling a plain-carbon eutectoid steel, the transformation from austenite to pearlite occurs over a range of temperatures rather than a single isothermal temperature. As a result, the final microstructure after continuous cooling will be complex since it is formed over a range of temperatures which changes the reaction kinetics.

The isothermal transformation diagram cannot therefore be used directly to predict what products will be formed by continuous cooling. Experimentally it has been found that, for an eutectoid steel, the continuous-cooling diagram has been shifted to slightly lower temperatures and longer times in relation to the isothermal diagram. This displacement is illustrated by the continuous-cooling diagram for an eutectoid plain-carbon steel shown in Fig. 2-9.

Different rates of cooling from the austenitizing temperature region are represented on the continuous-cooling diagram of an eutectoid plain-carbon steel shown in Fig. 2-10. Consider the continuous cooling of thin samples at the rates A to E shown in Fig. 2-10. Cooling curve A represents the slow cooling of a steel such as would be obtained by shutting the power off of a furnace and allowing the steel to cool as the furnace cools. The microstructure after cooling to room temperature would be lamellar coarse pearlite. Cooling curve B represents more rapid cooling such as would be obtained by removing the steel from the furnace at the austenitizing temperature and allowing it to cool in still air. A fine pearlite structure would be obtained in this case.

Cooling along curve C starts with the formation of a fine pearlite, but there is insufficient time to complete the austenite-to-pearlite transformation. Thus the

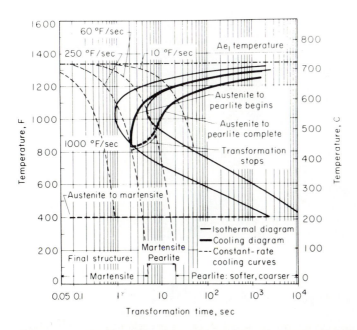

Figure 2-9 Continuous cooling diagram for a plain-carbon eutectoid steel. *(After R. A. Grange and J. M. Kiefer, Trans. ASM 29(1941):85, as adapted in E. C. Bain and H. W. Paxton, "Alloying Elements in Steel," 2nd ed., American Society for Metals, 1966, p. 254.)*

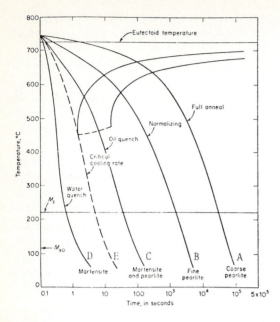

Figure 2-10 Variation in the microstructure of an eutectoid plain-carbon steel by continuously cooling at different rates. *(From R. E. Reed-Hill, "Physical Metallurgy Principles," 2nd ed., © 1973 by Litton Educational Publishing, Inc. Reprinted by permission of D. Van Nostrand Co.)*

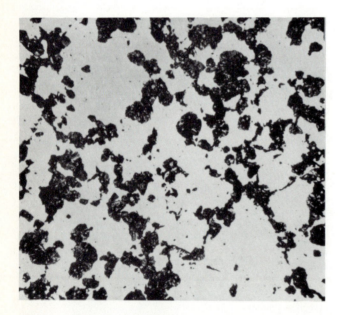

Figure 2-11 Microstructure obtained by continuously cooling an eutectoid steel at a rate to cause a split transformation. (Black is pearlite and white is martensite.) *(From D. S. Clark and W. R. Varney, "Physical Metallurgy for Engineers," 2nd ed., © 1962 by Litton Educational Publishing, Inc. Reprinted by permission of D. Van Nostrand Company.)*

remaining austenite that does not transform to pearlite at the upper temperatures transforms to martensite at lower temperatures beginning at about 200°C. Since the transformation in this case takes place in two stages, it is called a *split transformation*. The structure of the steel is thus a mixture of fine pearlite and martensite, as shown in Fig. 2-11. A rapid rate of cooling, such as represented by curve D, produces an entirely martensitic structure. A slightly slower cooling rate, as indicated by curve E, represents the slowest rate of cooling without obtaining pearlite in the structure. This cooling rate is called *critical cooling rate*. Cooling at a rate slower than curve E will not produce a fully hardened (martensitic) steel.

Another important point to be noted on the continuous-cooling-transformation diagram of an eutectoid plain-carbon steel is that the pearlitic transformation lines extend over and above the bainite start-to-finish transformation lines. Thus a martensitic or pearlitic structure will form, but not a bainitic one. Small amounts of bainite can be formed, however, by cooling at a rate which produces a split transformation. In order to get an all-bainitic structure, the eutectoid steel would first have to be rapidly cooled to some temperature above the M_s and then be isothermally transformed to bainite.

Continuous-Cooling Transformation for a Hypoeutectoid Plain-Carbon Steel

The continuous-cooling diagram for a 0.38% C hypoeutectoid plain-carbon steel is shown in Figs. 2-12*a* and *b*. In Fig. 2-12*b*, the hardness of the transformation products at various cooling rates is shown. The microstructures of this alloy, after cooling at each of the rates where the hardness is indicated, are shown in Fig. 2-13.

Cooling at the slowest rate indicated by the curve with the DPH[1] value of 139 produces the softest structure, which is a mixture of proeutectoid ferrite and pearlite in almost equal amounts (Fig. 2-13*a*). This structure is similar to that obtained by slow (furnace) cooling this type of steel. Increasing the cooling rate slightly produces finer pearlite and slightly less ferrite, with the hardness increasing (Fig. 2-13*b*). Increasing the cooling rate still further drastically reduces the amount of proeutectoid ferrite. The proeutectoid ferrite now outlines the former austenitic grain boundaries, with some Widmanstätten[2] ferrite being formed (Fig. 2-13*c*).

Increasing the cooling rate still further causes a split transformation to occur (Figs. 2-12*b* and 2-13*d*). The rate of cooling is so fast that very little proeutectoid ferrite is formed. Instead, pearlite (the dark phase) outlines the former

---

[1]   DPH = diamond pyramid hardness.

[2]   Widmanstätten structure—a structure characterized by a geometrical pattern resulting from the formation of a new phase along certain crystallographic planes of the parent solid. The orientation of the lattice in the new phase is related crystallographically to the orientation of the lattice in the parent phase. In steels exhibiting this structure, ferrite delineates the octahedral planes of austenite.

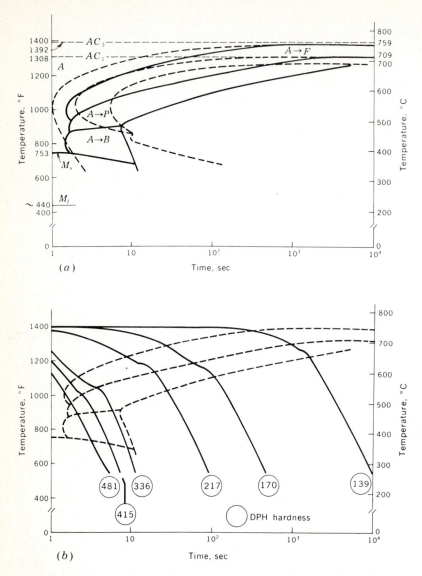

Figure 2-12 (*a*) Continuous cooling diagram for a 0.3% plain-carbon steel (0.70% Mn, 0.25% Si). The isothermal diagram for the steel is shown in dashed lines. (*b*) C-T diagram with selected cooling rates decreasing from left to right. DPH hardness values are indicated inside circles for each cooling curve. The microstructures for this steel cooled at the rates indicated are shown in Fig. 2-13 *a* to *f*. (*After C. Zurlipe and D. Grozier, Metal Progress, December, 1967, as presented by P. Shewmon, "Transformations in Metals," McGraw-Hill, New York, 1969.*)

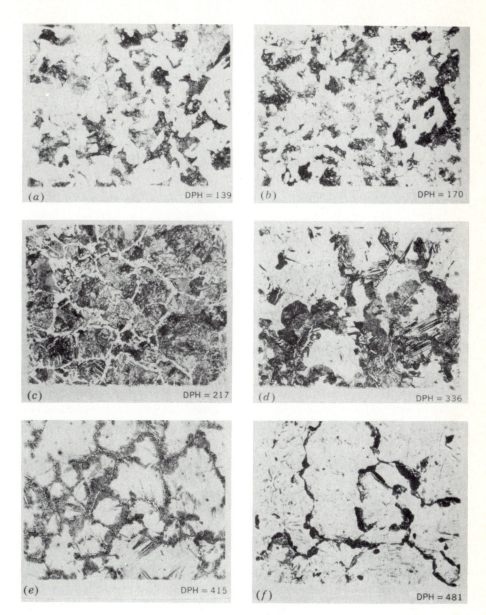

Figure 2-13 Microstructures of 0.38% C steel samples after continuously cooling at rates indicated in Fig. 2-12. The cooling rate is increased from *a* to *f* and each sample is identified by the DPH number. See text for explanation of structures. *(After C. Zurlipe and D. Grozier as presented in P. Shewmon, "Transformations in Metals," McGraw-Hill, New York, 1969, p. 250.)*

austenitic grain boundaries. Some bainite is formed, as indicated by the acicular structure. The white areas are martensite, which was formed when some of the austenite remained untransformed until the M_s temperature was reached.

Increasing the cooling rate even more increases the amount of martensite formed, and still gives a split transformation (Fig. 2-13e). Some proeutectoid ferrite and pearlite are formed at the former austenitic grain boundaries. Small amounts of acicular bainite are also observed. Note that the hardness of the sample has increased markedly due to the large percentage of martensite. Finally, in the last microstructure (Fig. 2-13f), the structure is almost completely martensitic. However, a small amount of pearlite and bainite are observed at the former austenitic grain boundaries.

2-4 ANNEALING AND NORMALIZING PLAIN-CARBON STEELS

Cold Working and Annealing

Most useful engineering alloys must possess an appropriate combination of strength and ductility. Ductility in metals and alloys allows them to be deformed plastically by various fabrication processes into the desired shape without fracturing. During plastic deformation or cold working, the main reason for the increase in strength is due to the increased generation and rearrangement of dislocations.

In order to make cold-worked metals ductile, they are annealed at appropriate temperatures. During annealing, highly distorted cold-worked structure is partly or completely returned to a softer more ductile structure containing fewer dislocations. The two most common types of annealing processes that are applied to commercial steel are *full annealing* and *process annealing*.

Full annealing In full annealing, hypoeutectoid and eutectoid steels are heated about 25°C above their A_{c_3} (upper critical) temperature, held for the necessary time for annealing, and then cooled slowly to room temperature. The most common type of full annealing utilizes a "box furnace" in which large coils of steel sheet are heated and cooled slowly. Figure 2-14 indicates the temperature range in the Fe-Fe$_3$C diagram commonly used for full annealing. In some newer steel mills, some kinds of steel strip are rapidly heated and annealed continuously. This continuous-annealing process for steel strip is more economical for some grades of steel, but it produces a finer-grain size and different mechanical properties than "box-annealed" steel. This difference in microstructure will be discussed later.

Process annealing Process annealing is usually applied to hypoeutectoid steels with up to about 0.3% C. The steel is heated to a temperature below the lower critical temperature, usually about 550 to 650°C, held for the necessary time,

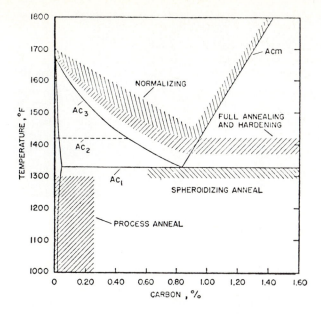

Figure 2-14 Commonly used temperature ranges for annealing plain-carbon steels. *(After T. G. Digges et al., "Heat Treatment and Properties of Iron and Steel," NBS Monograph 88, 1966, p. 10.)*

and then cooled at the desired rate (Fig. 2-14). Process annealing is frequently referred to as a "stress-relief" or "recovery" treatment since it partially softens cold-worked low-carbon steels by relieving internal stresses from cold working.

Microstructural Changes that Occur during Annealing

During annealing the changes in microstructure that occur can be subdivided into the following major processes:

1. *Recovery.* In this process, the cold-worked metal is heated to a temperature so that dislocations can be rearranged into lower-energy configurations.
2. *Recrystallization.* When a cold-worked metal is heated to a high enough temperature, termed the *recrystallization temperature*, new strain-free grains are formed by the migration of large-angle boundaries of high mobility.
3. *Grain growth.* Continued annealing of a recrystallized structure promotes the formation of a more stable grain structure. In this process, larger grains grow at the expense of the smaller ones.

   Since recovery and recrystallization processes are particularly important in the processing of flat-rolled sheet-mill products such as low-carbon sheet and strip, the treatment of this subject will be restricted to these low-carbon Fe-C alloys.

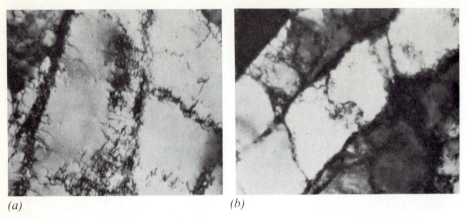

(a)   *(b)*

Figure 2-15 Dense tangles of dislocations formed in annealed iron by deformation at room temperature. (*a*) Strained 9% and (*b*) strained 20%. Note that the average spacing between the cell walls is decreased as the strain was increased. (Electron micrographs X20,000.) *(After J. T. Michalak, Metals Handbook, 8th ed., vol. 8, American Society for Metals, 1973, p. 219.)*

The Cold-Worked State

Cold working increases the strength of iron and steel by increasing the dislocation density and by rearranging the dislocations. When an annealed-iron or low-carbon steel is plastically deformed about 5 percent, dislocation tangles begin to form cell walls of deformation cells (Fig. 2-15*a*). As the plastic deformation is continued, the dislocation density increases, leading to an increased thickness of the cell walls and a decrease in their volume (Fig. 2-15*b*).

After about 65 percent deformation, a dislocation density of about 10^{10} to 10^{11} dislocations per square centimeter is reached, with a cell diameter of several microns. This high dislocation density is associated with a highly cold-worked and strained metal. Figure 2-16*a* shows how a preferred orientation is created in the rolling direction of a low-carbon steel cold rolled 65 percent. Figure 2-16*b* shows the dislocation substructure of the same steel at high magnification in a thin foil.

Recovery

During recovery or stress-relief heat treatment, the change in mechanical and physical properties introduced by cold working start to return to the values of the metal before cold working. In the case of iron and low-carbon steel, during recovery, internal stresses are relieved and the electrical resistivity is decreased due to the elimination of some crystal imperfections. However, changes in the mechanical properties are slight.

As the recovery process proceeds, dislocations climb and rearrange themselves in a more orderly manner, as shown in Figs. 2-17*a* and *b*. The climb

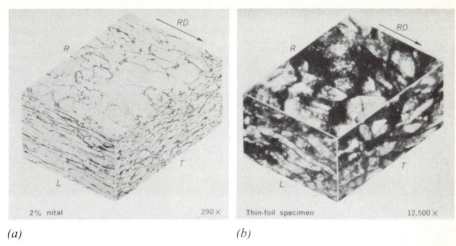

(a) (b)

Figure 2-16 (*a*) Low-carbon steel cold-rolled 65%, showing the grain boundaries in the rolling plane *R*, the longitudinal plane *L*, and the transverse plane *T*. (RD = rolling direction). (*b*) Thin foil electron micrograph of the same cold-rolled low-carbon steel as in (*a*) showing the dislocation substructure in the rolling, longitudinal and transverse planes. *(After J. T. Michalak, Metals Handbook, 8th ed., vol. 8, American Society for Metals, 1973, p. 220.)*

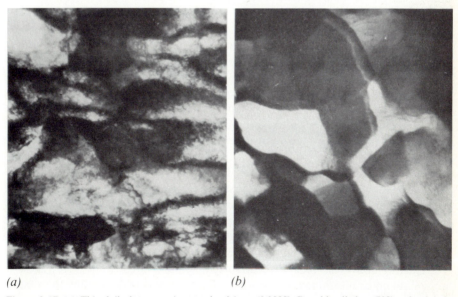

(a) (b)

Figure 2-17 (*a*) Thin foil electron micrograph of iron (0.002% C) cold-rolled to 70% reduction in thickness. Structure shows deformation cell walls of high-dislocation density. In the dark regions, many of the dislocations are not resolved due to their high density. (X12,500.) (*b*) Same cold-worked iron after recovery heat treatment. The structure consists of dislocation arrays and subgains which are typical of heavily deformed and recovered ferrite. (X20,000.) *(After W. Jolley and D. A. Witmer, Metals Handbook, 8th ed., vol. 8, American Society for Metals, 1973, p. 225.)*

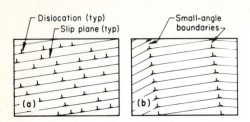

Figure 2-18 Schematic representation of the polygonization process. (*a*) Deformed crystal showing dislocations piled up on slip planes. (*b*) Recovered structure showing the dislocations having formed low-angle subgrain boundaries, producing a more stable arrangement. (*After L. E. Tanner and I. S. Servi, Metals Handbook, 8th ed., vol. 8, American Society for Metals, 1973, p. 222.*)

mechanism enables dislocations to form walls of new cells called *subgrains* (Fig. 2-17*b*). The formation of low-angle grain boundaries to form subgrains is called *polygonization*, and proceeds as shown schematically in Fig. 2-18. The formation of subgrains is a spontaneous process since the subgrain structure has a lower energy dislocation configuration than the original deformation cell structure (Fig. 2-17*a*). For each recovery temperature, an equilibrium cell size eventually is attained which corresponds to the fully recovered condition for that temperature.

Recrystallization and Grain Growth

Recrystallization, as noted earlier, is a process in which new strain-free grains are formed in the cold-worked metal. During this process, the mechanical and physical property changes caused by the cold work are returned approximately to their level before cold working.

If recovery has preceded recrystallization, the new grains can be nucleated from the subgrains developed in the recovery process. Recrystallization is essentially a nucleation and growth-type process. The driving force for the process is the decrease in volume free energy that results from the decrease in dislocation density. The new recrystallized grains are much larger than the subgrains and, even though energy must be provided for the new high-angle grain boundaries, more than enough energy is released due to the loss of the subgrain boundaries.

Figure 2-19 shows a low-carbon (0.06% C) steel that was partially recrystallized after cold working, and Fig. 2-20 shows the recrystallized grain structure of a capped steel (0.06% C) after box annealing. It should be noted that the new recrystallized grain structure contains approximately equiaxed grains, whereas in the 65 percent cold-worked structure of Fig. 2-16*a* the grains are preferentially aligned in the rolling direction and are elongated.

Heating above the recrystallization temperature range causes the grains to grow until an equilibrium size is reached. Larger grains will consume the smaller ones since the large grains are more thermodynamically stable. Grain growth occurs spontaneously because of the decrease in free energy due to the reduction in grain boundary area. Some further softening of the metal will occur with a larger grain size since the grain boundaries impede dislocation movement during plastic deformation. Grain growth will stop when grain boundary movement is prevented by impurities and substructure.

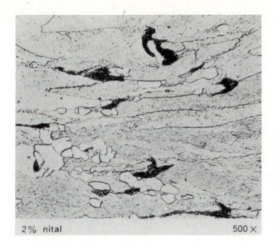

2% nital   500×

Figure 2-19 Low-carbon (0.06% C) steel sheet that was partially recrystallized after rolling, showing new grains (light) that were nucleated at carbide particles. (2% nital; X500.) *(After W. Jolley and D. A. Witmer, Metals Handbook, 8th ed., vol. 8, American Society for Metals, 1973, p. 227.)*

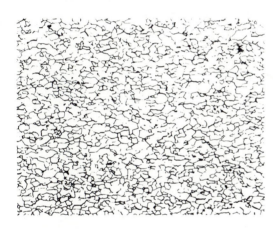

Figure 2-20 Low-carbon capped steel (0.06% C, 0.30% Mn), cold-rolled and then recrystallized by box annealing. (2% nital; X100.) *(After W. Jolley and D. A. Witmer, Metals Handbook, 8th ed., vol. 8, American Society for Metals, 1973, p. 228.)*

Normalizing

Normalizing is the process whereby a steel is heated about 40°C above the A_{c_3} or A_{cm}, held for the desired time, and then cooled in still air. Figure 2-14 indicates the commonly used temperature ranges for normalizing plain-carbon steels. The main purposes of normalizing are

1. To refine the grain structure or to insure a homogeneous austenite when a steel is reheated for quench hardening or full annealing
2. To reduce segregation in castings or forgings and thus provide a more uniform structure
3. To harden the steel slightly

The structures produced by normalizing are pearlite for eutectoid steels and pearlite with excess ferrite or cementite for hypo- and hypereutectoid steels, respectively. Because of the high diffusion rates at the higher temperatures of normalizing as compared to full annealing, segregation in the cast structures can be greatly reduced. The increase in grain size due to grain growth during normalizing can be reduced by a second heat treatment at a lower temperature.

2-5 QUENCH HARDENING PLAIN-CARBON STEELS

In order to obtain maximum hardening in a plain-carbon steel, it must be quenched from the austenitizing temperature (Fig. 2-14) at such a rate that a martensitic structure is produced throughout the entire sample (Fig. 2-10). To obtain a fully martensitic structure, the steel must be quenched at a rate equal to or greater than the critical cooling rate to achieve maximum hardening.

Since the critical cooling rate for some plain-carbon steels is so fast (e.g., an eutectoid 0.8% C plain-carbon steel), only thin-steel shapes can be made fully martensitic. If thick sections are rapidly cooled, the surface will cool more rapidly than the center, as is indicated in Fig. 2-21a for a 1-in cylindrical bar. Thus, using conventional quenchants, fully martensitic structures cannot be

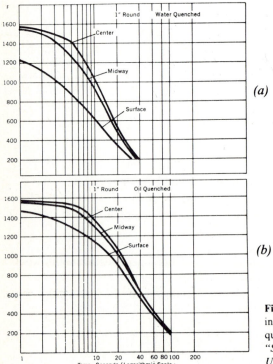

(a)

(b)

Figure 2-21 Cooling rates during quenching 1-in round steel bars. (*a*) Water quenching, (*b*) oil quenching. (*From "Suiting the Heat Treatment to the Job," United States Steel Co., 1968, p. 25.*)

obtained in plain-carbon steels with thick sections no matter how fast the cooling.

In order to obtain fully martensitic structures in thick sections of steels, alloying elements such as nickel, chromium, and molybdenum are added to plain-carbon steels thus increasing the time during quenching before the critical cooling rate is reached. This subject will be treated in Chap. 4, Alloy Steels.

The quenching media most commonly used for steels are water, water brine, and oils. In general brine (water plus various percentages of sodium chloride or calcium chloride) provides a faster cooling rate than water. Oil, on the other hand, cools at a slower rate than water (Fig. 2-21b). In general, agitation of the quench media increases the cooling rate.

Residual stresses When a steel bar of, say, 1-in thickness is cooled rapidly, the center of the sample remains at higher temperatures during the cooling than the surface. This temperature gradient by itself causes high stresses to be created inside the steel which can lead to cracking and distortion of the steel piece being heat-treated. Also, an expansion of the iron lattice occurs during the transformation from the FCC to BCC crystal structure. Thus two processes occur during the quenching to martensite in a piece of steel: (1) normal contraction due to cooling and (2) expansion due to the change from FCC to BCC iron. As a result, complex residual stresses are created in the steel which can lead to the formation of quench cracks and even rupture. As the carbon content of the steel is increased, the problem of cracking during quenching becomes more important since the BCC lattice is distorted more into a BCT type (Fig. 1-32).

In order to minimize the quench-cracking problem, the steel should be reheated to relieve the stresses as soon as possible. By the addition of alloying elements, slower (oil) quenching can be used to reduce quench stresses and distortion. Another possibility to reduce quench cracking and distortion is to use special heat treatments such as martempering and austempering which utilize an intermediate isothermal transformation during quenching (see Sec. 2-8).

2-6 TEMPERING OF PLAIN CARBON STEELS

The Tempering Process

Tempering is the process of heating a martensitic steel to a temperature below the transformation range to make it softer and more ductile. Figure 2-22 schematically illustrates the quenching and tempering process. As shown in this diagram, the steel is first austenitized above the A_{e_1}, then quenched at a rate fast enough to miss the nose of the IT diagram to form martensite. The steel is then reheated at an elevated temperature below the A_{e_1} to produce the desired tempered hardness.

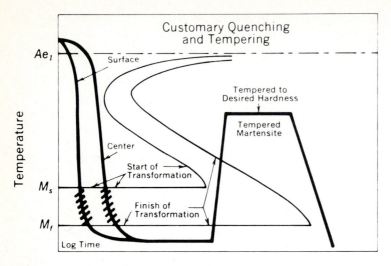

Figure 2-22 Schematic diagram to illustrate the customary quenching and tempering process for a plain-carbon steel. *(From "Suiting the Heat Treatment to the Job," United States Steel Co., 1968, p. 34.)*

Microstructural Changes in Plain-Carbon Steels which Occur during Tempering

During the tempering process, a number of solid-state reactions occur. The most important of these are:

1. Segregation of carbon atoms
2. Precipitation of carbides
3. Decomposition of retained austenite
4. Recovery and recrystallization of the ferrite matrix

These reactions do not all take place at the same temperature and over the same time period. Many of them overlap and occur simultaneously, so that the resultant microstructures are very complex.

Carbon segregation In the lath martensite of low-carbon steels, there is a high density of individual dislocations and many cell walls. The interstitial lattice sites near the dislocations provide lower-energy sites for carbon atoms than regular interstitial lattice positions. Therefore, when low-carbon martensitic steels are first tempered at 25 to 100°C, the carbon atoms will redistribute themselves to these lower-energy sites. Actually, much redistribution of the carbon atoms takes place during the quenching through the temperature range where martensite forms. For carbon contents less than 0.2 percent, Speich (Ref. 14), using electrical resistivity measurements, calculated that nearly 90 percent of the carbon segregated to lattice defects (mostly dislocations) during quenching. It is possible that the absence of tetragonality in the BCC lattice in martensitic plain-carbon steels with less than 0.2% C can be attributed to this type of segregation.

In high-carbon plain-carbon steels, the martensite formed is mainly plate type (see Sec. 1-7) and has an internally twinned structure. The main mode of carbon redistribution for these steels is by preprecipitation clustering. The driving force for this reaction is the lowering of the total elastic energy of the lattice. The number of low energy dislocation sites is much less in the high-carbon steels; therefore carbon segregation by this mechanism is substantially reduced.

Carbide precipitation In plain-carbon steels, three types of carbides which differ in chemical composition and crystal structure have been identified. These are ϵ carbide [$Fe_{2.3}C$, hexagonal close-packed (HCP)], Hägg carbide (Fe_5C_2, monoclinic), and cementite (Fe_3C, orthorhombic).

ϵ *carbide* ($Fe_{2.3}C$, HCP) When plain-carbon steels containing more than about 0.2% C are tempered in the 100 to 200°C range, ϵ carbide precipitates. In low-carbon steels with less than 0.2% C, no ϵ carbide precipitates. Presumably, at low-carbon concentrations, the carbon atoms can lower their energies more at dislocation sites than by precipitating as ϵ carbide. ϵ carbide is metastable, and at higher temperatures dissolves when Hägg carbide and cementite are formed. Precipitates of ϵ carbide formed in martensite in an Fe–24% Ni–0.5% C alloy tempered at 205°C for 1 h are shown in Fig. 2-23.

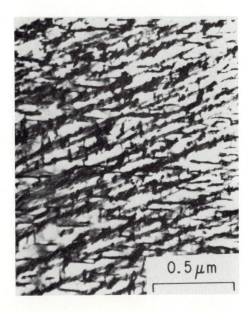

Figure 2-23 Precipitate of epsilon carbide (dark phase) in martensite in Fe-24% Ni-0.5%C steel during tempering for 1 h at 205°C. (Electron micrograph.) [*After M. G. Wells, Acta Met. 12(1964):389.*]

Figure 2-24 Precipitation of Fe_3C in Fe-0.39% C martensite tempered 1 h at 300°C. (Electron micrograph.) [*After G. R. Speich and W. C. Leslie, Met. Trans. 3(1972):1043.*]

Hägg carbide (Fe_5C_2, ***monoclinic***) This carbide has been identified by Mössbauer studies[1] (γ-ray absorption) and is formed in some high-carbon steels tempered at 200 to 300°C. It is a metastable carbide with a composition intermediate between ϵ carbide and cementite. There is some doubt as to whether Hägg carbide is part of the tempering sequence in low-carbon steels.

Cementite (Fe_3C, ***orthorhombic***) This carbide forms when plain-carbon steels are tempered between 250 to 700°C. The initial shape of cementite is needlelike (Fig. 2-24), either when formed by tempering between 200 to 300°C or when formed during the quenching of large steel sections. It is nucleated at martensite lath boundaries at low temperatures and at ferrite grain boundaries at higher temperatures. From 400 to 600°C, the lath-like carbides coalesce to form spheroidite, which reduces the overall surface energy (Fig. 2-25). From 600 to 700°C, spheroidite coarsens even more with the smaller particles dissolving (Fig. 2-26). Again, the driving force for the coalescence is the reduction of the overall surface energy of the cementite in the ferrite matrix.

Decomposition of retained austenite Retained austenite is present only in plain-carbon steels with a carbon content in excess of about 0.4 percent. It is therefore important in medium- and high-carbon steels. The decomposition of austenite occurs at tempering temperatures between 200 to 300°C, with the austenite being transformed into bainite.

[1]   J. Genin and P. A. Flinn, *Trans. AIME 242*(1968):1419.

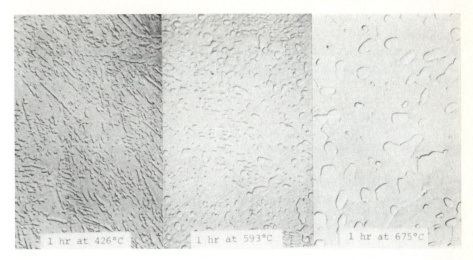

1 hr at 426°C   1 hr at 593°C   1 hr at 675°C

Figure 2-25 Structure of tempered martensite in a 0.75% C eutectoid steel. Note the progressive agglomeration of the cementite. (Nitrocellulose negative replicas, X15,000; electron micrographs, reduced 1/3.) [*After A. M. Turkalo and J. R. Low, Trans. AIME 212(1958):750.*]

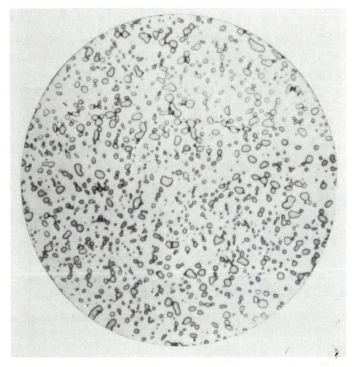

Figure 2-26 Spheroidite in a 1.1% C hypereutectoid steel. *(After J. Vilella, E. C. Bain and H. W. Paxton, "Alloying Elements in Steel," 2nd ed., American Society for Metals, 1966, p. 101.)*

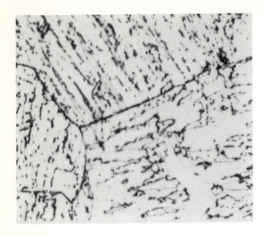

Figure 2-27 Recovered structure of Fe-0.18% C martensite after tempering 10 min at 600°C; optical micrograph. (2% nital; X1000.) [*After G. R. Speich and W. C. Leslie, Met. Trans. 3(1972):1043.*]

Recovery and recrystallization It is difficult to determine when the recovery of the defect structure of martensite begins during tempering, but it definitely affects the tempering process above 400°C (Ref. 15). During recovery, the cell boundaries and the random dislocations between them are annihilated, and a fine-grain structure is developed. The structure of a recovered Fe–0.18% C martensite after 10 min at 600°C is shown at low magnification in Fig. 2-27 and at high magnification in Fig. 2-28.

After a long time at 600°C, the recovered martensite recrystallizes and produces an equiaxed α-ferrite structure in which large particles of spheroidal Fe₃C are embedded. Figure 2-29 shows the partial recrystallized structure of an 0.18% C martensite after 10 min at 600°C. The structure obtained after tempering the 0.18% C steel 8 h at 700°C consists of coarse α ferrite with spheroidal cementite particles at the grain boundaries and within the grains (Fig. 2-30).

Figure 2-28 Recovered structure of Fe-0.18% C martensite after tempering 10 min at 600°C; electron transmission micrograph. (Thin-foil specimen; X30,000.) [*After G. R. Speich and W. C. Leslie, Met. Trans. 3(1972):1043.*]

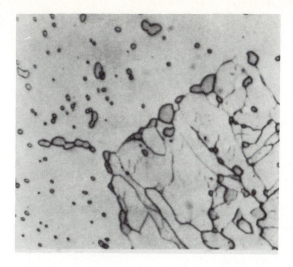

Figure 2-29 Partial recrystalliza-tion in Fe-0.18% C martensite tempered 96 h at 600°C. (2% nital; X1000.) [*After G. R. Speich and W. C. Leslie, Met. Trans. 3(1972): 1043.*]

Effect of Tempering on the Hardness of Plain-Carbon Steels

The effect of increasing tempering temperature (for 1 h) on decreasing the hardness of plain-carbon steels is shown in Figs. 2-31 and 2-32. For the low Fe-C martensites with 0.1% C, very little change in hardness occurs until the tempering temperature reaches about 200°C (Fig. 2-31). Above 200°C, the hardness gradually decreases as the tempering temperature is increased to 723°C. When the carbon content is increased to 0.4 percent, the hardness decreases steadily upon tempering at temperatures in the 150 to 723°C range.

For the high-carbon plain-carbon steels (Fig. 2-32), there is a slight increase in hardness in the 100 to 150°C tempering range. This hardness increase is

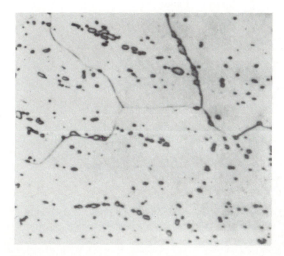

Figure 2-30 Completely recrystallized structure in Fe-0.18% martensite tem-pered 8 h at 700°C. (2% nital; X500.) [*After G. R. Speich and W. C. Leslie, Met. Trans. 3(1972):1043.*]

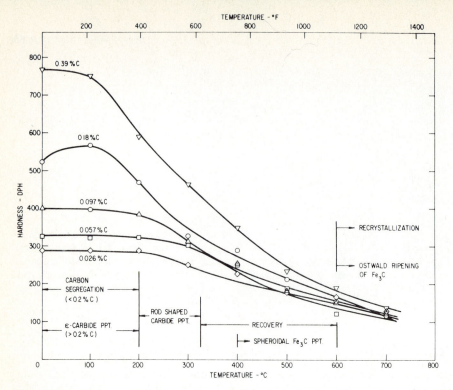

Figure 2-31 Hardness of iron-carbon martensites (0.026 to 0.39% C) tempered 1 h at 100 to 700°C. [*After G. R. Speich, Trans. AIME 245(1969):2559.*]

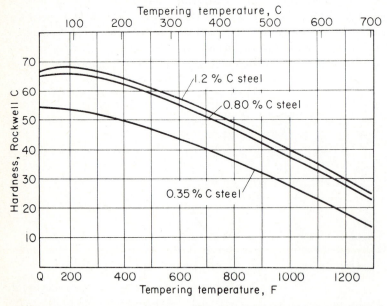

Figure 2-32 Hardness of iron-carbon martensites (0.35 to 1.2% C) tempered 1 h at indicated temperatures. *(After E. C. Bain and H. W. Paxton, "Alloying Elements in Steel," 2nd ed., American Society for Metals, 1966, p. 38.)*

attributed to the precipitation of the ϵ carbide. In this temperature range, two processes occur simultaneously. In one, softening occurs due to the loss of carbon from the martensite, and in the other hardening occurs due to the precipitation of the ϵ carbide. Above 150°C, the hardness decreases steadily with increasing tempering temperature from 150 to 723°C.

2-7 GRAIN-SIZE EFFECTS

Grain-Size Designation

In studying structural and property changes in steels and other metals, it is sometimes necessary to specify an "average grain size." The most common method of designating grain size in the United States is the ASTM[1] grain size number, N. The ASTM grain size number is related to the number of grains according to the following equation:

$$n = 2^{N-1}$$

where n is the number of grains per square inch at a magnification of $100 \times$.

Table 2-1 Grain-size number as related to grain count†

Timken- ASTM No.	Grains per square inch of image at $100 \times$			Grains per sq millimeter (mean actual)
	Maximum	Minimum	Mean	
−3	0.088	0.044	0.06	1
−2	0.176	0.088	0.125	2
−1	0.35	0.176	0.25	4
0	0.71	0.35	0.50	8
1	1.41	0.71	1.0	16
2	2.83	1.41	2.0	32
3	5.66	2.83	4.0	64
4	11.3	5.66	8.0	128
5	22.6	11.3	16	256
6	45.2	22.6	32	512
7	90.5	45.2	64	1024
8	181	90.5	128	2048
9	362	181	256	4096
10	724	362	512	8200
11	1448	724	1024	16400
12	2896	1448	2048	32800

Usual range (bracket spanning rows 1–8)

† After Ref. 9.

[1] American Society for Testing and Materials.

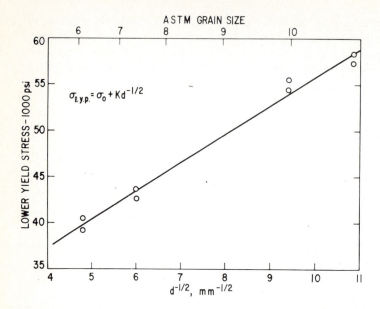

Figure 2-33 Effect of ferritic grain size on the strength of annealed mild steel. [*After W. B. Morrison, J. Iron Steel Inst. 201(1963):317, as adapted by R. A. Grange, ASM Trans. 59(1966):26.*]

The ASTM grain size numbers with the equivalent numbers of grains are listed in Table 2-1.

Effect of Grain Size on the Mechanical Properties of Low-Carbon Steels

The ferritic grain size has a large effect on the yield strength of low-carbon steels. Finer-grained low-carbon steels for the same carbon content and heat treatment have higher strengths than coarse-grained steels. Figure 2-33 shows how the yield strength of an annealed mild steel increased from 40 ksi at an ASTM grain size of 6 to 58 ksi for an ASTM grain size of 11.

The reason for this great increase in strength is that grain boundaries at lower temperatures act as barriers to dislocation movement. A quantitative relationship between yield stress and grain size has been proposed by Hall[1] and Petch[2] as follows:

$$\sigma_y = \sigma_i + kd^{-\frac{1}{2}}$$

where σ_y is the yield stress, σ_i is a lattice friction factor, k is a constant, and d is the grain diameter.

[1] E. O. Hall, *Proc. Phys. Soc.* B64(1951):747.
[2] N. J. Petch, *J. Iron Steel Inst.* 174(1953):25.

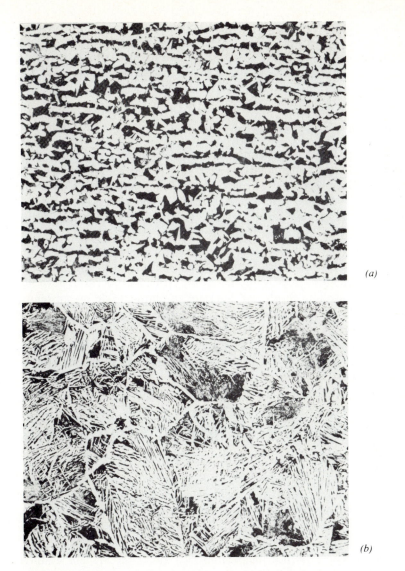

(a)

(b)

Figure 2-34 Effect of austenitic grain size on the proeutectoid ferrite distribution in a hypoeutectoid steel containing 0.23% C and 1.2% Mn after air cooling from (*a*) 900°C (small austenitic grain size) and (*b*) 1150°C (large austenitic grain size). *(After R. Yoe, U. S. Steel Co., as presented in "Transformations in Metals," McGraw-Hill, New York, 1969, p. 223.)*

When a metal or alloy is under stress, dislocations are not forced through grain boundaries, but pile up or concentrate at the boundaries. New sources of slip are created by the applied stress in neighboring grains. Consequently, a high density of grain boundaries will produce a higher yield stress in the metal or alloy.

Effect of Austenitic Grain Size on the Proeutectoid Ferrite Morphology

If a hypoeutectoid steel with a fine-grain size is fairly rapidly cooled (air-cooled) from the austenitic temperature range, proeutectoid ferrite will nucleate at the austenitic grain boundaries and reject carbon by diffusion into the centers of the grains until the transformation temperature is reached, whereupon the pearlite is produced from the remaining austenite. The resultant microstructure in an air-cooled 0.23% C hypoeutectoid carbon steel is shown in Fig. 2-34a.

However, if the austenitic grain size is relatively large compared to the size of the growing proeutectoid ferrite, during cooling the centers of the austenitic grains will become supersaturated with respect to ferrite. In order to relieve this supersaturation, proeutectoid ferrite will nucleate and grow inside the austenitic grains even though the activation energy is higher there. As a result, Widmanstätten plates of proeutectoid ferrite will form inside the austenitic grains, as shown in Fig. 2-34b. This type of structure is common in coarse-grained hypoeutectoid steels cooled near critical rates such as encountered in welds and steel castings.

2-8 AUSTEMPERING AND MARTEMPERING (MARQUENCHING)

Austempering

Austempering is an isothermal heat treatment process which produces a bainitic structure in plain-carbon steels. The process provides an alternative procedure to quenching and tempering for optimizing strength and toughness of some steels for certain hardness levels. For austempering (Fig. 2-35), the steel is austenitized, quenched into a hot salt bath at a temperature just above the M_s temperature, held isothermally, and then cooled to room temperature in air.

Austempering is usually substituted for conventional quenching and tempering (1) to obtain improved ductility and impact strength for a particular hardness and (2) to decrease cracking and distortion quenching. Austempering is particularly advantageous for the heat treatment of thin sections of plain-carbon steels to produce excellent toughness and ductility at a hardness of about Rockwell C50. From Table 2-2 it can be seen how austempering increases the impact strength and ductility in a 1095 steel. The reason for the increase in these properties is attributed to the favorable iron-carbide distribution in the bainitic structure.

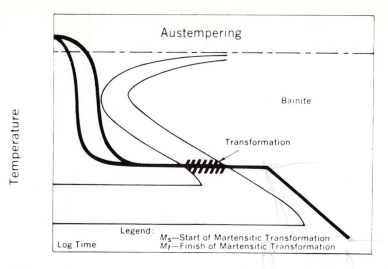

Figure 2-35 Cooling curves for austempering an eutectoid plain-carbon steel. The structure resulting from this treatment is bainite. An advantage of this heat treatment is that tempering is unnecessary. Compare with the customary quenching and tempering process shown in Fig. 2-22. *(From "Suiting the Heat Treatment to the Job," United States Steel Co., 1968, p. 34.)*

The austempering process, however, has its limitations and is impractical to use for some steels. In order to obtain a uniform structure and hence uniform mechanical properties, the entire cross section of the steel must be cooled rapidly enough to miss the nose of the IT curve. In plain-carbon steels, only relatively thin sections (about 3/8 in maximum) can be austempered since the time to start the austenite-to-bainite transformation near the nose of the IT diagram for plain-carbon steels is so short. With some alloy steels, larger cross sections can be austempered since the time to start the transformation is much longer. However, if the time required to complete the transformation becomes too long, the process becomes too time consuming to be practical.

Table 2-2 Mechanical properties of 1095 (0.95% C) steel heat-treated by austempering and conventional quenching†

Heat treatment	Rockwell C hardness	Impact, ft · lb	Elongation in 1 in, %
Austemper	52.0	45	11
Water quench and temper	53.0	12	0

† After Ref. 7.

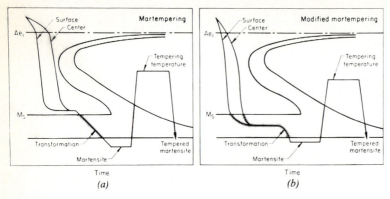

Figure 2-36 Cooling curves for (a) martempering and (b) modified martempering superimposed on eutectoid plain-carbon steel IT Diagrams. Tempering usually follows the martempering process. *(After Metals Handbook, 8th ed., vol. 2, American Society for Metals, 1964, p. 37.)*

Martempering (Marquenching)

Martempering (marquenching) is a modified quenching procedure used for steels primarily to minimize distortion of the heat-treated material. The martempering process consists of (1) austenitizing the steel, (2) quenching it in hot oil or molten salt at a temperature just slightly above (or slightly below) the martensite start temperature, (3) holding it in the quenching medium until the temperature is uniform throughout the steel (the isothermal treatment is normally stopped before the transformation of the austenite-to-bainite reaction begins), and (4) cooling at a moderate rate to prevent drastic temperature differences between the surface and center of the steel.

Figure 2-36 shows cooling curves for martempering and modified martempering of eutectoid plain-carbon steels. Martempered steels are usually tempered later to toughen the steel. The term "martempering" is a misnomer since in the martempering process the steel is not tempered. A more suitable name is *marquenching*.

Table 2-3   Mechanical properties of 1095 (0.95% C) steel heat-treated by martempering and conventional quenching†

Heat treatment	Rockwell C hardness	Impact, ft · lb	Elongation in 1 in, %
Water quench and temper	53.0	12	0
Martemper and temper	53.0	28	0

† After Ref. 7.

   In martempering, by allowing the martensitic transformation to take place at higher temperatures than used for conventional quenching, distortion and residual stresses in the workpiece are reduced. Table 2-3 compares the mechanical properties of 1095 steel after martempering and tempering with those after conventional quenching and tempering at a Rockwell hardness of about C50. The major difference appears in the increased impact resistance of the martempered and tempered material.

PROBLEMS

1. Describe the cooling path necessary to produce bainite in an eutectoid plain-carbon steel using an IT diagram.

2. How can bainite in eutectoid plain-carbon steels be defined in terms of its microstructure?

3. Describe the microstructural differences between upper and lower bainite in eutectoid plain-carbon steels.

4. How is the bainitic reaction in eutectoid plain-carbon steels similar to (*a*) the pearlitic reaction and (*b*) the martensitic reaction? In what ways does it differ?

5. How does the IT diagram of a hypoeutectoid plain-carbon steel differ from that of an eutectoid steel?

6. How does a continuous-cooling diagram of an eutectoid plain-carbon steel differ from the IT diagram? Why are continuous-cooling curves of more practical value to the engineer?

7. Show cooling curves on a CCT diagram for an eutectoid plain-carbon steel that produce (*a*) a martensitic structure, (*b*) split transformation of martensite and pearlite, and (*c*) coarse pearlite.

8. What is the critical cooling rate on a continuous cooling diagram of an eutectoid carbon steel?

9. Define a Widmanstätten structure in Fe-C alloys.

10. How does a CCT diagram differ from an IT diagram for a hypoeutectoid (0.38% C) plain-carbon steel?

11. Explain why, in a 0.38% C steel, the volume percent pearlite is increased by increasing the cooling rate during continuous cooling from austenitic temperatures. Assume cooling rates that form only pearlite and proeutectoid ferrite.

12. How can the decrease in weight percent proeutectoid ferrite in question 11 be accounted for from a materials balance standpoint?

13. Describe the difference between a full anneal and a process anneal using an Fe-Fe$_3$C diagram.

14. Briefly define the following processes that a cold-worked metal goes through upon annealing: (*a*) recovery, (*b*) recrystallization, and (*c*) grain growth.

15. Describe the general dislocation arrangement in a sheet of cold-worked (65 percent) iron sheet.

16. Describe what microstructural changes occur during recovery. What are subgrains? How are they formed? What is polygonization?

17. Describe the recrystallization process. What is the driving force for recrystallization? How can the higher-angle grain boundaries which are formed spontaneously in this process be accounted for energetically?

18. How is the overall energy of a recrystallized low-carbon steel decreased by grain growth?

19. Since grain growth is a spontaneous process, why does it not occur until single crystals are produced?

20. What is the normalizing process? What are it purposes? What type of structure is usually produced by this process in plain-carbon steels?

21. What is the origin of residual stresses in quenched plain-carbon steels? How can these be alleviated?

22. What is the tempering process for a steel? During the tempering of a plain-carbon steel, what important processes take place which change the microstructure?

23. Describe the microstructural changes which occur in a 0.39% C steel upon tempering in the 100 to 700°C range.

24. Describe the types of iron carbides that are formed upon tempering a plain-carbon steel.

25. What is retained austenite? Why is it sometimes important to know how much of it is produced?

26. Describe the microstructural changes which occur in a 0.18% C steel during (*a*) recovery and (*b*) recrystallization while tempering.

27. How does tempering affect the hardness of plain-carbon steels?

28. How is the grain size of a metal measured in the United States?

29. How does the grain size of an annealed low-carbon steel affect its yield stress?

30. How does the austenitic grain size affect the proeutectoid ferrite morphology in hypoeutectoid steels when cooled near critical rates?

31. Describe the austempering process used for plain-carbon steels. What are its advantages and disadvantages?

32 Describe the martempering (marquenching) process. What are its advantages and disadvantages?

33. During the isothermal tempering of a Fe–1.2% C alloy at 200°C, microcracking in the plate martensite is rapidly reduced in the "first stage" of tempering (that is, in the stage during which ϵ carbide precipitates and before the "second stage" when the retained austenite transforms). What action could account for the rapid decrease in microcrack severity? [See T. A. Balliett and George Krauss, *Met. Trans.* 7A(1976):81.]

REFERENCES

1. E. C. Bain and H. W. Paxton: "Alloying Elements in Steel," 2d ed., American Society for Metals, Metals Park, OH, 1966.
2. H. E. McGannon (ed.): "The Making, Shaping, and Treating of Steel," 9th ed., United States Steel Corporation, Pittsburgh, 1971.
3. R. E. Reed-Hill: "Physical Metallurgy Principles," 2d ed., Van Nostrand, New York, 1973.
4. R. E. Smallman: "Modern Physical Metallurgy," 3d ed., Butterworths, London, 1970.
5. P. G. Shewmon: "Transformations in Metals," McGraw-Hill, New York, 1969.
6. "Suiting the Heat Treatment to the Job," United States Steel Corporation, Pittsburgh, 1968.
7. *Metals Handbook*, vol. 2: "Heat Treatment, Cleaning, and Finishing," American Society for Metals, Metals Park, OH, 1964.
8. *Metals Handbook*, vol. 8: "Metallography, Structures and Phase Diagrams," American Society for Metals, Metals Park, OH, 1973.
9. M. A. Grossman and E. C. Bain: "Principles of Heat Treatment," 5th ed., American Society for Metals, Metals Park, OH, 1964.
10. R. F. Hehemann, K. R. Kinsman, and H. I. Aaronson: "A Debate on the Bainite Reaction," *Met. Trans.* 3(1972):1077.
11. D. M. Shackleton and P. M. Kelly: "Morphology of Bainite." In *Physical Properties of Martensite and Bainite*, Special Report No. 93, The Iron and Steel Institute, London, 1965.
12. G. R. Speich: "Growth Kinetics of Bainite in a Three Percent Chromium Steel." In V. F. Zackay and H. I. Aaronson (eds.): *Decomposition of Austenite by Diffusional Processes*, p. 353, Interscience Publishers, New York, 1962.
13. D. A. Witmer and G. Krauss: "Effect of Thermal History on the Recrystallization Behavior of Low-Carbon 0.30% Mn Steels Containing Oxygen and Sulfur," *Trans. ASM* 62(1969):447.

14. G. R. Speich: "Tempering of Low-Carbon Martensite," *Trans. AIME* 245(1969):2553.
15. G. R. Speich and W. C. Leslie: "Tempering of Steel," *Met. Trans.* 3(1972):1043.
16. A. M. Turkalo and J. R. Low, "The Effect of Carbide Dispersion on the Strength of Tempered Martensite," *Trans. AIME* 212(1958):750.
17. R. A. Grange, P. T. Kilhefner, and T. P. Bittner: "Austenite Transformation and Incubation in an Alloy Steel of Eutectoid Carbon Content," *Trans. ASM* 51(1959):495.

THREE

CARBON STEELS

Plain-carbon steels are the most important group of engineering alloys, and accounted for 85.3 percent of the steel produced in the United States in 1978. The relatively low cost and wide range of properties of plain-carbon steels make them of prime importance as engineering materials. The applications of plain-carbon steels are innumerable. Some of the major product forms of plain-carbon steels are sheet, strip, bar, wire, tubular products, structural shapes, forgings, plate, and castings.

3-1 MODERN STEELMAKING PROCESSES

Principal Steps in the Production of Finished Steel-Mill Products

A general diagram for the production of steel from raw materials to the finished mill products is shown in Fig. 3-1. The following is a brief description of the basic steps involved in the steelmaking process.

1. *Reduction of iron compounds* (*chiefly iron oxides*) *to molten iron* (*pig iron*). In this process, coke (carbon) acts as a reducing agent in a blast furnace (Fig. 3-2) to produce iron containing from 3 to 4.5% C according to the following typical reaction:

$$Fe_2O_3 + 3CO \rightarrow 2Fe + 3CO_2$$

Since most steel used today contains less than 1% C, the excess carbon must be removed from the pig iron to convert it to steel.

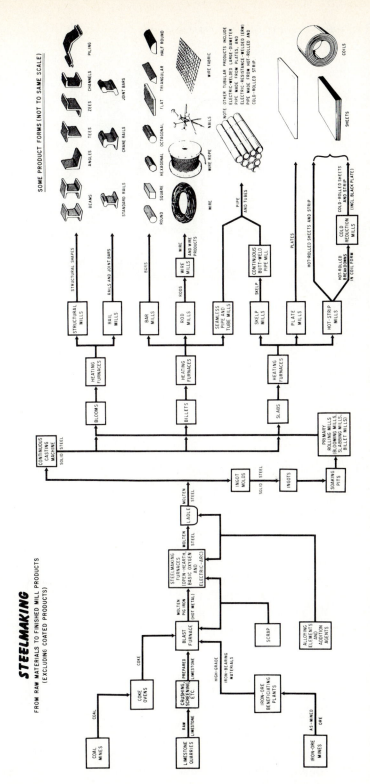

Figure 3-1 Flow diagram showing the principal process steps involved in converting raw materials into the major product forms, excluding coated products. [*After H. E. McGannon (ed.), "The Making, Shaping, and Treating of Steel," 9th ed., United States Steel Corporation, Pittsburgh, 1971, p. 2.*]

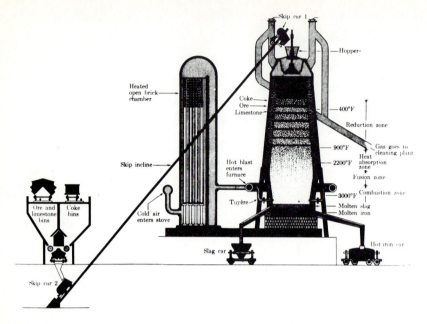

Figure 3-2 Cross section of the general operation of a modern blast furnace. *(After A. G. Guy, "Elements of Physical Metallurgy," 2d ed.,* © *1959, Addison-Wesley, Reading, Massachusetts, Fig. 2-5, p. 21.)*

2. *Process steelmaking.* In furnace process steelmaking, the excess carbon in the steel is reduced to the desired level by controlled oxidation of mixtures of pig iron and iron and/or steel scrap. The three principal furnace steelmaking processes used are (*a*) basic-oxygen furnace, (*b*) open-hearth furnace, and (*c*) electric-arc furnace. Alloy steels are made by adding elemental or alloy manganese, chromium, molybdenum, nickel, vanadium, etc., to the molten steel during or after the carbon-removal process.

3. *Casting.* After the steel has reached its desired composition, it is tapped or poured from the steelmaking furnace into a large container or *ladle*. Sometimes alloying elements or deoxidizing agents such as aluminum or ferrosilicon are added to the molten steel in the ladle to further adjust the chemical composition of the steel or to remove gaseous oxygen. The steel is then poured (*teemed*) into tall rectangular stationary ingot molds or tapped into a *tundish* (reservoir) for continuous casting of the steel.

4. *Rolling (forging).* Most ingots are reheated to a high temperature (below the melting point of the lowest-melting constituents in the steel) and held for a sufficient time so that the ingots will be homogeneously heated throughout. The reheated or soaked ingots are then hot-rolled or forged into the desired shape. Continuously cast steel can be directly cast into the semifinished shape desired.

5. *Mechanical treatment.* The semifinished products are further worked by hot rolling, cold rolling, forging, extruding or drawing, etc., to produce the

finished steel products such as plate, sheet, bars, tubular products, structural shapes, etc.

6. *Heat treatment*. In order to produce the finished steel product in the desired strength, it is sometimes necessary to heat-treat the steel. Heat treatment allows a certain degree of control over the structure and properties of the steel. The heat treatment of steel will be emphasized in this book since it greatly influences the structure and properties of steels.

Steelmaking Processes

Basic-oxygen process (BOP) In the past years, more and more of the steel produced in the United States has been made using the basic-oxygen process (Fig. 3-3). In this process, liquid pig iron and up to 30 percent scrap are charged into a barrel-shaped refractory-lined converter into which an oxygen lance is inserted from the top (Fig. 3-4). Pure oxygen from the lance reacts with the liquid bath to form iron oxide. Carbon in the steel reacts with the iron oxide to form carbon monoxide:

$$FeO + C \rightarrow Fe + CO$$

Immediately after the oxygen reaction starts, slag-forming fluxes (chiefly lime) are added in controlled amounts. The reaction proceeds rapidly and requires no external flame. In about 45 min, 200 tons of steel can be produced, which is much faster than the 6 to 10 h required by the open-hearth process to produce about as much steel. A schematic diagram of the reduction in concentration of the carbon, manganese, silicon, sulfur, and phosphorus is shown in Fig. 3-5.

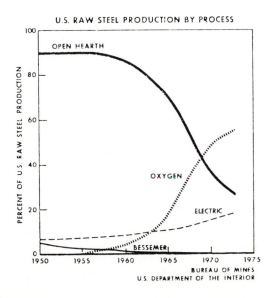

Figure 3-3 Percentage of raw steel produced in the United States by process. *Source:* American Iron and Steel Institute. *(After Minerals and Facts, 1975 edition, Bureau of Mines Bulletin 667, p. 556.)*

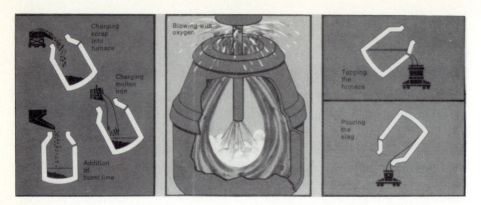

Figure 3-4 Steelmaking in a basic oxygen furnace. *(Courtesy of Inland Steel Company.)*

The chemical composition of the steel produced by the basic-oxygen process is superior to that of the basic open-hearth process in the following ways:

1. Sulfur contamination from external fuel (e.g., oil) is avoided, since no external fuel is required.
2. Since practically no nitrogen is present in the pure oxygen used for refining, the nitrogen content of BOP steels is very low, usually 0.004 percent.
3. Residual oxygen levels in BOP steels are lower, requiring lesser amounts of deoxidizing agents.

Other advantages of BOP steels are lower impurity levels due to less scrap used and easy adaptation for vacuum degassing and continuous casting.

Basic open-hearth process This process was the dominant steelmaking method in the United States from 1908 to 1969. In this process, a relatively shallow bath of steel is heated by a flame which passes over the bath from burners at one end of the furnace (Fig. 3-6). The bath is covered with a chemically basic slag to

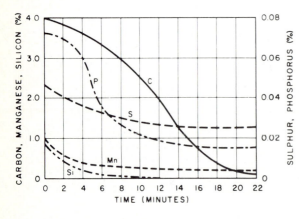

Figure 3-5 Schematic representation of progress of refining in a top-blown basic-lined vessel. [*After H. E. McGannon (ed.), "The Making, Shaping, and Treating of Steel," 9th ed., United States Steel Corporation, Pittsburgh, 1971, p. 494.*]

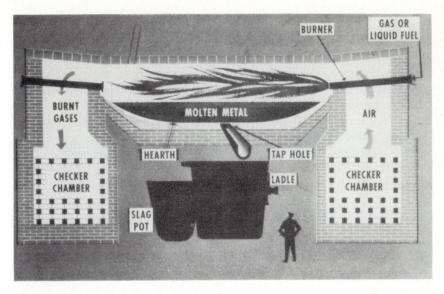

Figure 3-6 Operation of an open-hearth steelmaking process. *(Courtesy of Bethlehem Steel Co.)*

remove phosphorus and sulfur impurities, while carbon is removed from the steel by the addition of iron ore or by using oxygen. The main disadvantage of the process is it requires 6 to 10 h to make a batch of steel. This steelmaking process is being phased out since in most cases it is more economical to use the basic-oxygen process.

Electric-arc process In this process, adjustable electrodes are lowered to a point just above a charge of cold steel scrap. An electric arc is struck between the electrodes and the steel scrap, resulting in the melting of the steel (Fig. 3-7). Since about 1945, this process has been used increasingly for melting plain-carbon steel scrap. Since the electric-arc furnace has a relatively low capital

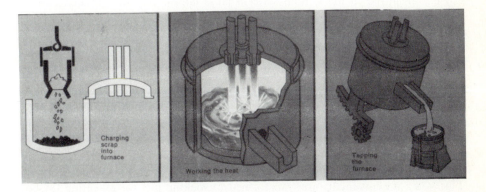

Figure 3-7 Steelmaking in an electric-arc furnace. *(Courtesy of Inland Steel Company.)*

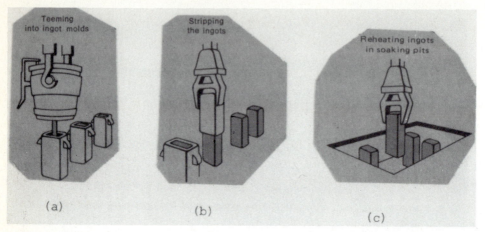

Figure 3-8 Individual ingot casting. (*a*) Teeming into ingot molds. (*b*) Stripping the ingots. (*c*) Reheating ingots in soaking pits. (*Courtesy of Inland Steel Company.*)

investment cost and can remelt steel scrap, this process is used where local supplies of steel scrap are available.

The electric-arc furnace is also used for producing special alloy steels which contain an appreciable amount of easily oxidized alloying elements such as chromium, tungsten, and molybdenum. It is also used when very low sulfur and phosphorus levels are necessary in some alloy steels. Special slag covers are used to lower sulfur and phosphorus levels and give protection against oxidation of the alloying elements. Careful temperature control is also possibie with this process.

3-2 INGOT CASTING

In general, most steel is produced by the basic-oxygen or open hearth processes and is transferred in the molten state in large ladles. The steel is then cast or "teemed" into large stationary ingot molds, or cast continuously into long ingots.

Individual Ingot Casting

Casting the steel into individual ingot molds is the conventional method of producing steel ingots for hot working, and most steel is cast in this way since the method is so versatile. In this process, the full ladle of steel is moved by overhead crane so that it can be tapped (or teemed) into individual molds standing upright on rail cars (Fig. 3-8*a*). The ingot molds are slightly tapered for easy removal after solidification of the steel. After stripping the ingot molds (Fig. 3-8*b*), the hot ingots are transformed to soaking pits for reheating for hot rolling (Fig. 3-8*c*). About 85 percent of the steel cast today is still cast in individual molds.

Continuous Casting

In continuous casting, the ladle of molten steel is transported to an elevated casting platform above a casting machine (Fig. 3-9). The molten steel is discharged into a rectangular trough, called a *tundish*, which acts as a reservoir for the steel. From a spout in the bottom of the tundish, the molten steel is poured into a water-cooled mold with a movable bottom, which is slowly lowered. As the molten steel enters the mold, the metal at the mold surface solidifies, forming a thin skin. This skin thickens as the metal passes through the mold. The remaining metal in the center of the ingot is solidified by cold water sprayed onto the ingot as it leaves the mold.

The solid metal billet is pulled by rollers so that a long continuous steel slab is produced, as shown in Fig. 3-9. For many types of steel, this process is more economical than stationary casting into individual molds. More steel can be cast into slab form in a shorter time than with the individual molds. The metal does not have to be cooled down so the molds can be stripped off. The trend in the steel industry today is to use continuous casting wherever possible since it can in many cases produce a higher quality product at lower cost.

3-3 TYPES OF INGOT STRUCTURES

When steel is cast or teemed into individual stationary molds, various types of structures can be produced depending on how the steel is allowed to solidify (Fig. 3-10), while the decision as to type of ingot structure to be produced

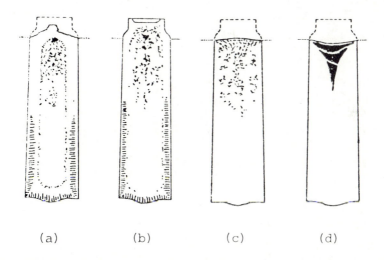

(a)   (b)   (c)   (d)

Figure 3-10 Types of ingot structure. (*a*) Rimmed; (*b*) capped; (*c*) semikilled; (*d*) killed. Note the distribution of blow holes in the rimmed steel and the pipe cavity in the killed steel. [*After H. E. McGannon (ed.), "The Making, Shaping, and Treating of Steel," 9th ed., United States Steel Corporation, 1971, p. 597.*]

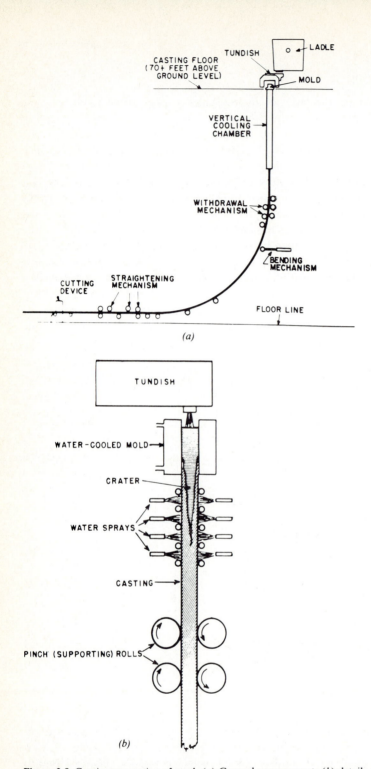

Figure 3-9 Continuous casting of steel. (*a*) General arrangement; (*b*) details of mold region. [*After H. E. McGannon (ed.), "Making, Shaping, and Treating of Steel," 9th ed., United States Steel Corporation, Pittsburgh, 1971, pp. 707–708.*]

depends on the final use of the steel. During the solidification of the steel, excess gases are expelled from the metal. Oxygen in the molten steel in the form of FeO reacts with carbon in the steel to produce carbon monoxide according to the following reaction:

$$FeO + C = Fe + CO(gas)$$

Since the steel solidifies over a range of temperatures, gases evolved from the unsolidified liquid can be trapped at the solid-liquid interfaces, producing *blowholes*. The amount of oxygen dissolved in the liquid steel just before casting can be controlled by the addition of deoxidizing agents such as aluminum and ferrosilicon. Depending on the amount of gases, mainly oxygen, remaining in the liquid steel during the solidification process, the following types of ingot structures can be produced: rimmed, capped, semikilled, and killed (Fig. 3-10).

Rimmed Ingot Structure

In a rimmed steel, the reaction of the dissolved oxygen and carbon to form carbon oxide gases (CO and CO_2) is allowed to progress until a heavy rim of relatively pure iron free from voids is produced (Fig. 3-10a). The interior of the ingot contains gas porosity in the form of blowholes of various sizes and shapes. During hot rolling, these voids are welded together to produce sheet and plate with good surface quality. The rimming action lowers the carbon content of steel and causes segregation of carbon, sulfur, and phosphorus toward the center and top of the ingot. The segregation of the metalloids causes a variation in composition and mechanical properties from sheet to sheet. Rimmed steel is cheaper to produce since the top part of the ingot does not have a large pipe cavity, which means the yield is higher.

Capped Ingot Structure

In capped steels, the rimming action is stopped chemically or mechanically (Fig. 3-10b). Chemically capped steel is poured into the mold and allowed to rim for 1 to 3 min, and then the reaction is stopped by the addition of shot aluminum or ferrosilicon to the top. In mechanical capping, a heavy cast-iron cap is used to close the top opening as soon as pouring is stopped. The gas evolution is stopped due to the increased pressure when the metal strikes the cap.

Semikilled Ingot Structure

A typical semikilled ingot structure is shown in Fig. 3-10c. In this type of ingot, only a slight amount of gas is allowed to evolve during solidification. Only a sufficient number of blowholes are allowed to form so that the volume contraction due to solidification can be compensated for.

Killed Ingot Structure

Fully killed steels evolve no gas and form a pipe cavity at the top of the ingot (Fig. 3-10d) since the addition of aluminum or silicon to the molten steel in the ladle or mold stops the gas reaction. Aluminum-killed steels are widely used for cold-rolled sheet that will be used for severe forming or deep drawing and also for sheet that will be stored for long periods before being used. These steels show minimum strain aging and have a fine-grain size. (Strain aging will be discussed in Sec. 3-8.) The composition of killed steels is more uniform than rimmed steels because there is no gas reaction.

3-4 CLASSIFICATION OF PLAIN-CARBON STEELS

Plain-carbon steels are classified by several different systems, depending on the type of steel and its application. There is thus no one classification system that applies to all plain-carbon steels. The two most commonly used systems are the AISI-SAE[1] system and the ASTM[2] classification.

AISI-SAE Classification System for Plain-Carbon Steels

This system is applied to hot-rolled and cold-finished bars, wire, rod, and seamless tubing, and semifinished products for forging. Since the carbon content of plain-carbon steels essentially determines their strength, this system uses the percent carbon to designate the different steels. A four-digit number is used, with the first two digits being 10 to designate a plain-carbon steel. The second two digits indicate the hundreths of percent carbon. For example, the number 1020 indicates a plain-carbon steel with 0.20 nominal percent carbon. As will be seen in the next chapter on alloy steels, this system is also used for alloy steels, with the first two digits altered to indicate other major alloying elements. Table 3-1 lists some selected grades of plain-carbon steels.

ASTM System

In the ASTM system, standards are written for various alloys to meet special requirements. In addition to establishing chemical compositions, the ASTM standards also set mechanical property levels and often specify fabrication procedures and heat treatments. Plate steels, for example, are mainly classified according to ASTM standards.

Other Systems

Special standards are often set for special products. For example, many low-carbon steel products such as tin plate and special automotive sheet are

---

[1] AISI: American Iron and Steel Institute; SAE: Society of Automotive Engineers.
[2] American Society for Testing and Materials.

Table 3-1 AISI-SAE carbon-steel compositions

AISI-SAE No.	% C	% Mn
1006	0.08 max.	0.25–0.40
1010	0.08–0.13	0.30–0.60
1015	0.13–0.18	0.30–0.60
1020	0.18–0.23	0.30–0.60
1025	0.22–0.28	0.30–0.60
1030	0.28–0.34	0.60–0.90
1035	0.32–0.38	0.60–0.90
1040	0.37–0.44	0.60–0.90
1045	0.43–0.50	0.60–0.90
1050	0.48–0.55	0.60–0.90
1055	0.50–0.60	0.60–0.90
1065	0.60–0.70	0.60–0.90
1070	0.65–0.75	0.60–0.90
1075	0.70–0.80	0.40–0.70
1080	0.75–0.88	0.60–0.90
1085	0.80–0.93	0.70–1.00
1090	0.85–0.98	0.60–0.90
1095	0.90–1.03	0.30–0.50

P, 0.040 max; S, 0.05 max.

produced according to special specifications, and so there is no general numbering system for these steels.

3-5 EFFECTS OF OTHER ELEMENTS IN PLAIN-CARBON STEELS

In addition to carbon, plain-carbon steels contain the following other elements:

Manganese up to 1.0 percent
Sulfur up to 0.05 percent
Phosphorus up to 0.04 percent
Silicon up to 0.30 percent

The effects of each of these elements in plain-carbon steels are summarized in the following subsections.

Manganese

Manganese in plain carbon steels ranges from 0.35 percent maximum in AISI 1005 steel to 1.0 percent maximum in AISI 1085 steel. Manganese combines with the sulfur present in the steel to produce manganese sulfide (MnS), which exists as soft gray inclusions in the steel. The MnS inclusions are scattered in the grain bodies and are elongated in the direction of working (Fig. 3-11a). Figure

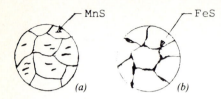

Figure 3-11 Schematic distribution of (*a*) manganese sulfide and (*b*) iron sulfide in plain-carbon steels.

3-12 shows an inclusion in a rimmed low-carbon steel which consists of mixed sulfides of iron and manganese. MnS is preferable to iron sulfide (FeS) in the steel since FeS is a brittle, low-melting compound which forms at the grain boundaries (Fig. 3-11*b*). Manganese also raises the yield strength of plain-carbon steels by refining the pearlite and by solid-solution strengthening of the ferrite. Manganese increases the depth of hardening during quenching from austenite, but in large amounts it also increases the tendency toward cracking and distortion during quenching.

Sulfur

Sulfur is present up to a maximum of 0.05 percent in plain-carbon steels. It usually is combined with manganese in the steel to produce MnS inclusions, as indicated in Fig. 3-11*a*. However, if the sulfur combines with iron, it forms FeS, which usually occurs as a grain-boundary precipitate (Fig. 3-11*b*). Since FeS is hard and has a low melting point, it can cause cracking during hot and cold working of the steel. Thus in order to avoid the FeS inclusions, the manganese to sulfur ratio in these steels is usually about 5:1.

Phosphorus

Phosphorus is limited to a maximum of 0.04 percent in plain-carbon steels since it forms a compound, Fe_3P, which is extremely brittle and segregates in the steel.

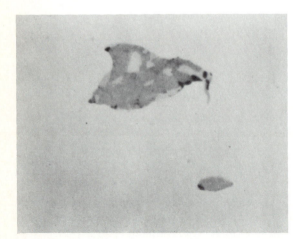

Figure 3-12 Mixed sulfides of iron and manganese containing a few small oxide spots. (Rimmed low-carbon steel.) (As polished: $1000 \times$.) (*After Metals Handbook, 8th ed., vol. 7, American Society for Metals, 1972, p. 16.*)

Silicon

The amount of silicon in plain-carbon steels varies from about 0.1 to 0.3 percent. Silicon is used as a deoxidizer, and forms SiO_2 or silicate inclusions. Otherwise silicon has little effect on the mechanical properties of plain-carbon steels, since it dissolves in ferrite.

3-6 HOT AND COLD WORKING OF CARBON STEELS

Primary Rolling

Reheated ingots are removed from the preheating furnaces (soaking pits) at about 1370°C and are hot-worked by primary rolling mills into slabs, billets, and/or blooms (Fig. 3-13). In the slabbing mill, the ingot is rolled into a flat *slab* (Fig. 3-13*a*), which is later further rolled into plate and sheet. In the blooming mill, the ingot is rolled into a rectangular shape called a *bloom* (Fig. 3-13*b*), which is subsequently rolled into structural shapes and rails. In the billet mill, the ingot is rolled into a smaller rectangular shape than a bloom, called a *billet* (Fig. 3-13*c*), which is later rolled into bars, rods, and seamless pipe and tube stock.

By continuously casting, slabs, billets, and blooms may be cast directly so that the primary working stage can be circumvented. Wherever economically and technically feasible, therefore, continuous casting is used, although most steel today is still cast as individual ingots. Figure 3-1 shows the steel flow diagrams and indicates the stages by which the various steel products are processed from slabs, billets, and blooms, but in this book the emphasis will be on the processing of *sheet products* since these are economically the most important steel products.

Hot Rolling

To produce hot-rolled flat steel strip, slabs are reheated to about 1315°C and are reduced from about 10 in in thickness to about 0.1 in by a series of reductions in

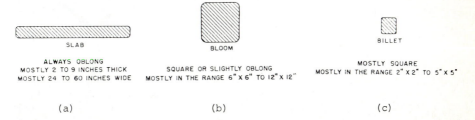

Figure 3-13 Typical cross sections and dimensional characteristics of shapes of steel products after primary rolling. (*a*) Slab, (*b*) bloom, (*c*) billet. [*After H. E. McGannon (ed.), "Making, Shaping, and Treating of Steel," 9th ed., United States Steel Corporation, Pittsburgh, 1971, p. 675.*]

HOT STRIP ROLLING MILL

TYPICAL REDUCTIONS PER PASS IN FINISHING STANDS

(THIS DRAWING IS ENTIRELY SCHEMATIC AND NOT TO SCALE)

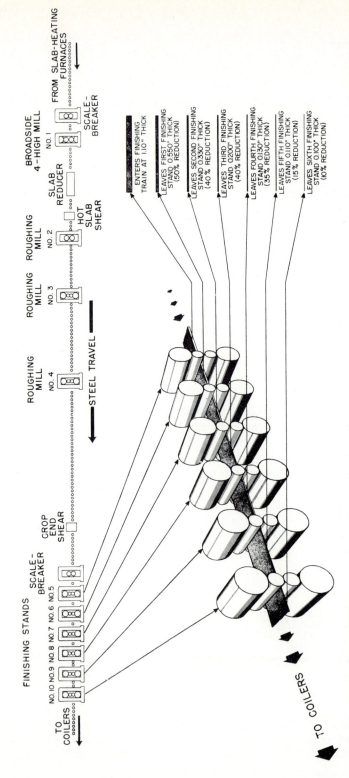

Figure 3-14 Typical reductions in the finishing stands of a hot-strip mill equipped with four roughing stands and six finishing stands. [*After H. E. McGannon (ed.), "Making, Shaping, and Treating of Steel," 9th ed., United States Steel Corporation, Pittsburgh, 1971, p. 937.*]

a line of hot-rolling mills (Fig. 3-14). A series of roughing mills reduces the thickness of the slab to about 1 in. This slab is then rolled to a final hot-strip thickness of about 0.1 in in a series of closely spaced finishing mills, and then is coiled.

If a deep-drawing quality low-carbon killed steel is being rolled, it is important to keep the reheating temperature high enough so that aluminum nitride (AlN) will be taken into solid solution. The temperature of the slab must also be kept high enough so that iron oxide (scale) formed on the strip surface can be removed by high-pressure water sprays at each roughing stand. If the scale is not removed, it can be detrimental to the surface of the final cold-rolled sheet.

The temperature most closely controlled during hot rolling is the temperature of the strip after it leaves the last finishing stand. The temperature of the strip is usually controlled by water sprays located between the finishing stands and the coiler.

Hot rolling is carried out above the recrystallization temperature so that the grain structure is reformed after working. The temperature of hot working must not be too high, however, or excessive grain growth will occur. Hot working should be finished at a temperature slightly above the recrystallization temperature so that a fine-grain size will be obtained upon cooling below the recrystallization temperature.

The effects of hot rolling steel ingots can be summarized as follows:

1. Hot rolling breaks down the coarse columnar structure of the cast ingots.
2. Hot rolling homogenizes the dendritic segregation which occurs during casting.
3. In rimming steels, the blowholes are welded together. In all steels, porosity is healed.
4. Nonmetallic inclusions are broken up and elongated in the direction of rolling. This leads to direction properties in the rolled product. The strength is increased in the direction of rolling.
5. If the finishing temperature is close to the recrystallization temperature, grain refinement will be obtained.

Pickling

Most hot-rolled strip which is to be cold-reduced is acid-cleaned, or *pickled*, to remove the scale from the hot-rolling operation (Fig. 3-15*a*). In this process, which is usually continuous, the strip is immersed in an acid bath (HCl or H_2SO_4) at approximately 82°C. The pickled strip is then rinsed with water, air-dried, oiled, and coiled.

Cold Reduction

In order to produce cold-rolled sheet products, the pickled hot-mill strip is cold-reduced from 40 to 70 percent. A minimum amount of cold reduction is

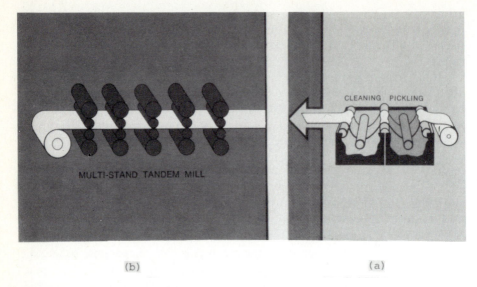

Figure 3-15 (*a*) Pickling and cleaning of low-carbon sheet steel. (*b*) Cold-rolling of low-carbon sheet steel in multitandem mill. Note that *a* and *b* are two separate batch operations and that the sheet steel is coiled up after cleaning and then uncoiled again at the start of the cold-rolling operation. (*Courtesy of Inland Steel Company.*)

necessary to ensure recrystallization of the cold-worked sheet during subsequent annealing. Figure 3-15*b* shows the schematic arrangement of a multistand tandem mill for cold rolling. Modern cold-rolling mills produce cold-rolled sheet with a high-quality surface, good shape, and close gauge control.

3-7 NON-HEAT-TREATABLE LOW-CARBON SHEET STEEL

Low-carbon sheet steel is used in large tonnages primarily for consumer products such as automobile body stock, tin plate, and sheet steel for porcelain enameling. These mass-produced materials, which are relatively low in cost, must have special properties, some of which are

1. Ease of fabrication (formability and weldability)
2. Sufficient strength after fabrication
3. Attractive appearance before and after fabrication
4. Compatibility with other materials and for various coatings

In order to produce low-carbon sheet steel which meets some or all of the above requirements, the chemical composition, fabrication practices, and heat treatment procedures are varied as is necessary.

Chemical Composition

About 80 percent of the amount of low-carbon sheet steel and strip has a composition in the following ranges:

Element	% Composition
Carbon	0.03 to 0.12
Manganese	0.20 to 0.60
Silicon	0.02 to 0.15
Phosphorus	0.04 max.
Sulfur	0.04 max.

The normal carbon content of low-carbon sheet is from 0.06 to 0.12 percent. However, for porcelain enameling, the carbon may range from 0.04 to as low as 0.002 percent. In order to achieve very-low-carbon contents, the sheet must be decarburized by a special process. For sheet for deep drawing, the phosphorus and sulfur contents are kept as low as possible.

Deoxidation Practice

Low-carbon sheet steel is produced from ingots of rimmed, capped, semikilled, or killed steel. See Sec. 3-3 for the details of these types of ingot structures and their deoxidation procedures.

Heat Treatment and Microstructure of Low-Carbon Sheet Steel

Rimmed steel Rimmed steel is hot-rolled at as high a temperature as possible to produce a refined grain structure for subsequent cold rolling and annealing. After hot rolling, the strip is pickled and cold-rolled from about 40 to 65 percent reduction depending on the final use for the sheet. Figure 3-16 shows the structure of cold rolled (65 percent reduction) rimmed steel.

The cold-reduced coils of sheet are then softened by annealing if the steel is to be used for deep drawing or forming. The most common method of annealing cold-reduced coils of sheet steel is by *box annealing* (Fig. 3-17). In this batch process, coils are stacked three or four high and placed under a cover. They are heated to the desired temperature and held for the necessary time using a special reducing atmosphere that prevents decarburizing of the surface. Rimmed sheet steel is annealed just under the A_1 temperature at about 705°C for sufficient time, and then slow-cooled to about 90°C. Figure 3-18a shows the typical equiaxed recrystallized grain structure of annealed rimmed steel (0.06% C) at $100 \times$ after box annealing at 705°C. Figure 3-18b shows some of the spheroidized iron carbides in the interior of a grain at $1000 \times$.

Figure 3-16 Cold-worked structure of rimmed sheet steel (0.06% C) after 65 percent reduction. (Etch: 2% nital; 100 × .) *(Courtesy of Inland Steel Company.)*

Killed steel Aluminum-killed steel is coiled at the end of the hot-strip mill at a temperature just below 600°C to keep the AlN in solution. During box annealing, the AlN precipitates, producing an elongated grain shape which has high formability. Since the AlN inhibits recrystallization, box-annealed aluminum-killed steels have to be annealed at higher temperatures than rimmed steels. It is common practice to anneal aluminum-killed steel at about 730°C, which is between the A_1 and A_3 (intracritical anneal).

During box annealing, which involves a slow heating rate, AlN precipitates in the subboundaries of the unrecrystallized matrix (Ref. 14). By controlling the extent of AlN precipitation at low temperatures during the recovery, polygonization, and coalescence stages, recrystallization at higher temperatures results in a grain size that has an elongated grain structure (Fig. 3-19a). This elongated

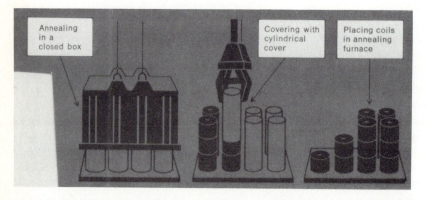

Figure 3-17 Box-annealing sheet steel coils. (A reducing atmosphere is used to prevent decarburizing of the sheet steel surface.) *(Courtesy of Inland Steel Company.)*

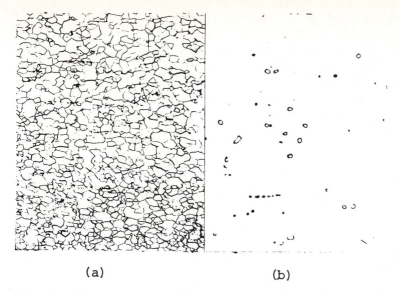

(a)　　　　　　　　　　　(b)

Figure 3-18 Rimmed steel sheet (0.06% C) after box annealing at 705°C. (*a*) Equiaxed recrystallized grain structure at 100 × . (*b*) Spheroidized iron carbide within a grain at 1000 × . Etch: 2% nital. *(Courtesy of Inland Steel Company.)*

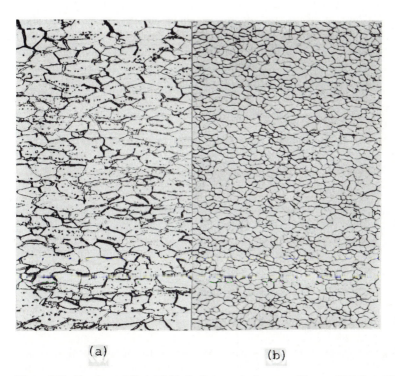

(a)　　　　　　　　　　　(b)

Figure 3-19 Aluminum killed 0.06% C sheet steel after annealing at 715°C. (*a*) Elongated grain structure; (*b*) equiaxed grain structure. (Etch: 2% nital; 100 × .) *(Courtesy of Inland Steel Company.)*

grain structure has a special crystallographic texture which makes it desirable for deep drawing and high formability. Rapidly heating aluminum-killed steels to the annealing temperature results in a finer-grain size (Fig. 3-19b) and higher mechanical properties, while it prevents the development of the elongated structure. This difference is attributed to the prevention of the precipitation of finely dispersed AlN.

Continuous Annealing of Low-Carbon Sheet Steels

Continuous annealing with its rapid heating rate results in a finer-grain size than that developed by the box-annealing process with its slow heating rate (Fig. 3-20). Continuously annealed steels thus have higher strengths and lower ductilities than similar box-annealed steels. Thus, the formability of box-annealed special-killed sheet steels will be higher than those that are continuously annealed.

Mechanical Properties of Low-Carbon Sheet Steels

After annealing, the cold-rolled rimmed steel is usually temper-rolled. This light (approximately 1 percent) cold-rolled reduction retards strain aging in rimmed steels. The temper rolling, while decreasing the strain-aging effect in rimmed steels, increases their strength and decreases their ductility and formability. Figure 3-21 gives typical ranges for the mechanical properties of rimmed and killed low-carbon sheet steels. It is noted that the special-killed steels have lower strength and higher ductility.

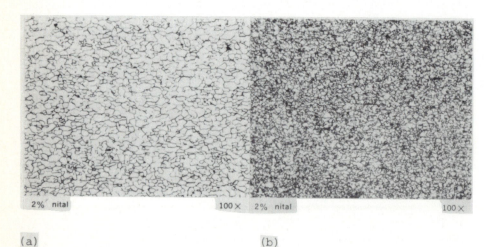

2% nital   100×   2% nital   100×

(a)   (b)

Figure 3-20 Low-carbon capped steel (0.06% C, 0.30% Mn); cold-rolled and then (a) recrystallized by box annealing and (b) recrystallized by continuous annealing. Note the finer grain size of the continuously annealed steel. (*After Metals Handbook, 8th ed., vol. 8, American Society for Metals, 1973, p. 228.*)

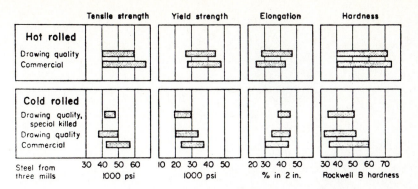

Figure 3-21 Typical range of mechanical properties of low-carbon steel from three mills. Hot-rolled sheet thicknesses from 0.598 to 0.135 in; cold-rolled from 0.029 to 0.0598 in. All cold-rolled grades include a temper pass. All grades were rolled from rimmed steel except the one labeled special killed. *(After Metals Handbook, 8th ed., vol. 1, American Society for Metals, 1961, p. 81.)*

3-8 QUENCH AGING AND STRAIN AGING OF CARBON STEELS

Quench Aging

Aging in low-carbon steels can be divided into two types: quench aging and strain aging. Quench aging is caused by the precipitation of carbon, nitrogen, or both from supersaturated solid solution. The solubilities of both these elements decreases sharply with decreasing temperature. The interstitial solubility of carbon in ferrite decreases from about 0.02 percent at 723°C (eutectoid temperature) to as low as 10^{-7} percent at room temperature. The solubility of nitrogen also decreases rapidly with decreasing temperature to an exceedingly low value.

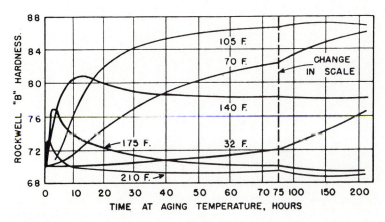

Figure 3-22 Hardness changes in a 0.06% C steel quenched from 720°C and aged at indicated temperatures. *(After Metals Handbook, American Society for Metals, 1948, p. 439.)*

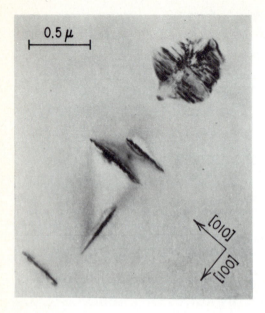

Figure 3-23 Metastable carbide precipitation on $\{100\}_\alpha$ planes in an Fe–0.013% C alloy quenched from 700°C and aged 6 h at 200°C. [*After H. W. Wagenblast and R. Glenn, Met. Trans 1(1970):2299.*]

If low-carbon sheet steels containing about 0.1% C are rapidly cooled after annealing, the carbon and nitrogen will be retained in interstitial supersaturated solid solution. Upon subsequent aging at room temperature or slightly above it, finely dispersed precipitates of ε carbide will be produced. These precipitates cause an increase in hardness and strength of the steel, as shown in Fig. 3-22.

The principal hardening agent in quench aging is carbon, since carbon steels contain much more carbon than nitrogen. This carbide precipitate, which is

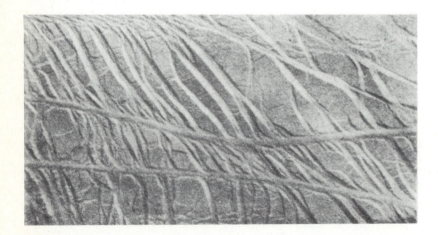

Figure 3-24 Stretcher strains in a sheet steel part (three-fourths actual size). *(After Metals Handbook, 8th ed., vol. 1, American Society for Metals, 1961, p. 325.)*

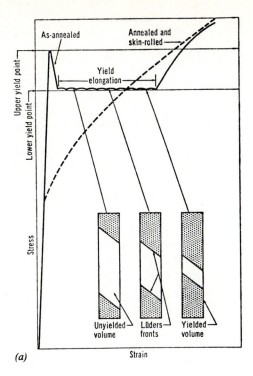

(a)

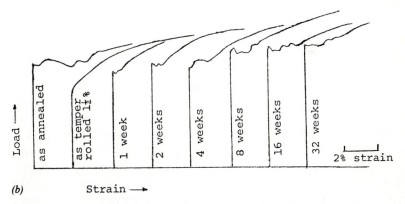

(b)

Figure 3-25 (*a*) Yield point behavior in low-carbon steel sheet. Unless skin rolled (dotted line), annealed sheet has definite upper and lower yield points. Once the steel yields in a sheet tensile specimen, it elongates for a period at the lower yield point. Stretcher strains (Lüders bands) develop during the yield elongation. (*After D. J. Blickwede, Metal Progress, vol. 95, no. 6, June, 1969, p. 12.*) (*b*) Temper rolling of annealed sheet eliminates the yield point, according to stress-strain tests. As the sheet ages, for the indicated periods, the yield point gradually returns. (*After M. R. Baren and P. G. Nelson, Metal Progress, Dec. 1970, p. 87.*)

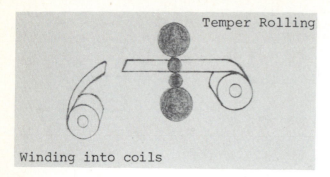

Figure 3-26 Temper rolling of low-carbon sheet steel. *(Courtesy of Inland Steel Company.)*

formed on aging, has been identified as ϵ carbide,[1] and is shown in Fig. 3-23 as precipitates on the $\{100\}_\alpha$ planes in an Fe–0.013% C alloy. The maximum hardness achieved by quench aging in steel is at about room temperature. Aging at slightly elevated temperatures leads to a rapid rise in hardness (although not as high), and then to a decrease caused by *overaging* and coarsening of the precipitates (Fig. 3-22). The maximum hardness is due to an optimum size and interparticle spacing of the carbide precipitates.

Strain Aging

Strain aging is industrially more important for low-carbon sheet steels than quench aging, since it can cause unsightly stretcher strains during deep drawing, as shown in Fig. 3-24. Strain aging manifests itself during plastic deformation as a sharp upper yield point, which is followed by a lower-yield-point elongation (Fig. 3-25a). Cold working during the lower-yield-point elongation can lead to the formation of stretcher strains, especially during deep drawing. A small amount of cold working just after annealing can eliminate the problem.

The cause of stretcher strains is the segregation of interstitial solute atoms (mainly nitrogen and carbon) to the strain fields of dislocations in the α-iron lattice. When this occurs, dislocations are anchored in place. The upper yield point is attributed to the extra stress required either to tear the dislocations loose from their "atmospheres" of interstitial solute atoms or to initiate new sources of dislocations. By temper rolling 1 to 3 percent (Fig. 3-26) before deep drawing, the lower-yield-point elongation is eliminated.

If temper-rolled rimmed steel is allowed to age before deep drawing, the yield point will gradually come back (Fig. 3-25b). It is believed that the nitrogen interstitial atoms have time to diffuse back to dislocations and thus cause a new upper yield point. Aluminum-killed low-carbon sheet steels do not show this phenomenon since the nitrogen is combined with aluminum as AlN.

3-9 HARDENABLE CARBON STEELS

Hardenable plain-carbon steels can be divided into the following groups according to their carbon contents: (1) low-carbon steels with 0.10 to 0.25% C, (2)

[1] F. W. Langer, *Met. Sci. J.* 2(1968):59.

medium-carbon steels with 0.25 to 0.55% C and (3) high-carbon steels with 0.55 to 1.00% C.

Low-Carbon Steels with 0.10 to 0.25% C

Steels in this group have increased strength and hardness and reduced cold formability compared to non-heat-treatable 0.06 to 0.10% C low-carbon steels.

Table 3-2 Mechanical properties of selected hardenable plain-carbon steels†

		Hot-rolled, normalized, and annealed					
AISI No.	Treatment	Yield strength psi	Tensile strength psi	Elonga-tion, %	Reduc-tion in area, %	Hard-ness, Bhn	Impact strength (Izod), ft · lb
1015	As-rolled	45,500	61,000	39.0	61.0	126	81.5
	Normalized (1700°F)	47,000	61,500	37.0	69.6	121	85.2
	Annealed (1600°F)	41,250	56,000	37.0	69.7	111	84.8
1020	As-rolled	48,000	65,000	36.0	59.0	143	64.0
	Normalized (1600°F)	50,250	64,000	35.8	67.9	131	86.8
	Annealed (1600°F)	42,750	57,250	36.5	66.0	111	91.0
1022	As-rolled	52,000	73,000	35.0	67.0	149	60.0
	Normalized (1700°F)	52,000	70,000	34.0	67.5	143	86.5
	Annealed (1600°F)	46,000	65,250	35.0	63.6	137	89.0
1030	As-rolled	50,000	80,000	32.0	57.0	179	55.0
	Normalized (1700°F)	50,000	75,500	32.0	60.8	149	69.0
	Annealed (1550°F)	49,500	67,250	31.2	57.9	126	51.2
1040	As-rolled	60,000	90,000	25.0	50.0	201	36.0
	Normalized (1650°F)	54,250	85,500	28.0	54.9	170	48.0
	Annealed (1450°F)	51,250	75,250	30.2	57.2	149	32.7
1050	As-rolled	60,000	105,000	20.0	40.0	229	23.0
	Normalized (1650°F)	62,000	108,500	20.0	39.4	217	20.0
	Annealed (1450°F)	53,000	92,250	23.7	39.9	187	12.5
1060	As-rolled	70,000	118,000	17.0	34.0	241	13.0
	Normalized (1650°F)	61,000	112,500	18.0	37.2	229	9.7
	Annealed (1450°F)	54,000	90,750	22.5	38.2	179	8.3
1080	As-rolled	85,000	140,000	12.0	17.0	293	5.0
	Normalized (1650°F)	76,000	146,500	11.0	20.6	293	5.0
	Annealed (1450°F)	54,500	89,250	24.7	45.0	174	4.5
1095	As-rolled	83,000	140,000	9.0	18.0	293	3.0
	Normalized (1650°F)	72,500	147,000	9.5	13.5	293	4.0
	Annealed (1450°F)	55,000	95,250	13.0	20.6	192	2.0

Table 3-2 –continued

		Quenched and tempered				
AISI No.	Tempering temperature, °F	Tensile strength, psi	Yield strength, psi	Elongation, %	Reduction in area, %	Hardness, Bhn
1030‡	400	123,000	94,000	17	47	495
	600	116,000	90,000	19	53	401
	800	106,000	84,000	23	60	302
	1000	97,000	75,000	28	65	255
	1200	85,000	64,000	32	70	207
1040‡	400	130,000	96,000	16	45	514
	600	129,000	94,000	18	52	444
	800	122,000	92,000	21	57	352
	1000	113,000	86,000	23	61	269
	1200	97,000	72,000	28	68	201
1040	400	113,000	86,000	19	48	262
	600	113,000	86,000	20	53	255
	800	110,000	80,000	21	54	241
	1000	104,000	71,000	26	57	212
	1200	92,000	63,000	29	65	192
1050‡	400	163,000	117,000	9	27	514
	600	158,000	115,000	13	36	444
	800	145,000	110,000	19	48	375
	1000	125,000	95,000	23	58	293
	1200	104,000	78,000	28	65	235
1050	400	—	—	—	—	—
	600	142,000	105,000	14	47	321
	800	136,000	95,000	20	50	277
	1000	127,000	84,000	23	53	262
	1200	107,000	68,000	29	60	223
1060	400	160,000	113,000	13	40	321
	600	160,000	113,000	13	40	321
	800	156,000	111,000	14	41	311
	1000	140,000	97,000	17	45	277
	1200	116,000	76,000	23	54	229
1080	400	190,000	142,000	12	35	388
	600	189,000	142,000	12	35	388
	800	187,000	138,000	13	36	375
	1000	164,000	117,000	16	40	321
	1200	129,000	87,000	21	50	255
1095‡	400	216,000	152,000	10	31	601
	600	212,000	150,000	11	33	534
	800	199,000	139,000	13	35	388
	1000	165,000	110,000	15	40	293
	1200	122,000	85,000	20	47	235
1095	400	187,000	120,000	10	30	401
	600	183,000	118,000	10	30	375
	800	176,000	112,000	12	32	363
	1000	158,000	98,000	15	37	321
	1200	130,000	80,000	21	47	269

† After Ref. 10.   ‡ Water quenched.

Although carbon steels of this type can be quenched and tempered for increased strength, it is not usually economical. For heat-treating purposes, these steels are carburized or case hardened. For carburizing applications, AISI 1016, 1018, and 1019 steels are commonly used for thin sections while AISI 1022 and 1024 steels are used for heavier sections. Typical mechanical properties of AISI 1015, 1020, and 1022 steels in the as-rolled, normalized, and annealed conditions are listed in Table 3-2.

Medium-Carbon Steels with 0.25 to 0.55% C

Plain-carbon steels of the medium-carbon type are usually strengthened by quenching and tempering because of their higher carbon content. These grades

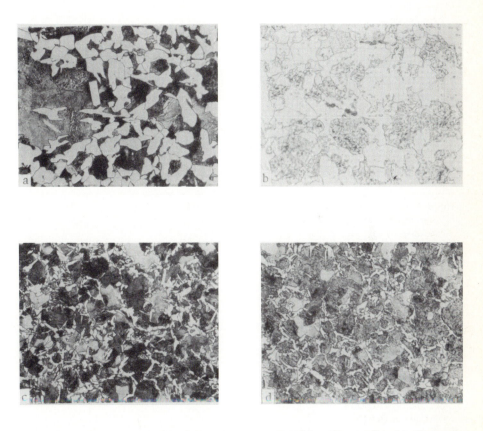

Figure 3-27 Microstructure of AISI 1040 steel after (*a*) normalizing, (*b*) annealing, (*c*) oil quenching, (*d*) oil quenching and tempering. (*a*) Normalized at 871°C 1 h and air-cooled; structure shows ferrite and pearlite; coarse pearlite was caused by coarsening of austenite during normalizing. (*b*) Normalized at 871°C 1 h and air-cooled; annealed at 691°C for 24 h; structure shows ferrite + spheroidized carbides. (*c*) Austenitized at 843°C for 1 h and oil quench; structure shows fine pearlite with ferrite outlining former austenitic grain boundaries. (*d*) Same as (*c*) plus tempering 4 h at 538°C; structure is similar to (*d*); little effect of tempering on structure. Etchant: 2% nital 400 × . *(Courtesy of Republic Steel Co.)*

Table 3-3 Usage of carbon steels in automobiles†

Use	Type of steel
Body	
Body sheet metal	1006 and 1008
Hood springs	1065
Suspension and steering	
Struts	1040 and 1030
Pitman and idler arms	1038, 1040, and 1041
Torsion bar housing	1021
Steering knuckles	1046
Tie rod ends	1040
Sway bars	1090
Tie rod studs	1018
Ball joint studs	1041
Center link	1040
Engine	
Fan	1020 and 1025
Pulley	1010
Crankshaft	1046 and 1049 modified
Connecting rod	1041
Piston pin	1016
Rocker-arm shaft	1030
Rocker arm	1040
Intake valve	1041
Oil pan	1010
Torque converter; flex plate	1020
Cover, turbine and impeller shell, and values	1006 and 1010
Overrunning clutch cam	1060 modified
Overrunning clutch hub	1060
Impeller hub	1137
Reaction shaft	1030
Transmission	
Input shaft	1024
Output shaft	1024 and 1036
Kickdown and reverse bands	1040
Sun gear driving shell	1010
Planet pinion shaft	1041
Annulus support plates	1027
Clutch disks and plates	1050
Overrunning clutch cam	1060

† From Ref. 6.

are normally produced as killed steels. By proper selection of quenching medium and temperature, a wide range of mechanical properties (85 to 160 ksi) can be obtained, as indicated in Table 3-2 for AISI 1030, 1040, and 1050 steels. When the section size is relatively small or if the properties required after heat treatment are not too high, oil quenching instead of water quenching is used since this treatment eliminates the cracking problem and reduces distortion. Figure 3-27 shows the optical microstructures of AISI 1040 steel after various heat treatments.

The medium-carbon steels are the most versatile of the three groups of hardenable plain-carbon steels, and are used for a wide range of applications. Many parts for automobiles are made from medium-carbon steels, such as parts for engines, transmissions, suspensions, and steering (Table 3-3).

High-Carbon Steels with 0.55 to 1.00% C

Steels in this group are more restricted in application than the medium-carbon steels since they are more costly to make, and have poor formability and weldability. These steels have more carbon than is needed to attain maximum as-quenched hardness, and consequently, have lower ductility than the medium-carbon steels. Table 3-2 lists the mechanical properties of AISI 1060, 1080, and 1095 steels. The ultimate tensile strengths of these steels range from 90 to 216 ksi, while their elongations range from 9 to 25 percent.

Figure 3-28 shows the microstructure of an AISI 1060 steel rod which was air-cooled after hot rolling, producing a fine pearlitic structure. Figure 3-29

Figure 3-28 Microstructure of an AISI 1060 steel rod, 1/4 in in diameter, which was cooled from hot-rolling using a high-velocity air blast. Structure is mostly unresolved pearlite with some lamellar pearlite visible; some white areas of ferrite partly outlining the prior austenite grain boundaries. (Etch: picral; 1000 × .) *(Courtesy of United States Steel Corporation.)*

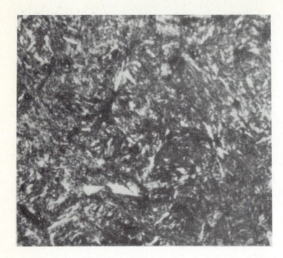

Figure 3-29 Microstructure of an AISI 1070 valve-spring steel wire in the quenched and tempered condition; steel was austenitized at 871°C, oil-quenched, and tempered at 454°C; structure is mainly tempered martensite with some free ferrite, which are the white regions. (Etch: 2% nital; ×1000.) *(Courtesy of United States Steel Corporation.)*

shows the microstructure of a 1070 steel valve spring wire that was quenched and tempered, producing a tempered martensitic structure. In most cases the high-carbon steels are heat-treated by oil quenching and tempering. Water quenching is used for heavier sections and when cutting edges are required.

3-10 MICROALLOYED STEELS

In recent years, the microalloying of plain-carbon steels with small amounts (rarely exceeding about 0.1 wt%) of strong carbide- and nitride-forming elements such as Cb, Ti, and V has achieved a great improvement in their mechanical properties. The addition of small amounts of Cb, Ti, and V in conjunction with controlled rolling practices has produced low-carbon (0.05 to 0.1% C) plain-carbon steels at low cost with yield stresses of 50 to 80 ksi and good toughness qualities. These improvements in mechanical properties are a result of many factors, the most important of which are

1. Refinement of the ferrite grain size by the formation of a fine-subgrain structure
2. Strain-induced precipitation of the carbides and nitrides of the strong carbide- and nitride-forming elements
3. Precipitation strengthening of the ferrite

Precipitation Mechanisms in Hot-Rolled Microalloyed Steels

Before the hot-rolling operation, the steel ingots are preheated (soaked) at temperatures above 1230°C and, as a result, a significant amount of the carbonitrides are dissolved. As the temperature decreases during hot rolling, the

Figure 3-30 Fine CbC (columbium carbide) precipitates formed in austenite during hot rolling. Small areas of retained austenite are indicated by the arrows. (Dark field illumination.) (Electron micrograph; 10,000 × .) *(After A. T. Davenport, L. C. Brossard, and R. E. Miner, J. Metals, June 1975, p. 21.)*

carbonitrides become insoluble and precipitate out in the austenite during hot rolling (Fig. 3-30).

In the initial stages of hot rolling, the coarse austenite grains produced by preheating are progressively reduced in size by recrystallization induced by the deformation of each reduction. The carbonitride particles which are induced by the deformation reduce the size of the recrystallized grains by "pinning" the grain boundaries. Second, in the final stages of hot rolling, these precipitates retard recrystallization because they prevent the substructure from changing by the processes of dislocation and subgrain boundary migration. The net effect of these processes is to progressively flatten the austenitic grains so that a "pancake" structure is produced which has a higher austenitic grain boundary area per unit volume than normally would be obtained.

Since the ferrite nucleates mainly in the austenitic grain boundaries, the increased grain boundary area will provide more nuclei for ferrite and hence lead to a finer ferrite grain size. Finally, any microalloying element left unprecipitated during hot rolling will precipitate in the ferrite either during cooling to room temperature (plate steels) or during the coiling operation (strip steels). The precipitation in the ferrite will provide additional strength to the microalloyed steels.

Precipitation of Cb, Ti, and V Carbides and Nitrides

M(C, N) precipitation in austenite The principal compound which precipitates in microalloyed steels is the FCC-type phase (NaCl) of the general formula

Figure 3-31 Fine CbC particles nucleated in the ferrite matrix in a quenched and tempered Cb-microalloyed steel. (110,000 × .) *(After A. T. Davenport, L. C. Brossard, and R. E. Miner, J. Metals, June 1975, p. 21.)*

M(C, N). This phase precipitates in the austenite according to the following relationship:

$$(100)_{M(C, N)} \| (100)_\gamma$$

Figure 3-30 shows fine CbC particles formed in austenite when a 0.1% C–1.3% Mn–0.1% Cb steel was preheated at 1288°C, hot-rolled at 870°C, and aged 15 min at the rolling temperature.

M(C, N) precipitates in ferrite M(C, N) precipitation can also occur in the ferrite both in the matrix and at the γ/α interface boundaries. The following relationship has been determined for the precipitation of M(C, N) in supersaturated ferrite:

$$(100)_{MC} \| (100)_\alpha$$

Figure 3-32 A colony of VC particles formed by interphase precipitation in a normalized V-bearing steel. (25,000 × .) *(After A. T. Davenport, L. C. Brossard, and R. E. Miner, J. Metals, June 1975, p. 21.)*

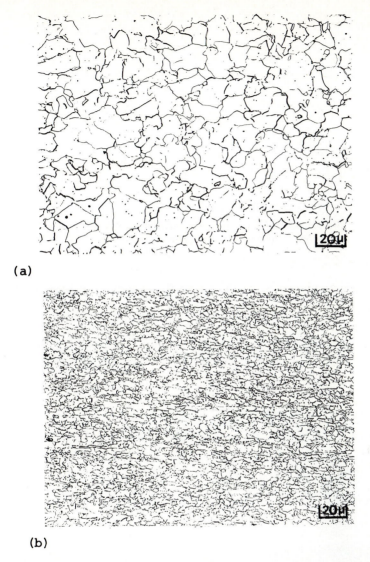

(a)

(b)

Figure 3-33 Optical microstructures of 50 and 80 ksi microalloyed columbium steels. *(After P. L. Mangonon and W. E. Heitmann, Microalloying 75, Union Carbide Co., New York, 1977, p. 59.)*

Figure 3-31 shows fine CbC particles nucleated in the ferrite in a quenched and tempered Cb low-carbon steel.

M(C, N) precipitates nucleated at the γ/α interphase boundaries form with the same ferrite nucleation orientation as described above and lie in sheetlike arrays, the plane of which denotes the position of the interphase boundary at the time of nucleation. These interphase precipitates sometimes appear as precipitate rows (Fig. 3-32), which are formed by VC in a normalized microalloyed steel.

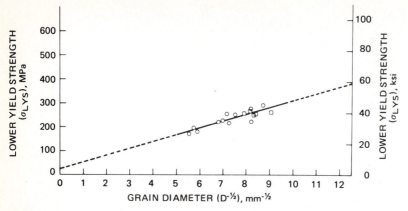

Figure 3-34 Relationship between grain size and lower yield strength in AISI 1005 rimmed steel sheet. *(After P. L. Mangonon and W. E. Heitmann, Microalloying 75, Union Carbide Co., New York, 1977, p. 59.)*

Strengthening of Microalloyed Steels by Grain Refinement and Subgrains

Microalloyed steels are also strengthened to some degree by a fine-grain size and fine-subgrain structure. Figure 3-33 compares the grain size of 50- and 80-ksi columbium steels. Although grain size is only one factor contributing to the increased strength of microalloyed steels, there is some increase in lower yield strength due to a finer-grain size, as shown in Fig. 3-34.

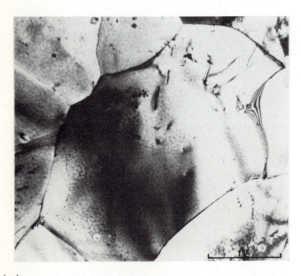

(a)

Figure 3-35 Substructure of (*a*) 50 ksi and (*b*) 80 ksi microalloyed columbium steels; electron transmission micrographs. *(After P. L. Mangonon and W. E. Heitmann, Microalloying 75, Union Carbide Co., New York, 1977, p. 59.)*

(b)

Figure 3-35 Continued.

A far greater structural contribution to increased strength, however, is due to a fine subgrain structure. Figure 3-35 shows the observed microscopic difference in substructure between the 50- and 80-ksi microalloyed steels. The highly refined substructure of the 80-ksi steel is quite noticeable. Figure 3-36 shows the incremental increase in yield strength due to the subgrain refinement, which is found to depend on the subgrain size and the volume fraction of grains with subgrains.

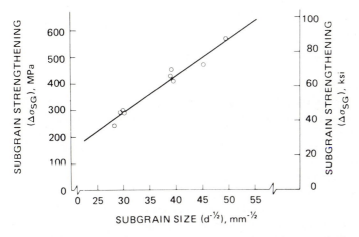

Figure 3-36 Regression line between subgrain size (d) and its strengthening effect ($\Delta\sigma_{SG}$) for AISI 1005 rimmed steel. *(After P. L. Mangonon, and W. E. Heitmann, Microalloying 75, Union Carbide Co., New York, 1977, p. 59.)*

Thus microalloyed steels are strengthened by a combination of grain refinement, subgrain formation, and precipitation hardening. The amount of strengthening from columbium carbonitride precipitates depends upon the amount of columbium added, finishing rolling temperatures, and amount of deformation.

PROBLEMS

1. Briefly describe the following steelmaking processes: (*a*) basic oxygen, (*b*) basic open hearth, and (*c*) electric arc.

2. Why has the basic-oxygen process become the most important steelmaking process in the United States?

3. What are the advantages of the basic-oxygen process?

4. Why is most of the steel in the United States still cast into individual molds?

5. Describe the continuous-casting process for steel. What are its advantages? Disadvantages?

6. Describe the following types of steel ingot structures and the processes used to produce each of them: (*a*) rimmed, (*b*) capped, (*c*) semikilled, and (*d*) killed

7. What are the advantages and disadvantages of rimmed steel ingots? Of aluminum-killed ingots?

8. How would it be possible to produce a composite ingot with a rimmed outer zone and an aluminum-killed core?

9. What is the AISI-SAE classification system for plain-carbon steels? Why can it not be used for all plain-carbon steels for all purposes?

10. Describe the effects of the following elements in plain-carbon steels: (*a*) manganese, (*b*) sulfur, (*c*) phosphorus, and (*d*) silicon.

11. Why are manganese sulfide inclusions in steel preferable to iron sulfide ones?

12. When manganese sulfide inclusions are rolled, they are elongated in the direction of rolling. What would be the disadvantage of these inclusions in rolled plate?

13. Describe the typical cross sections of (*a*) slabs, (*b*) blooms, and (*c*) billets.

14. Describe the effects of hot rolling on the structure of steel strip.

15. Describe the pickling process which is used for hot-rolled steel strip before it is cold-rolled.

16. What properties are desirable in non-heat-treatable low-carbon sheet steel?

17. What is the chemical composition of about 80 percent of the non-heat-treatable low-carbon sheet steel?

18. Describe the box-annealing process. What are its advantages and disadvantages?

19. What is the effect of AlN in the recrystallization of low-carbon sheet steels?

20. Describe the type of structure of killed low-carbon sheet steels which have especially high formability.

21. How does the grain structure of low-carbon sheet steel which has been continuously annealed differ from that which has been box-annealed?

22. What causes the quench-aging effect in low-carbon sheet steel?

23. Describe the strain-aging phenomenon in low-carbon sheet steels.

24. How can strain aging be avoided in rimmed low-carbon sheet steels?

25. Why is strain aging not encountered in killed low-carbon sheet steels?

26. Describe the three major groups of hardenable plain-carbon steels and some of their applications.

27. What are microalloyed steels? What are the principal elements that are added to produce microalloyed steels?

28. Describe the precipitation mechanisms which strengthen microalloyed steels.

29. Describe how microalloyed steels are strengthened by grain refinement and subgrain structure.

30. What processing factors affect the amount of strengthening obtained in microalloyed steels?

31. In the rolling of sheet, plate, and strip in modern steel mills, deformation takes place mainly in the longitudinal direction. During this operation, manganese sulfides are deformed plastically into longitudinal stringers.

(*a*) What difference in strength with respect to the longitudinal, transverse, and thickness directions would be expected?

(*b*) Cerium and calcium additions modify the sulfide inclusions to isolated globules with decreased plasticity. How would the strength properties of plate or sheet be changed in the different directions? [See T. M. Banks and T. Gladman, *Met. Tech.* 6(1979):81.]

32. At present only about 15 percent of the steel produced in the United States is continuously cast. It is estimated that about 85 percent of the steel could be continuously cast. What difficulties and reasons account for this difference?

REFERENCES

1. H. E. McGannon (ed.): "The Making, Shaping, and Treating of Steel," 9th ed., United States Steel Corporation, Pittsburgh, 1971.
2. E. C. Bain and H. W. Paxton: "Alloying Elements in Steel," 2d ed., American Society for Metals, Metals Park, OH, 1966.
3. *Metals Handbook*, vol. 1: "Properties," American Society for Metals, Metals Park, OH, 1961; vol. 2: "Heat Treatment," ASM, 1964; vol. 8: "Metallography, Structures, and Phase Diagrams, ASM, 1973.
4. *Microalloying 75*, Union Carbide Co, New York, 1977.
5. R. W. K. Honeycombe: "Transformation from Austenite in Alloy Steels," *Met. Trans.* 7A(1976):915.
6. C. R. Weymueller, "Steels in the Auto Industry," *Metal Progress*, Oct. 1965, p. 128.
7. L. V. Gallagher and B. S. Olds, "The Continuous Casting of Steel," *Scientific American*, vol. 209, Dec. 1963, p. 75.
8. A. T. Davenport, L. C. Brossard, and R. E. Miner, "Precipitation in Microalloyed High-Strength Low-Alloy Steels, *J. Metals*, June 1975, p. 21.
9. W. C. Leslie and A. S. Keh: "Aging of Flat-Rolled Steel Products as Investigated by Electron Microscopy." Volume 26 in T. G. Bradbury (ed.): *Mechanical Working of Steel 2, Metallurgical Society Conferences*, Gordon and Breach, New York, 1965.
10. "ASM Databook," published in *Metal Progress*, vol. 112, no. 1, mid-June 1977.
11. J. K. Stone: "Oxygen in Steelmaking," *Scientific American*, vol. 218, April 1968, p. 24.
12. J. B. Ballance (ed.): "The Hot Deformation of Austenite," American Institute of Mining, Metallurgical, and Petroleum Engineers, New York, 1977.
13. J. M. Gray (ed.): "Processing and Properties of Low-Carbon Steel," American Institute of Mining, Metallurgical, and Petroleum Engineers, New York, 1973.
14. R. H. Goodenow: "Recrystallization and Grain Structure in Rimmed and Aluminum Killed Low-Carbon Steel," *Trans. ASM* 59(1966):804.
15. R. A. Grange: "Strengthening Steel by Austenite Grain Refinement," *Trans. ASM* 59(1966):26.
16. M. R. Baren and P. G. Nelson: "Strain Aging of Low-Carbon Sheet Steel," *Metal Progress*, Dec. 1970.
17. P. L. Mangonon and W. E. Heitmann: "Subgrain and Precipitation-Strengthening Effects in Hot-Rolled, Columbium-Bearing Steels." In *Microalloying 75*, Union Carbide Co., New York, 1977.
18. H. E. Chandler: "A Look at American Steel Technology from the User Point of View," *Metal Progress*, Oct. 1978, p. 40.
19. R. W. K. Honeycombe: "Transformation from Austenite in Alloy Steels," *Met. Trans.* 7A(1976):915.

FOUR

ALLOY STEELS

Although plain-carbon steels can be produced in a great range of strengths at a relatively low cost, their properties are not always adequate for all engineering applications of steel. In general plain-carbon steels have the following limitations:

1. They cannot be strengthened beyond about 100,000 psi without significant loss in toughness (impact resistance) and ductility.
2. Large sections cannot be made with a martensitic structure throughout, and thus are not deep-hardenable.
3. Rapid quench rates are necessary for full hardening in medium-carbon plain-carbon steels to produce a martensitic structure. This rapid quenching leads to shape distortion and cracking of heat-treated steel.
4. Plain-carbon steels have poor impact resistance at low temperatures.
5. Plain-carbon steels have poor corrosion resistance for many engineering environments.
6. Plain-carbon steels oxidize readily at elevated temperatures.

For these and other reasons, *alloy steels* have been developed which, although they cost more, are more economical for many uses. In some applications, alloy steels are the only materials that are able to meet engineering requirements. The principal elements that are added to make alloy steels are nickel, chromium, molybdenum, manganese, silicon, and vanadium. Other elements sometimes added are cobalt, copper, and lead.

In this chapter, engineering alloys which are used for construction and automotive applications are mainly considered. Stainless steels are treated in Chap. 7, and tool steels in Chap. 9.

4-1 CLASSIFICATION OF ALLOY STEELS

In a general sense, alloy steels may contain up to about 50 percent of alloying elements and still be called alloy steels. However, in a technical sense, the term *alloy steels* will be used in this text to refer to heat-treatable construction and automotive alloy steels which contain from about 1 to 4 percent alloying elements.

Alloy steels in the United States are usually referred to by the AISI-SAE system, which uses four digits to designate each alloy steel. The first two digits

Table 4-1 Principal types of standard alloy steels†

13xx	Manganese 1.75
40xx	Molybdenum 0.20 or 0.25; or molybdenum 0.25 and sulfur 0.042
41xx	Chromium 0.50, 0.80, or 0.95, molybdenum 0.12, 0.20, or 0.30
43xx	Nickel 1.83, chromium 0.50 or 0.80, molybdenum 0.25
44xx	Molybdenum 0.53
46xx	Nickel 0.85 or 1.83, molybdenum 0.20 or 0.25
47xx	Nickel 1.05, chromium 0.45, molybdenum 0.20 or 0.35
48xx	Nickel 3.50, molybdenum 0.25
50xx	Chromium 0.40
51xx	Chromium 0.80, 0.88, 0.93, 0.95, or 1.00
51xxx	Chromium 1.03
52xxx	Chromium 1.45
61xx	Chromium 0.60 or 0.95, vanadium 0.13 or min 0.15
86xx	Nickel 0.55, chromium 0.50, molybdenum 0.20
87xx	Nickel 0.55, chromium 0.50, molybdenum 0.25
88xx	Nickel 0.55, chromium 0.50, molybdenum 0.35
92xx	Silicon 2.00; or silicon 1.40 and chromium 0.70
50Bxx	Chromium 0.28 or 0.50
51Bxx	Chromium 0.80
81Bxx	Nickel 0.30, chromium 0.45, molybdenum 0.12
94Bxx	Nickel 0.45, chromium 0.40, molybdenum 0.12

Note: B denotes boron steel.
† After Ref. 13.

Table 4-2 Nominal compositions and typical applications of selected standard alloy steels

AISI-SAE No.	% C	% Mn	% Cr	% Mo	% Ni	% Si†	Typical applications
Manganese steels							
1330	0.30	1.75					High-strength bolts
1340	0.40	1.75					
Chromium steels							
5120	0.20	0.80	0.80				Carburizing steel
5130	0.30	0.80	0.95				Steering parts
5140	0.40	0.80	0.80				
5160	0.60	0.88	0.80				Spring steels
E52100	1.04	0.35	1.45				Ball and roller bearings
Molybdenum steels							
4023	0.23	0.80		0.25			Carburizing steel
4037	0.37	0.80		0.25			
4047	0.47	0.80	0.25				
Chromium-molybdenum steels							
4118	0.18	0.80	0.50	0.13			
4130	0.30	0.50	0.95	0.20			Pressure vessels, aircraft structural parts, auto
4140	0.40	0.88	0.95	0.20			axles, steering knuckles
Chromium-vanadium steels							
6150	0.50	0.80	0.95				0.15V; valves and springs
Nickel-molybdenum steels							
4620	0.20	0.55		0.25	1.83		Transmission gears, chain pins, shafts, roller
4820	0.20	0.60		0.25	3.50		bearings
Nickel (1.83%)-chromium-molybdenum steels							
4320	0.20	0.55	0.50	0.25	1.83		Carburizing steel
4340 (E)	0.40	0.70	0.80	0.25	1.83		Heavy sections, landing gears, truck parts
Nickel (0.55%)-chromium-molybdenum steels							
8620	0.20	0.80	0.50	0.20	0.55		Carburizing steel
8640	0.40	0.88	0.50	0.20	0.55		Auto springs, small machine axles, shafts
8660	0.60	0.88	0.50	0.20	0.55		
Silicon steels							
9260	0.60	0.88				2.0	Leaf springs

† All steels contain 0.28% min Si except 9260; all steels contain 0.035% max P and 0.040% max S except electric furnace steels (E), which have 0.025% max P and 0.025% max S.

indicate the principal alloying element or group of alloying elements, such as those listed in Table 4-1. The last two digits indicate the approximate nominal carbon content of the alloy. Table 4-2 lists the nominal composition of some selected standard alloy steels.

4-2 EFFECTS OF ALLOYING ELEMENTS IN ALLOY STEELS

General Effects of Alloying Elements in Steel

Alloying elements are added to plain-carbon steels for many purposes. Some of the most important of these are:

1. To improve mechanical properties by increasing the depth to which a steel can be hardened[1]
2. To allow higher tempering temperatures while maintaining high strength and good ductility
3. To improve mechanical properties at high and low temperatures
4. To improve corrosion resistance and elevated-temperature oxidation
5. To improve special properties such as abrasion resistance and fatigue behavior

Items 1 and 2 are particularly important. By increasing the depth of hardening of plain-carbon steels (item 1), larger sections can be made martensitic throughout, and thus the strength and toughness advantage of a tempered martensitic structure can be obtained. Also, by increasing the depth of hardening in a steel, a slower quench rate can be used, and thus cooling stresses can be lessened. Oil or air quenching reduces thermal gradients which can lead to distortion and cracking of steels.

By increasing the resistance to softening during tempering (item 2), the alloy steels are able to resist softening at higher tempering temperatures. A lower carbon content may therefore be used to obtain the same tempered hardness as in a higher-carbon-containing plain-carbon steel. Since a steel with a lower carbon content is in general tougher than one containing more carbon, the lower-carbon alloy steel will have increased toughness. Likewise, the toughness of an alloy steel can be increased over that of a plain-carbon steel of the same carbon content by tempering at a higher temperature, which allows greater relaxation of stresses while maintaining the same hardness.

Distribution of Alloying Elements in Alloy Steels

The distribution of the alloying elements in an alloy steel depends on its composition. Many complex interactions can occur and, as the number and amount of alloying elements are increased, the complexity of the interactions also increases. However, there are basic trends in the distribution of the alloying elements which can be observed and which are listed in Table 4-3 for annealed alloy steels at room temperature.

---

[1] The property of steels that determines the depth and degree of their hardening is called *hardenability*, which will be discussed in detail in Sec. 4-3.

Table 4-3 Approximate distribution of alloying elements in alloy steels† *

Element	Dissolved in ferrite	Combined in carbide	Combined as carbide	Compound	Elemental
Nickel	Ni			Ni₃Al	
Silicon	Si				SiO₂·M_xO_y
Manganese	Mn ◄─────► Mn		(Fe,Mn)₃C	MnS; MnO·SiO₂	
Chromium	Cr ◄─────► Cr		(Fe,Cr)₃C		
			Cr₇C₃		
			Cr₂₃C₆		
Molybdenum	Mo ◄─────► Mo		Mo₂C		
Tungsten	W ◄─────► W		W₂C		
Vanadium	V ◄─────► V		V₄C₃		
Titanium	Ti ◄─────► Ti		TiC		
Columbium	Cb ◄─────► Cb		CbC		
Aluminum	Al			Al₂O₃; AlN	
Copper	Cu (small amount)				
Lead	Pb				Pb

† From Ref. 1.
* The arrows indicate the relative tendencies of the elements listed to dissolve in the ferrite or combine in carbides.

Nickel has less of a carbide-forming tendency than iron and dissolves in the α ferrite. Silicon combines to some extent with oxygen to form nonmetallic inclusions, but otherwise dissolves in the ferrite. Much of the manganese in alloy steels dissolves in the α ferrite regardless of the carbon content. Manganese is only moderately more carbide-forming than iron, and the manganese that does form carbides in steels usually enters the cementite as (Fe,Mn)₃C.

Chromium partitions between the ferrite and carbide phases. The distribution of chromium depends on the amount of carbon and other carbide-forming elements present in the steel. Tungsten and molybdenum combine with carbon to form carbides if sufficient carbon is present and if other stronger carbide-forming elements such as titanium and columbium are absent. Vanadium, titanium, and columbium are strong carbide-forming elements and will be found in steels mainly as carbides. If sufficient nitrogen is present, some columbium nitride will also form. Aluminum combines with oxygen and nitrogen to form the compounds Al₂O₃ and AlN, respectively.

Effects of Alloying Elements on the Eutectoid Point of Steels

All common substitutional alloying elements in steel such as nickel, chromium, silicon, manganese, tungsten, molybdenum, and titanium lower the eutectoid carbon content, as is shown graphically in Fig. 4-1a. Titanium, tungsten, and molybdenum are the most effective, whereas nickel and chromium are the least

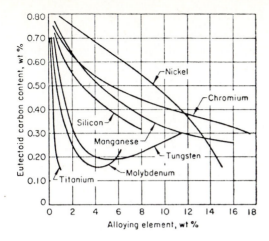

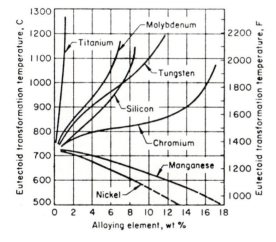

Figure 4-1 The effect of percentage substitutional elements in steel on (*a*) the carbon content of the eutectoid point and (*b*) the temperature of the eutectoid transformation point. *(After Metals Handbook, 9th ed., vol. 8, American Society for Metals, 1973, p. 191.)*

effective. For example, a steel containing 5% Cr has its eutectoid carbon content reduced from 0.8 to 0.5 percent.

Some elements lower the eutectoid temperature of steels and others raise it, as shown in Fig. 4-1*b*. Manganese and nickel both lower the eutectoid temperature, and are thus considered *austenite-stabilizing elements*. The effect of increasing additions of manganese from 0.35 to 9 percent in enlarging the austenitic region in carbon steels is shown in Fig. 4-2*a*. Nickel behaves in a similar way as manganese, and enlarges the austenitic region. In some steels with sufficient amounts of nickel or manganese, austenite may be retained at room temperature.

The carbide-forming elements such as tungsten, molybdenum, silicon, and titanium shift the eutectoid temperature to higher values and reduce the austenitic phase field. These elements are thus termed *ferrite-stabilizing elements*.

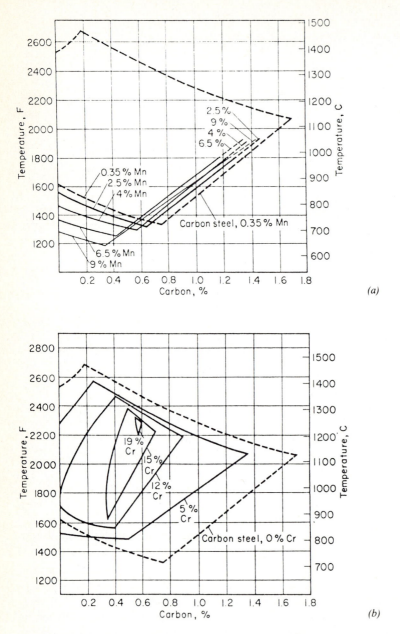

Figure 4-2 The effect of (*a*) manganese and (*b*) chromium additions on the austenite phase region in carbon steels. (*After E. C. Bain and H. W. Paxton, "Alloying Elements in Steel," 2d ed., American Society for Metals, 1966, pp. 104–105.*)

Figure 4-2*b* shows how increasing the chromium content from 0 to 19 percent decreases the austenitic phase field in carbon steels. With over about 12 percent chromium, the α-ferrite and δ-ferrite regions merge, as indicated in Fig. 4-2*b*.

4-3 HARDENABILITY

Definition

The *hardenability* of a steel is defined as that property which determines the depth and distribution of hardness induced by quenching. Hardenability is a characteristic of a steel and is principally determined by the following factors:

1. Chemical composition of the steel
2. Austenitic grain size
3. Structure of the steel before quenching

Hardenability should not be confused with the *hardness* of a steel, which is its resistance to plastic deformation. Hardness is usually measured by a hardness-testing machine that makes an indentation into the surface of the steel. Hardenability, on the other hand, is a measure of the depth of hardening of a steel upon quenching from austenite.

Determination of Hardenability by Grossmann's Method

Critical diameter of hardenable steel bar To determine the hardenability of a steel by Grossmann's method, a series of cylindrical steel bars of a specified steel of different diameters (i.e., 0.5 to 2.5 in) are hardened by quenching from austenitic temperatures to room temperature in a particular quenching medium. After a metallographic examination, the bar that has 50% martensite at its center is selected as the bar with the *critical diameter*, D_o (in inches). Thus, the critical diameter is the diameter of the largest bar whose cross section contains no unhardened core after etching. The critical diameter is also called the *actual critical diameter*.

Ideal critical diameter The critical diameter of a hardenable steel bar depends upon, in addition to its structure and composition, the medium in which it is quenched. Thus, the rate at which the steel bar is quenched from the austenitic temperature range will affect the value of the critical diameter of the bar. In order to eliminate the cooling rate variable, all hardenability measurements are referred to an "ideal quench." The ideal quench is obtained with an hypothetical cooling medium which is assumed to remove heat from the surface of the bar as soon as it can flow out from within the bar; that is, the surface of the quenched bar would be cooled instantly to the temperature of the quenching liquid. The critical diameter of the steel bar when using the ideal quench is called the *ideal critical diameter*, D_I (in inches).

No ideal quench exists. However, a comparison can be made of the ideal quench with ordinary quenching media such as brine, water, or oil. The cooling intensities of different cooling media are assigned H numbers, which represent coefficients of severity of the cooling media. The ideal quenching medium is assigned a value of infinity. The H values for oil, water, and brine quenching media are listed in Table 4-4.

The relationship among the actual critical diameter D_o, the ideal critical diameter D_I, and the severity of quench (H values) is shown graphically in Fig. 4-3. In practice, D_o values are determined using this graph from calculated values of D_I and appropriate H values.

Example problem The ideal critical diameter of a steel was calculated to be 2.2 in. What is its actual critical diameter D_o if the steel is subjected to an oil quench with moderate agitation?

ANSWER Referring to Table 4-4, a value of 0.40 will be taken as the H value. Using the chart in Fig. 4-3, with $H = 0.4$ and $D_I = 2.2$ in, a value of 0.9 in is obtained for D_o.

Effect of austenitic grain size on the hardenability of steels The effect of austenitic grain size on the hardenability of steels is explained by the heterogeneous nucleation of pearlite at the austenitic grain boundaries. During the transformation of austenite to pearlite, pearlite nucleates preferentially at the austenitic grain boundaries. Thus, the more grain boundary surface available for pearlitic nucleation, the easier it is for pearlite to form. The smaller the grain size, therefore, the lower the hardenability of the steel when all other factors are constant.

A coarse grain size is not a desirable structure for most steels since it leads to lower strengths and decreased ductility. Also, the tendency to crack increases

Table 4-4 Cooling intensities of different quenching media (H-factors)†

Agitation	Oil	Coefficient of severity of quench H, cooling medium Water	Brine
None	0.25–0.30	0.9–1.0	2.0
Mild	0.30–0.35	1.0–1.1	2.0–2.2
Moderate	0.35–0.40	1.2–1.3	
Good	0.4–0.5	1.4–1.5	
Strong	0.5–0.8	1.6–2.0	
Violent	0.8–1.1	4.0	5.0

† After Ref. 2.

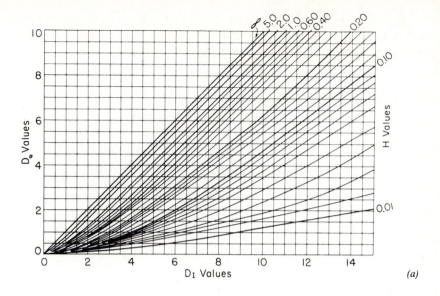

(a)

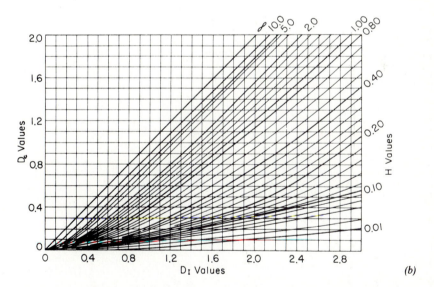

(b)

Figure 4-3 Relationships among ideal critical diameter D_I, actual critical diameter D, and severity of quench H. The lower diagram (*b*) is an enlargement of the lower left-hand section of the upper diagram (*a*). (*After M. A. Grossmann and E. C. Bain, "Principles of Heat Treatment," 5th ed., American Society for Metals, 1964, pp. 99–100.*)

in a coarse-grained steel. Increasing the grain size to increase the hardenability of a steel is therefore not a beneficial procedure overall and is not normally used. It is more efficient to add other alloying elements to increase the hardenability of steels.

Effect of carbon content on the hardenability of steels Increasing the carbon content of a steel greatly increases its hardenability. Since a high carbon content in a steel is not always desirable, a relatively low-carbon steel with other alloying additions to increase hardenability is the most common situation. The relationship between carbon content, austenitic grain size, and ideal critical diameter for plain-carbon steels is shown in Fig. 4-4. Using this chart, the ideal critical diameter of a plain-carbon steel can be determined for a particular austenitic grain size.

Example Problem Determine the ideal critical diameter of a 0.6% plain-carbon steel with an ASTM grain size of 8.

ANSWER Using the chart in Fig. 4-4, a value of 0.24 in is obtained for the ideal critical diameter.

This means that even with an ideal quench, the maximum hardenable diameter of this steel is about 0.25 in. An ordinary quench would not even harden a cylinder of this diameter. Commercial plain-carbon steels do not have

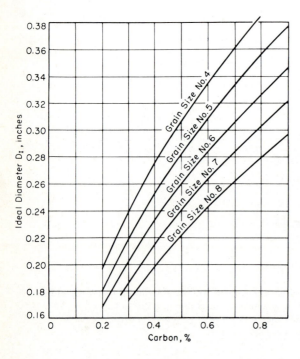

Figure 4-4 Ideal critical diameter D_I, as a function of carbon content and austenitic grain size for plain carbon steels. *(After M. A. Grossmann and E. C. Bain, "Principles of Heat Treatment," 5th ed., American Society for Metals, 1964, p. 122.)*

as low a hardenability as indicated for this steel since all commercial steels contain some manganese and other impurities which increase hardenability.

Effect of alloying elements on hardenability Each element in a steel has some effect on its hardenability. All common alloying elements except *cobalt* increase the hardenability of steel. Cobalt increases the rate of nucleation and growth of pearlite, and hence decreases hardenability. The relative effect of common alloying elements on hardenability is shown in Fig. 4-5, which gives the multiplying factors for each alloying element at various percentages in the steel. These multiplying factors make possible an approximate calculation of the hardenability of a steel when only its chemical composition and austenitic grain size are known. The following example problem shows how such calculations can be made.

> **Example problem** Calculate the approximate hardenability of an 8630 alloy steel which has an ASTM grain size of 7 and the following chemical composition: 0.3% C, 0.3% Si, 0.7% Mn, 0.5% Cr, 0.6% Ni, 0.2% Mo.
>
> ANSWER First, the base diameter D_I is looked for on Fig. 4-4, and is found to be 0.185 in. Next, the multiplying factors for each element are determined from Fig. 4-5. This is done by drawing a vertical line at the composition of the element in question and finding where it intersects the curve for that element. The value of the multiplying factor is determined by drawing a horizontal line from the point of intersection back to the ordinate value.

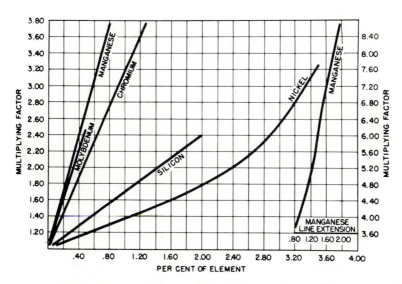

Figure 4-5 Multiplying factors for different alloying elements for hardenability calculations. [*After H. E. McGannon (ed.), "The Making, Shaping, and Treating of Steel," 9th ed. United States Steel Corporation., 1971, p. 1132.*]

Using this method, the following multiplying factors for this problem are found:

Percentage of alloying element	Multiplying factor
0.3 Si	1.2
0.7 Mn	3.4
0.5 Cr	2.1
0.6 Ni	1.2
0.2 Mo	1.6

Finally, the ideal critical diameter is found by multiplying the base diameter by the multiplying factors:

$$D_I = 0.185 \times 1.2 \times 3.4 \times 2.1 \times 1.2 \times 1.6 = 3.04 \text{ in}$$

If a mild water quench of $H = 1.0$ is used, the actual critical diameter D_o is reduced to 2.3 in.

In comparison, a 1030 plain-carbon steel with 0.7% Mn has an ideal critical diameter of 0.65 in. If a mild water quench is used, the actual critical diameter D_o is reduced to 0.2 in. Thus, the alloying element additions in the 8630 alloy steel increased the actual critical diameter of the 1030 steel from 0.2 to 2.3 in, which is a considerable increase in hardenability.

Determination of Hardenability by the Jominy Method

The Grossmann method of determining hardenability of steels is too complicated and costly to be of great practical importance commercially. The most common method for determining hardenability in industry is the Jominy method. In the Jominy test, a single specimen replaces the series of samples needed for the Grossmann test.

In the Jominy end-quench test, the specimen consists of a cylindrical bar with a diameter of 1 in and a length of 4 in (Fig. 4-6). Since prior structure has a strong effect on hardenability, the specimen should be normalized before testing. In the Jominy test, after the specimen has been austenitized, it is placed in a fixture, as shown in Fig. 4-6, and a jet of water is quickly showered on the end of the specimen. After cooling, two parallel flats are ground on opposite sides of the test bar and Rockwell C hardness tests are made along these surfaces. End-quench hardenability curves are made by plotting the hardness of the steel as a function of distance from the quenched end, as shown in Fig. 4-7 for an AISI 1050 plain-carbon steel.

A comparison of the hardenability of different steels can be made by plotting Jominy end-quench test curves together as shown in Fig. 4-8. The high hardenability of the 4340 alloy steel is shown by its ability to maintain a

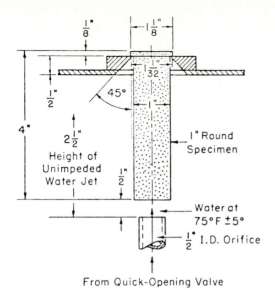

Figure 4-6 Specimen and fixture for end-quench hardenability test. *(After M. A. Grossmann and E. C. Bain, "Principles of Heat Treatment," 5th ed., American Society for Metals, 1964, p. 114.)*

Rockwell C40 hardness up to 2 in from the quenched end of the specimen. For the plain-carbon 1050 steel, its hardness falls to about Rockwell C35 at $\frac{3}{16}$ in from the quenched end (Fig. 4-7), and hence plain-carbon steels like this one have relatively low hardenability. The hardness change along the side of a Jominy end-quenched alloy specimen can be correlated with its continuous-cooling transformation diagram, as is indicated in Fig. 4-9 for a 1080 eutectoid steel. It is the simplicity of the Jominy end-quench test along with detailed hardenability data that make this test so widely used industrially.

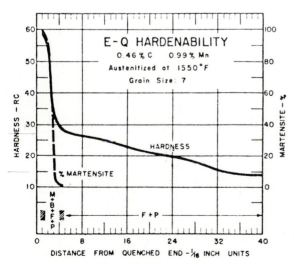

Figure 4-7 End-quench hardenability curve for an AISI 1050 steel. *(After "Isothermal Transformation Diagrams," United States Steel Corporation, 1963, p. 19.)*

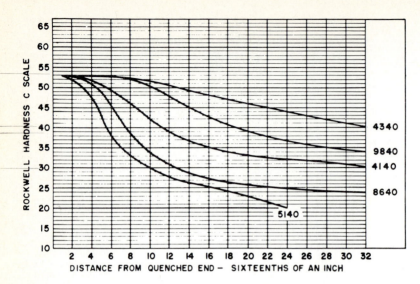

Figure 4-8 Comparative hardenability curves for 0.40% carbon alloy steels. [*After H. E. McGannon (ed.), "The Making, Shaping, and Treating of Steels," United States Steel Corporation, 1971, p. 1139.*]

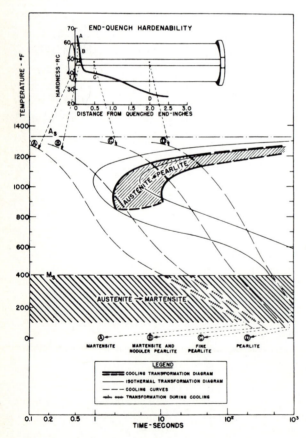

Figure 4-9 Correlation of continuous cooling transformation diagram and end-quench hardenability test data for eutectoid carbon steel. *(After "Isothermal Transformation Diagrams," United States Steel Corporation, 1963, p. 181.)*

4-4 MANGANESE STEELS

Chemical Compositions and Applications

Manganese is added to all commercial steels in the range of 0.25 to 1.00 percent to deoxidize it and to combine with sulfur to form globular MnS. Manganese is most effective when strength increase is considered in relationship to cost increase. Thus, when higher strength than mild steel is required combined with weldability, 1.6 to 1.9% Mn steels are widely used. The AISI 13xx series of manganese low-alloy steels have nominal levels of carbon from 0.30 to 0.45 percent, and 1.75% Mn. These 13xx steels have higher strengths and hardenabilities than their plain-carbon steel counterparts and are used for axles, shafts, gears, and tie rods for automobiles and farm implements.

Structure

The hardenability of 13xx alloy steels is slightly higher than the 10xx plain-carbon steels and is the result of the increase in manganese content to a nominal 1.75 percent in the 13xx alloys. The I-T diagram of 1340 alloy is shown in Fig. 4-10. As compared to the diagram for 1040 plain-carbon steel, the transformation boundaries in the 1340 alloy diagram are slightly shifted to the right. Manganese, by reducing diffusion rates, makes the transformation from

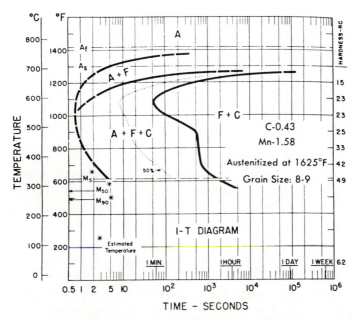

Figure 4-10 Isothermal transformation diagram of AISI 1340 steel. *(After "Isothermal Transformation Diagrams," United States Steel Corporation, 1963, p. 26.)*

Figure 4-11 Microstructure of AISI 1340 steel, containing 1.74% Mn and 0.40% C. Air-cooled from 828°C. Structure shows fine pearlite with some ferrite outlining prior austenite boundaries. (Etch: picral; 500 × .) *(Courtesy of R. M. Fisher, U. S. Steel Research Laboratories.)*

austenite to ferrite-pearlite more sluggish, thus increasing the hardenability of carbon steels. Manganese also refines the pearlite in carbon steels and thereby strengthens them. The pearlitic refinement action of manganese is clearly seen in the austenitized and air-cooled AISI 1340 alloy microstructure shown in Fig. 4-11.

When the manganese content of carbon steels exceeds about 2 percent, the steels become embrittled. However, if the manganese content is increased to about 12 percent and the carbon content to about 1.1 percent, the manganese steel becomes austenitic at room temperature if rapidly quenched from the austenitic state. This alloy, which is known as *Hadfield's manganese steel*, was developed in 1882 and was one of the first high-alloy steels. In the austenitic condition, it is particularly resistant to wear and abrasion under high-impact stresses, since it work-hardens at a very high rate.

Mechanical Properties

The effect of manganese in strengthening plain-carbon steels can be divided into the following three parts: solid-solution hardening, grain-size refinement, and increase in proportion of pearlite. Manganese is soluble in γ and α iron, and strengthens the ferrite in carbon steels by solid-solution strengthening. The extent of the strengthening for a 0.15% C steel as a function of manganese

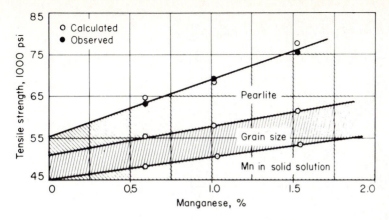

Figure 4-12 Factors contributing to the effect of manganese on the tensile strength of annealed 0.15% C steels. [*After K. J. Irving and F. B. Pickering, JISI, 201(1963):944, as presented in E. C. Bain and H. W. Paxton, "Alloying Elements in Steel," 2d ed., American Society for Metals, 1966, p. 270.*]

content up to 2 percent is shown in Fig. 4-12. By both refining and increasing the proportion of pearlite, manganese considerably strengthens low-carbon steels, as indicated in Fig. 4-12. The overall effect of 1.75% Mn in increasing the hardness upon tempering a 1340 alloy steel as compared to a 1040 carbon steel is shown in Fig. 4-13. Table 4-5 lists the mechanical properties of alloys 1330 and 1340 after quenching and tempering, and Table 4-6 gives the properties of alloy 1340 after normalizing and annealing.

Table 4-5 Mechanical properties of quenched and tempered low-alloy manganese steels†

AISI No.	Tempering temperature, °F	Tensile strength, psi	Yield strength, psi	Elongation, %	Reduction in area, %	Hardness, Bhn
1330	400	232,000	211,000	9	39	459
	600	207,000	186,000	9	44	402
	800	168,000	150,000	15	53	335
	1000	127,000	112,000	18	60	263
	1200	106,000	83,000	23	63	216
1340	400	262,000	231,000	11	35	505
	600	230,000	206,000	12	43	453
	800	183,000	167,000	14	51	375
	1000	140,000	120,000	17	58	295
	1200	116,000	90,000	22	66	252

† After "ASM Databook," published in *Metal Progress*, vol. 112, no. 1, mid-June 1977.

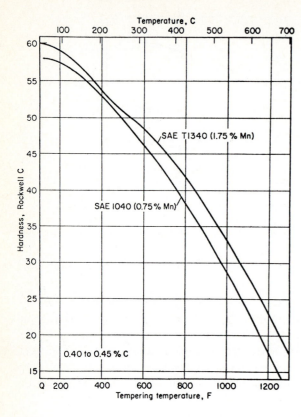

Figure 4-13 Softening, with increasing tempering temperature, of quenched 0.40 to 0.45% C steels as influenced by an increase of manganese from about 0.75 to 1.75%. *(After E. C. Bain and H. W. Paxton, "Alloying Elements in Steel," 2d ed., American Society for Metals, 1966, p. 194.)*

Table 4-6 Mechanical properties of normalized and annealed AISI 1340 low-alloy manganese steel†

AISI No.	Treatment	Yield strength, psi	Tensile strength, psi	Elongation, %	Reduction in area, %	Hardness, Bhn	Impact strength (Izod), ft · lb
1340	Normalized (1600°F)	81,000	121,250	22.0	62.9	248	68.2
	Annealed (1475°F)	63,250	102,000	25.5	57.3	207	52.0

† After "ASM Databook," published in *Metal Progress*, vol.112, no. 1, mid-June 1977.

4-5 LOW-ALLOY CHROMIUM STEELS

Chemical Compositions and Typical Applications

Chromium is added to plain-carbon steels to improve hardenability, strength, and wear resistance. Chromium has a BCC crystal structure and hence is a strong ferrite stabilizer. Chromium also combines with carbon in iron to form carbides (Table 4-3). Since the chromium content of low-alloy steels is less than 2 percent, the chromium atoms replace iron atoms in Fe_3C to produce the complex carbide, $(Fe,Cr)_3C$.

Table 4-7 lists the chemical compositions and typical applications of the low-alloy chromium steels. Alloy steels of the 51xx series contain from 0.20 to 0.60% C and from 0.8 to 0.9% Cr. The low-carbon grades of this series are used for producing very hard, wear-resistant surfaces, but they lack a tough core. Higher-carbon grades are used for spring steels where high strength and wear resistance are required. The 52100 steel, which contains about 1% C and 1.5% Cr, is used for ball and roller bearings where very high wear resistance and strength are required. These steels, however, are susceptible to temper embrittlement and care must be taken with their heat treatment.

Structure

Continuous-cooling transformation kinetics The introduction of 0.9% Cr to a 0.4% plain-carbon steel shifts the diffusion-controlled austenite → ferrite + pearlite reaction to the right and downward in the continuous-cooling transformation diagram (Fig. 4-14). In the 5140 low-alloy steel, some bainitic products are possible with rapid quenching because of the increased hardenability caused by the presence of the 0.9% Cr.

Microstructure The microstructure of spring-steel alloy 5160 after hot rolling and air cooling consists of ferrite and unresolved pearlite (Fig. 4-15). If this alloy is austenitized and oil-quenched, a martensitic structure with some retained

Table 4-7 Chemical compositions and typical applications of low-alloy chromium steels

Alloy AISI No.	% C	% Mn	% Cr	Typical applications
5120	0.20	0.80	0.80	Carburizing grade
5130	0.30	0.80	0.95	Steering parts
5140	0.40	0.80	0.80	
5160	0.60	0.88	0.80	Spring steels
52100	1.04	0.35	1.45	Ball and roller bearings; races

† After "ASM Databook," published in *Metal Progress*, vol. 112, no. 1, mid-June 1977.

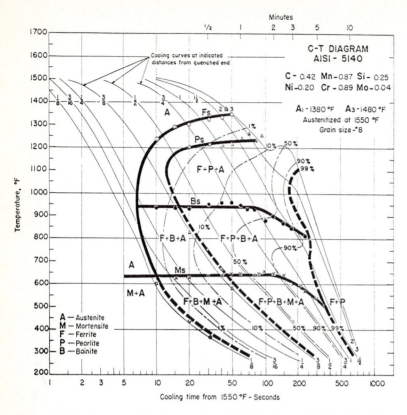

Figure 4-14 Continuous-cooling diagram for AISI 5140 alloy steel. *(After Metal Progress, Dec. 1965, p. 84.)*

2% nital   550 ×

Figure 4-15 Alloy 5160 hot-rolled, 0.635 in diameter, air-cooled from finish-rolling temperature of 982°C; structure consists of unresolved pearlite (dark) and ferrite (light). *(After Metals Handbook, 8th ed., vol. 7, American Society for Metals, 1972, p. 49.)*

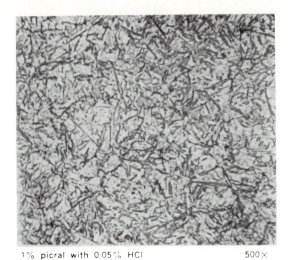

1% picral with 0.05% HCl   500×

Figure 4-16 Alloy 5160 hot-rolled coil-spring steel, austenitized at 871°C for 30 min and oil-quenched; structure consists of untempered martensite (dark constituent) and retained austenite (light constituent). *(After Metals Handbook, 8th ed., vol. 7, American Society for Metals, 1972, p. 49.)*

austenite is produced (Fig. 4-16). After tempering 1 h at 204°C, a structure consisting of tempered martensite is produced (Fig. 4-17). When 52100 alloy steel is austenitized at 843°C, oil-quenched, and tempered 1 h at 399°C, a structure consisting of tempered martensite is produced (Fig. 4-18). A dispersion of carbide particles which were not dissolved during austenitizing is also present. These carbides provide the extra-hard wear-resistant surfaces of this ball-bearing alloy.

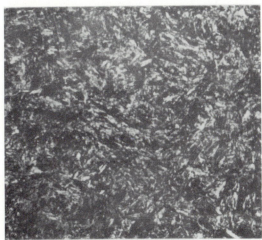

4% nital, 4% picral, mixed 1 to 1   1000×

Figure 4-17 Alloy 5160 hot-rolled coil-spring steel, austenitized at 871°C for 30 min, oil-quenched, and tempered 1 h at 204°C; structure consists of tempered martensite. *(After Metals Handbook, 8th ed., vol. 7, American Society for Metals, 1972, p. 49.)*

4% nital, 4% picral, mixed 1 to 1   500×

Figure 4-18 Alloy 52100 steel bar first spheroidized and then austenitized at 843°C for 0.5 h, oil-quenched and tempered 1 h at 399°C; structure consists of tempered martensite and a dispersion of carbide particles not dissolved during austenitizing. *(After Metals Handbook, 8th ed., vol. 7, American Society for Metals, 1972, p. 51.)*

Mechanical Properties

The mechanical properties of some of the 51xx series alloys are listed in Table 4-8 for the normalized and annealed conditions. Table 4-9 gives their properties when they are quenched and tempered. Of special note is the high strength and hardness of these alloy steels. Their ductility, however, is relatively low, and under some conditions they are susceptible to temper embrittlement.

4-6 MOLYBDENUM STEELS

Chemical Compositions and Typical Applications

Molybdenum is added in small amounts to plain-carbon steels to improve their *strength* and *hardenability*. Table 4-10 lists the chemical compositions and applications of the currently used 40xx series of molybdenum low-alloy steels. The amount of molybdenum added to these steels (and to almost all standard alloy steels) is restricted to about 0.25 percent since this amount has been found experimentally to be the optimum for increased toughness, hardenability, and strength properties.

The low-carbon alloy steels of the 40xx series are used primarily as carburizing grades in the auto industry. They are extensively used for rear-axle gears and automatic transmission components.

Structure

Alloy 4047 will be taken as an example for this series of alloy steels since it is the strongest and most hardenable.

Table 4-8 Mechanical properties of normalized and annealed low-alloy chromium steels†

AISI No.	Treatment	Yield strength, psi	Tensile strength, psi	Elonga-tion, %	Reduc-tion in area, %	Hard-ness, Bhn	Impact strength (Izod), ft · lb
5140	Normalized (1600°F)	68,500	115,000	22.7	59.2	229	28.0
	Annealed (1525°F)	42,500	83,000	28.6	57.3	167	30.0
5150	Normalized (1600°F)	76,750	126,250	20.7	58.7	255	23.2
	Annealed (1520°F)	51,750	98,000	22.0	43.7	197	18.5
5160	Normalized (1575°F)	77,000	138,750	17.5	44.8	269	8.0
	Annealed (1495°F)	40,000	104,750	17.2	30.6	197	7.4

† After "ASM Databook," published in *Metal Progress*, vol. 112, no. 1, mid-June 1977.

Table 4-9 Mechanical properties of quenched and tempered low-alloy chromium steels†

AISI No.	Tempering tempera-ture, °F	Tensile strength, psi	Yield strength, psi	Elonga-tion, %	Reduc-tion in area, %	Hard-ness, Bhn
5130	400	234,000	220,000	10	40	475
	600	217,000	204,000	10	46	440
	800	185,000	175,000	12	51	379
	1000	150,000	136,000	15	56	305
	1200	115,000	100,000	20	63	245
5140	400	260,000	238,000	9	38	490
	600	229,000	210,000	10	43	450
	800	190,000	170,000	13	50	365
	1000	145,000	125,000	17	58	280
	1200	110,000	96,000	25	66	235
5150	400	282,000	251,000	5	37	525
	600	252,000	230,000	6	40	475
	800	210,000	190,000	9	47	410
	1000	163,000	150,000	15	54	340
	1200	117,000	118,000	20	60	270
5160	400	322,000	260,000	4	10	627
	600	290,000	257,000	9	30	555
	800	233,000	212,000	10	37	461
	1000	169,000	151,000	12	47	341
	1200	130,000	116,000	20	56	269

† After "ASM Databook," published in *Metal Progress*, vol. 112, no. 1, mid-June 1977.

Table 4-10 Chemical compositions and typical applications of low-alloy molybdenum steel†

Alloy AISI-SAE No.	Chemical composition, nominal wt %				Typical applications
	C	Mn	Mo	Si	
4023	0.23	0.80	0.25	0.23	Carburizing grades:
4027	0.27	0.80	0.25	0.23	rear-axle drive pinions and gears; automatic transmission components
4037	0.37	0.80	0.25	0.23	
4047	0.47	0.80	0.25	0.23	

† After "ASM Databook," published in *Metal Progress*, vol. 112, no. 1, mid-June 1977.

Continuous-cooling transformation kinetics When an unalloyed 0.40% C steel is cooled from its austenitizing temperature, it normally decomposes into ferrite and pearlite. Only with very rapid cooling can intermediate (bainitic) structures be produced. The introduction of 0.25% Mo to a 0.47% C steel shifts the diffusion-controlled austenite → ferrite + pearlite transformation substantially to the right and downward in the continuous-cooling transformation diagram (Fig. 4-19). As a result, an increased amount of bainitic transformation products are produced.

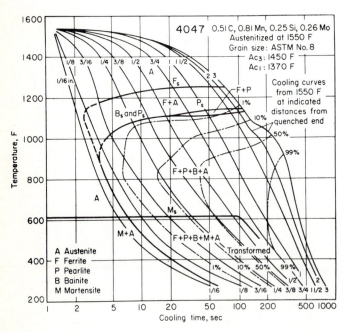

Figure 4-19 Continuous-cooling diagram for AISI 4047 alloy steel. *(After Metal Progress, Dec. 1963, p. 114.)*

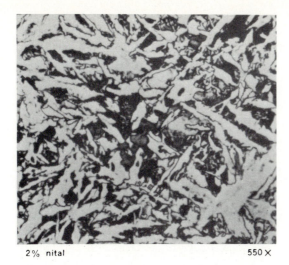

2% nital   550×

Figure 4-20 Alloy 4047 steel forging (1/2-in section thickness) air-cooled from forging temperature of 1204°C; longitudinal section; structure consists of plates of ferrite (white) and fine pearlite (dark). *(After Metals Handbook, 8th ed., vol. 7, American Society for Metals, 1972, p. 370.)*

Microstructure The microstructure of air-cooled alloy 4047 ($\frac{1}{2}$-in section) consists of proeutectoid ferrite and fine pearlite (Fig. 4-20). When the cooling rate from the austenitizing temperature for this alloy is slowed down as in furnace cooling, the pearlite becomes much coarser, as shown in Fig. 4-21.

Mechanical Properties

The addition of 0.25% Mo to 1040 plain-carbon steel retards the softening process during tempering to some extent, as indicated in Fig. 4-22. The large molybdenum atoms enter the Fe_3C and, by inhibiting diffusion, slow down the rate of coalescence of the Fe_3C. However, the small amount of molybdenum in

2% nital   500×

Figure 4-21 Alloy 4047 steel forging (5/8-in thick, longitudinal section) austenitized at 829°C, cooled to 663°C and held 6 h, furnace-cooled to 538°C, air-cooled. Ferrite (white) and lamellar pearlite (dark). *(After Metals Handbook, 8th ed., vol. 7, American Society for Metals, 1972, p. 37.)*

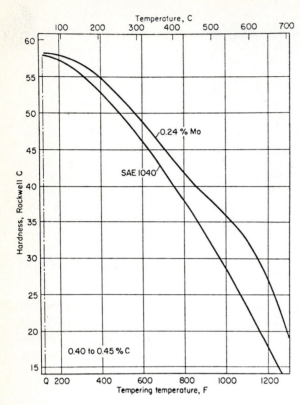

Figure 4-22 Comparison of the softening with increasing tempering temperature of alloy 1040 with the same alloy with 0.25% Mo. *(After E. C. Bain and H. W. Paxton, "Alloying Elements in Steel," 2d ed., American Society for Metals, 1966, p. 198.)*

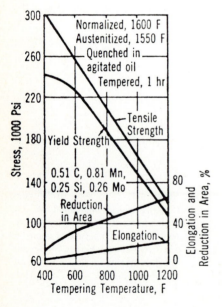

Figure 4-23 Effect of tempering temperature on the mechanical properties of alloy 4047. *(After Metal Progress, Dec. 1963, p. 114.)*

alloy 4047 does not greatly affect the rapid decrease in strength with increasing tempering temperature (Fig. 4-23). The hardenability of alloy 4047 is only slightly increased above that of the plain-carbon steel with the same carbon content.

4-7 CHROMIUM-MOLYBDENUM STEELS

Chemical Compositions and Typical Applications

Chromium (0.5 to 0.95 percent) is added, along with a small amount of molybdenum (0.13 to 0.20 percent), to make the 41xx series of alloy steels. The addition of chromium further increases the hardenability, strength, and wear resistance of the plain-carbon steels of the same carbon content. However, the addition of chromium to low-alloy structural steels tends to make them susceptible to temper embrittlement under some conditions. This subject is discussed in detail in Sec. 4-10. Table 4-11 lists the chemical compositions and typical applications of the most important 41xx alloy steels.

Low-alloy steels with chromium and molybdenum, because of their increased hardenability, can be oil-quenched to form martensite instead of being water-quenched. Since the slower oil quench reduces temperature gradients and internal stresses due to volume contraction and expansion during quenching, distortion and cracking tendencies can be minimized.

Structure

Alloy 4140 will be taken as an example from the 41xx series of alloy steels since it is one of the most commonly used alloy steels.

Continuous-cooling transformation kinetics The continuous-cooling transformation diagram of alloy 4140 is shown in Fig. 4-24. The effectiveness of molybdenum in modifying the phase transformation of a 0.40% C steel is

Table 4-11 Chemical compositions and typical applications of low-alloy chromium-molybdenum steels†

Alloy AISI-SAE No.	Chemical composition, nominal wt%				Typical applications
	C	Mn	Cr	Mo	
4118	0.18	0.80	0.50	0.13	
4130	0.30	0.50	0.55	0.20	Pressure vessels, aircraft
4140	0.40	0.88	0.95	0.20	structural parts, auto axles,
4150	0.50	0.88	0.95	0.20	steering knuckles

† After "ASM Databook," published in *Metal Progress*, vol. 112, no. 1, mid-June 1977.

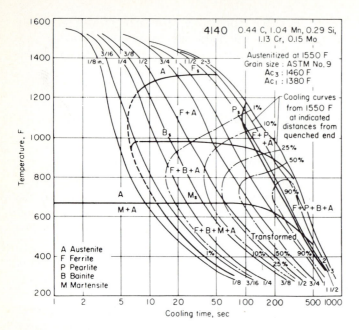

Figure 4-24 Continuous-cooling transformation diagram for AISI 4140 alloy steel. *(After Metal Progress, Jan. 1964, p. 100.)*

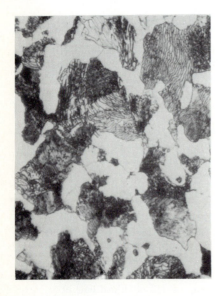

Figure 4-25 Alloy 4140 steel fully annealed 24 h at 691°C; structure consists of pearlite with ferrite. (Etch: 2% nital; 800 × .) *(Courtesy of Republic Steel Co.)*

enhanced by the addition of chromium, especially if in amounts of over 0.7 percent. The temperature and time range for the austenite-to-martensite and austenite-to-bainite transformations is widened and the B_s temperature is lowered with the chromium addition. Also the hardenability of the alloy steel increases with the chromium additions and there is a greater delay in the ferrite-to-pearlite transformation in the chromium-molybdenum alloy steels. Compare the 4140 CCT diagram of Fig. 4-24 with that of the 4047 diagram of Fig. 4-19.

Microstructure The microstructure of alloy 4140 after being fully annealed at 691°C consists of blocky ferrite and fine to coarse pearlite (Fig. 4-25). After austenitizing at 843°C and oil quenching, a martensitic structure is produced (Fig. 4-26) and, with subsequent tempering at 315°C, a fine-tempered martensitic structure is the result (Fig. 4-27). Unfortunately, very little of the fine structure of these alloys is shown in optical micrographs.

Recently Krauss, Materkowski, and Schupmann (Ref. 4, p. 240) have obtained more information about the fine structure of low-alloy steels using electron transmission microscopy. They have shown that the martensite in low alloy steels (e.g., 4130 alloy) consists of packets of fine units of martensite called *laths* that align themselves parallel to one another to form packets (Fig. 4-28). The orientation of the units or laths within a packet are limited, and frequently large volumes of laths within a packet have only one orientation. Therefore many of the boundaries within a packet are low-angle and as an approximation, the entire packet has essentially one orientation.

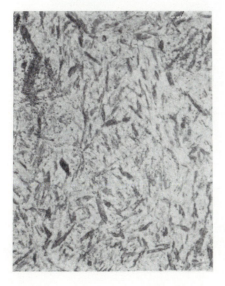

Figure 4-26 Alloy 4140 steel fully hardened; sample was austenitized at 843°C and oil-quenched; structure is martensite. (Etch: 2% nital; 800 × .) *(Courtesy of Republic Steel Co.)*

Figure 4-27 Alloy 4140 steel fully hardened and tempered; sample was austenitized at 843°C and oil-quenched tempered at 315°C; structure is tempered martensite. (Etch: 2% nital; 800 × .) *(Courtesy of Republic Steel Co.)*

Figure 4-28 Lath martensite in a 4130 alloy steel. [*After G. Krauss in D. V. Doane and J. S. Kirkaldy (eds.), "Hardenability Concepts with Applications to Steel," AIME, 1978.*]

Mechanical Properties

The mechanical properties of some of the 41xx series of alloys are listed in Table 4-12 for the normalized and annealed conditions. The effect of tempering temperature on the mechanical properties of these alloys is shown in Table 4-13. The degree of softening with increasing temperature in the Cr-Mo low-alloy steels is essentially the same as that shown by the molybdenum low-alloy steels.

Table 4-12 Mechanical properties of normalized and annealed low-alloy chromium-molybdenum steels†

AISI No.	Treatment	Yield strength, psi	Tensile strength, psi	Elonga-tion, %	Reduc-tion in area, %	Hard-ness, Bhn	Impact strength (Izod), ft · lb
4130	Normalized (1600°F)	63,250	97,000	25.5	59.5	197	63.7
	Annealed (1585°F)	52,250	81,250	28.2	55.6	156	45.5
4140	Normalized (1600°F)	95,000	148,000	17.7	46.8	302	16.7
	Annealed (1500°F)	60,500	95,000	25.7	56.9	197	40.2
4150	Normalized (1600°F)	106,500	167,500	11.7	30.8	321	8.5
	Annealed (1500°F)	55,000	105,750	20.2	40.2	197	18.2

† After "ASM Databook," published in *Metal Progress*, vol. 112, no. 1, mid-June 1977.

Table 4-13 Mechanical properties of quenched and tempered low-alloy chromium-molybdenum steels†

AISI No.	Tempering tempera-ture, °F	Tensile strength, psi	Yield strength, psi	Elonga-tion, %	Reduc-tion in area, %	Hard-ness Bhn
4130	400	236,000	212,000	10	41	467
	600	217,000	200,000	11	43	435
	800	186,000	173,000	13	49	380
	1000	150,000	132,000	17	57	315
	1200	118,000	102,000	22	64	245
4140	400	257,000	238,000	8	38	510
	600	225,000	208,000	9	43	445
	800	181,000	165,000	13	49	370
	1000	138,000	121,000	18	58	285
	1200	110,000	95,000	22	63	230
4150	400	280,000	250,000	10	39	530
	600	256,000	231,000	10	40	495
	800	220,000	200,000	12	45	440
	1000	175,000	160,000	15	52	370
	1200	139,000	122,000	19	60	290

† After "ASM Databook," published in *Metal Progress*, vol. 112, no. 1, mid-June 1977.

4-8 NICKEL-CHROMIUM-MOLYBDENUM STEELS

Chemical Compositions and Typical Applications

Low-alloy steels consisting of about 1.8% Ni, 0.5 to 0.8% Cr, and 0.20% Mo make up the 43xx alloy series. In the 86xx series, the nickel content is reduced to 0.55 percent. Table 4-14 lists the chemical compositions and typical applications for the low-alloy nickel-chromium-molybdenum steels.

Nickel in combination with chromium produces low-alloy steels with higher elastic limits, greater hardenability, and higher impact and fatigue resistance than the plain-carbon steels. The further addition of about 0.2% Mo increases hardenability still more and minimizes the susceptibility of these alloys to temper embrittlement. The 4320 and 4340 alloy steels are used for heavy-duty, high-strength parts such as gears and aircraft tubing. When slightly lower strengths are required, the 8620 and 8640 alloys with lower nickel levels are used. Both of these 86xx alloys are used for shafts and forgings requiring high strength.

Structure

Alloy 4340 will be taken as an example for the structural changes which take place in the 43xx nickel-chromium-molybdenum alloy steels.

Continuous-cooling transformation kinetics The continuous-cooling transformation diagram of alloy 4340 is shown in Fig. 4-29. The combination of nickel-chromium-molybdenum delays the ferrite-to-pearlite transformation to much longer times than in the case of the chromium-molybdenum alloys (Fig. 4-24). The temperature (M_s) for the beginning of the austenite-to-martensite transformation is decreased to about 290°C since nickel depresses the M_s temperature as well as the A_{c_3} and A_{c_1} temperatures. The time for the beginning of the

Table 4-14 Chemical compositions and typical applications of low-alloy nickel-chromium-molybdenum steels†

Alloy AISI-SAE No.	Chemical composition, nominal wt%					Typical applications
	C	Mn	Ni	Cr	Mo	
4320	0.20	0.55	1.83	0.50	0.25	Carburizing grade
4340	0.40	0.60	1.83	0.80	0.25	Heavy sections, landing gears
8620	0.20	0.80	0.55	0.50	0.20	Carburizing grade
8640	0.40	0.88	0.55	0.50	0.20	Auto springs, small machine axles, shafts
8660	0.60	0.88	0.55	0.50	0.20	

† After "ASM Databook," published in *Metal Progress*, vol. 112, no. 1, mid-June 1977.

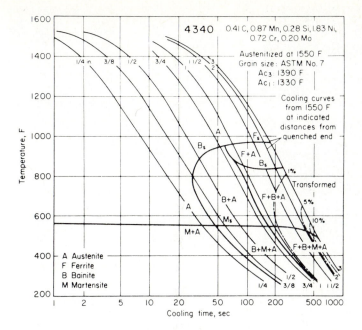

Figure 4-29 Continuous-cooling transformation diagram for AISI 4340 alloy steel. *(After Metal Progress*, Sept., 1964, p. 106.)

austenite-to-bainite transformation is also increased significantly when nickel, chromium, and molybdenum are all present.

Microstructures The microstructures of alloy 4340 resulting from various heat-treated conditions are shown in Figs. 4-30 to 4-33. Air cooling from austenitizing temperature produces a bainitic structure, as shown in Fig. 4-30. The bainitic structure is made possible because of the long delay in the austenite → ferrite + pearlite transformation (Fig. 4-29). Oil quenching from austenitizing produces a martensitic structure with some possible retained austenite (Fig. 4-32). Oil quenching can be used to obtain a martensitic structure because of the delayed austenite → ferrite + pearlite reaction. The martensite of the 4340 steel consists of many laths of about the same orientation within packets (Fig. 4-34).

Mechanical Properties

Nickel strengthens the 4340 alloys since it is soluble in both austenite and ferrite. The tensile strength of alloy 4340 in the quenched and 315°C tempered condition is about 250 ksi. The mechanical properties of some of the 43xx and 86xx alloy steels are listed in Table 4-15 for the normalized and annealed conditions and in Table 4-16 for the quenched and tempered states. Upon tempering, there is a steady decline in strength similar to the softening of the plain-carbon steels, but at higher strength levels.

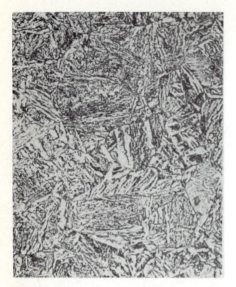

Figure 4-30 Alloy 4340 normalized at 871°C for 1 h and air-cooled; structure consists of upper bainite. (Etch: 2% nital; 400 × .) *(Courtesy of Republic Steel Co.)*

Figure 4-31 Alloy 4340 normalized at 871°C for 1 h, air-cooled, and annealed at 691°C for 24 h, tempered structure, tending toward spheroidization. (Etch: 2% nital; 400 × .) *(Courtesy of Republic Steel Co.)*

Figure 4-32 Alloy 4340 austenitized at 843°C for 1 h and oil-quenched; structure consists of martensite with some possible retained austenite. (Etch: 2% nital; 400 × .) *(Courtesy of Republic Steel Co.)*

Figure 4-33 Alloy 4340 austenitized at 843°C for 1 h, oil-quenched and tempered 4 h at 538°C; structure consists of tempered martensite. (Etch: 2% nital; 400 × .) *(Courtesy of Republic Steel Co.)*

Figure 4-34 Lath martensite in a 4340 alloy steel. [*After G. Krauss in D. V. Doane and J. S. Kirkaldy (eds.), "Hardenability Concepts with Applications to Steel," AIME, 1978, p. 240.*]

Table 4-15 Mechanical properties of normalized and annealed nickel-chromium-molybdenum alloy steels†

AISI No.	Treatment	Yield strength, psi	Tensile strength, psi	Elonga-tion, %	Reduc-tion in area, %	Hard-ness, Bhn	Impact strength (Izod), ft · lb
4320	Normalized (1640°F)	67,250	115,000	20.8	50.7	234	53.8
	Annealed (1560°F)	61,625	84,000	29.0	58.4	163	81.0
4340	Normalized (1600°F)	125,000	185,500	12.2	36.3	363	11.7
	Annealed (1490°F)	68,500	108,000	22.0	49.9	217	37.7
8620	Normalized (1675°F)	51,750	91,750	26.3	59.7	183	73.5
	Annealed (1600°F)	55,875	77,750	31.3	62.1	149	82.8
8630	Normalized (1600°F)	62,250	94,250	23.5	53.5	187	69.8
	Annealed (1550°F)	54,000	81,750	29.0	58.9	156	70.2
8650	Normalized (1600°F)	99,750	148,500	14.0	40.4	302	10.0
	Annealed (1465°F)	56,000	103,750	22.5	46.4	212	21.7
8740	Normalized (1600°F)	88,000	134,750	16.0	47.9	269	13.0
	Annealed (1500°F)	60,250	100,750	22.2	46.4	201	29.5

† After "ASM Databook," published in *Metal Progress*, vol. 112, no. 1, mid-June 1977.

Table 4-16 Mechanical properties of quenched and tempered nickel-chromium-molybdenum alloy steels†

AISI No.	Tempering temperature, °F	Tensile strength, psi	Yield strength, psi	Elongation, %	Reduction in area, %	Hardness, Bhn
4340	400	272,000	243,000	10	38	520
	600	250,000	230,000	10	40	486
	800	213,000	198,000	10	44	430
	1000	170,000	156,000	13	51	360
	1200	140,000	124,000	19	60	280
8630	400	238,000	218,000	9	38	465
	600	215,000	202,000	10	42	430
	800	185,000	170,000	13	47	375
	1000	150,000	130,000	17	54	310
	1200	112,000	100,000	23	63	240
8640	400	270,000	242,000	10	40	505
	600	240,000	220,000	10	41	460
	800	200,000	188,000	12	45	400
	1000	160,000	150,000	16	54	340
	1200	130,000	116,000	20	62	280
8650	400	281,000	243,000	10	38	525
	600	250,000	225,000	10	40	490
	800	210,000	192,000	12	45	420
	1000	170,000	153,000	15	51	340
	1200	140,000	120,000	20	58	280
8660	400	—	—	—	—	580
	600	—	—	—	—	535
	800	237,000	225,000	13	37	460
	1000	190,000	176,000	17	46	370
	1200	155,000	138,000	20	53	315
8740	400	290,000	240,000	10	41	578
	600	249,000	225,000	11	46	495
	800	208,000	197,000	13	50	415
	1000	175,000	165,000	15	55	363
	1200	143,000	131,000	20	60	302

† After "ASM Databook," published in *Metal Progress*, vol. 112, no. 1, mid-June 1977.

4-9 NICKEL-SILICON-CHROMIUM-MOLYBDENUM STEELS

The addition of about 2% Si to AISI 4340 alloy steel significantly increases its strength and toughness, as shown in Fig. 4-35. The increase in toughness of the 4340 + 2% Si steel is attributed to the silicon retarding the precipitation of cementite from retained austenite in the tempered martensite and to the stabilization of the ϵ carbide (Ref. 18). In quenched and tempered steels with low

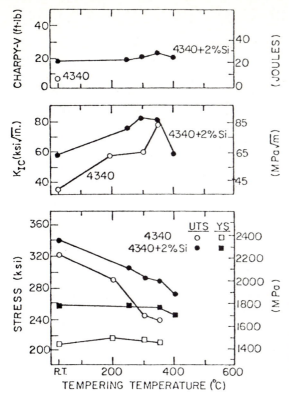

Figure 4-35 Plots showing the effects of tempering on the mechanical properties of AISI 4340 steel and AISI type 4340 steel to which 2% silicon has been added. Steels quenched from 870 and 950°C, respectively. [*After E. R. Parker, Met. Trans. 8A(1977):1025.*]

silicon contents, the retained austenite decomposes upon tempering in the 200 to 370°C range and cementite films form around it. This reaction contributes to martensitic embrittlement which is discussed further in Sec. 4-10. In the silicon-containing 4340 steel, the formation of cementite from retained austenite is suppressed, as well as the formation of cementite from ϵ carbide. As a result, the 4340 + 2% Si alloy is stronger and tougher in the tempered condition.

The 300M alloy steel utilizes the favorable effect of silicon on the 4340 alloy, and is presently used extensively for ultrahigh-strength steel for landing gears. The 300M steel has the nominal composition 0.40% C, 0.75% Mn, 1.6% Si, 0.8% Cr, 1.8% Ni, 0.40% Mo, 0.08% V, 0.015% max P, and 0.015% max S. Vanadium is added for grain refinement, and the sulfur and phosphorus levels are kept very low to reduce temper embrittlement and to increase toughness and transverse ductility. This alloy is vacuum-arc-remelted to lower the hydrogen and oxygen contents. The lower oxygen content minimizes the formation of oxide inclusions[1] and thus increases the toughness of the alloy, as shown in Fig. 4-36. A lower hydrogen content reduces susceptibility to flaking.[1]

---

[1] W. M. Imrie, *Roy. Soc. Lond. Phil. Trans.* A282(1976):91.

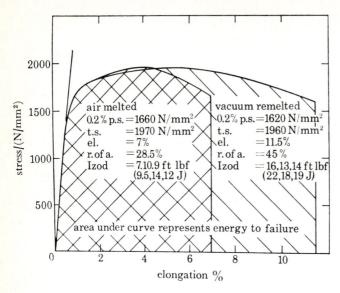

Figure 4-36 Energy absorbed in fracturing air and vacuum remelted 300M alloy steel (transverse properties). [*After W. M. Imrie, Royal Soc. London Phil. Trans. 282(1976):91 (series A).*]

4-10 TEMPER EMBRITTLEMENT IN LOW-ALLOY STEELS

In this section two types of temper embrittlement commonly exhibited by high-strength low-alloy steels will be discussed. These types have been designated *one-step temper embrittlement* and *two-step temper embrittlement* by Briant and Banerji (Ref. 14).

1. *One-step temper embrittlement*, commonly known as 350°C embrittlement, is often encountered in commercial high-strength low-alloy steels which have quenched and tempered martensitic microstructures. In this case, the alloy is austenitized, quenched, and tempered for a short time (about 1 h) at a relatively low temperature ($< 400°C$). This embrittlement can be recognized by an anomalous decrease in notched-bar energy when tempered in the 250 to 350°C range.
2. *Two-step temper embrittlement* refers to the decrease in notched toughness that is frequently observed when tempered alloy steels are isothermally aged in the temperature range 375 to 560°C. This same type of embrittlement can be obtained by slowly cooling a steel after tempering.

One-Step Embrittlement

The mechanisms causing one-step embrittlement are not completely understood at the present time (Ref. 14). Experimental evidence shows that it must be

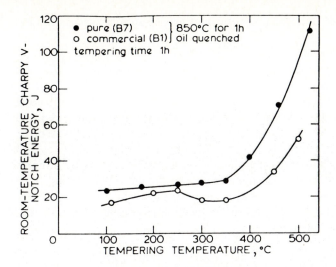

Figure 4-37 Comparison of one-step embrittlement curves pure 4340 (B1) and commercial 4340 (B7) alloy steels; note the absence of the embrittlement trough in the pure-base alloy. [*After S. Banerji, H. C. Feng, and C. J. McMahon, Met. Trans. 9A(1978):237.*]

caused by impurities in the steel since it is absent in pure low-alloy steels like 4340. As shown in Fig. 4-37, pure 4340 alloy does not show the embrittlement trough whereas commercial 4340 does. The mode of fracture for one-step temper embrittlement is principally intergranular. Figure 4-38 shows how the maximum amount of intergranular fracture coincides with the minimum in the impact-energy trough.

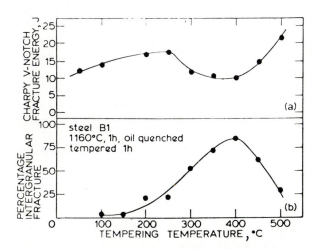

Figure 4-38 Change in fracture mode as a function of temperature for a commercial 4340 alloy steel (B7); note that the trough in the embrittlement curve corresponds to the maximum amount of intergranular fracture. [*After S. Banerji, H. C. Feng, and C. J. McMahon, Met. Trans. 9A(1978):237.*]

Important aspects of one-step embrittlement can be summarized as follows (Ref. 14):

1. The occurrence of the anomalous impact-energy trough coincides with the beginning of cementite precipitation.
2. Since one-step embrittlement causes an intergranular mode of fracture along prior austenitic grain boundaries, it is believed that the segregation of P, N, and possibly S to the austenitic grain boundaries is essential for this type of embrittlement.
3. Alloying elements such as manganese may have an indirect effect by promoting the segregation of the embrittling elements to the grain boundaries.
4. The presence of undissolved carbides at the prior austenitic grain boundaries is thought to accentuate the impurity-induced intergranular fracture, the carbides acting as slip barriers.

Two-Step Embrittlement

Two-step embrittlement occurs when some tempered alloy steels are isothermally aged in the temperature range of 375 to 560°C or are slowly cooled after tempering. This type of temper embrittlement is attributed to the segregation of impurity elements to the grain boundaries since, if the impurities are removed from the steel, it does not become embrittled during aging. When the impurities are segregated to the grain boundaries, the brittle fracture mode is intergranular, as shown in Fig. 4-39.

From the many studies that have been made on two-step temper embrittlement, the following general conclusions can be made (Ref. 14):

1. The ductile-brittle transition temperature is directly dependent on the grain boundary concentration of the impurities. This effect in a nickel-chromium steel doped with antimony, tin, and phosphorus is shown in Fig. 4-40. The relative effect of these impurities was found to be Sn > Sb > P.
2. Alloying elements sometimes cosegregate to the grain boundaries with the impurities. For example, nickel cosegregates with antimony.
3. The segregation of impurities to the grain boundaries appears to be an equilibrium phenomenon.
4. The equilibrium grain boundary concentration of impurities increases with decreasing aging temperature. Time also is important at lower temperatures. For example, Fig. 4-41 shows how increased aging time increases the concentration of antimony in a 3.5% Ni–1.7% Cr–0.008% C–0.06% Sb steel.

The rate and amount of impurity segregation, and hence the resulting intergranular embrittlement, depend on the total composition of the system. Nickel, chromium, and manganese increase two-step temper embrittlement caused by Sb, Sn, P, or As. Additions of molybdenum to the alloy steel retard

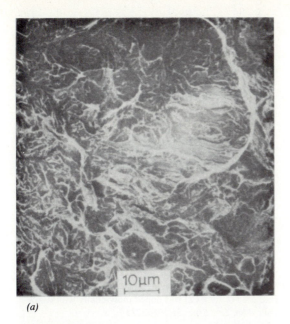

(a)

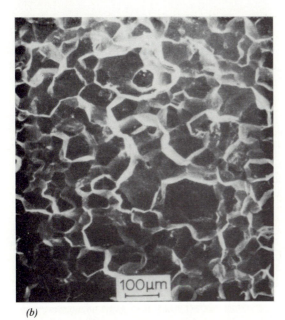

(b)

Figure 4-39 Comparison of (*a*) cleavage fracture in a quenched and tempered alloy steel and (*b*) intergranular fracture in a quenched and tempered, and aged alloy steel. (Steel is HY130:4.88% Ni, 0.57% Cr, 0.49% Mo, 0.88% Mn, 0.11% C.) [*After C. L. Briant, H. C. Feng, and C. J. McMahon, Met. Trans. 9A(1978):625.*]

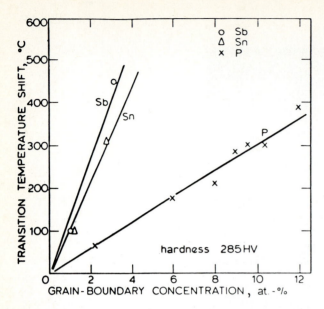

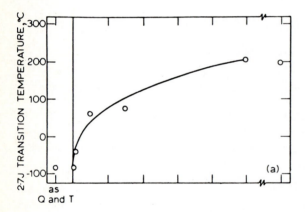

Figure 4-40 Change in ductile to brittle transition temperature as function of grain-boundary impurity concentration; the 3340 alloy steel (3.5% Ni, 1.7% Cr) was doped individually with 0.06% P, 0.06% Sb, or 0.06% Sn. [*After C. J. McMahon, Met. Sci. Engr. 25(1976):233.*]

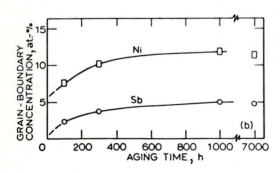

Figure 4-41 Two-step temper embrittlement of a 3.5% Ni–1.7% Cr–0.008% C–0.06% Sb steel showing that as aging time is increased the ductile–brittle transition temperature is raised and the amount of Ni and Sb segregated to the grain boundaries is increased. The alloy was austenitized, quenched, tempered, and then aged at 520°C. [*After H. Ohtani, H. C. Feng, and C. J. McMahon, Met. Trans. 7A(1976): 87.*]

temper embrittlement since molybdenum inhibits the segregation of the impurities. Molybdenum readily precipitates as phosphides in the matrix and hence inhibits segregation.

4-11 MARAGING STEELS

Composition

Maraging steels are a class of high-strength steels which are characterized by very-low-carbon contents and the use of substitutional elements to produce age hardening in iron-nickel martensites. The name maraging was coined from a combination of *mar*tensite and *age* harden*ing*.

Maraging steels containing 18% Ni along with Co, Mo, Ti, and Al additions have been established as ultrahigh-strength structural steels. The nominal yield strengths of these steels in the fully age-hardened condition are 200, 250, 300, and 350 ksi and the corresponding designations for them are 18Ni(200), 18Ni(250), 18Ni(300), and 18Ni(350). Table 4-17 lists the chemical composition of these maraging steels.

Martensitic Formation

The 18% Ni maraging steels transform to martensite on cooling from austenitic temperatures since their nickel content is so high. The M_s temperature for these alloys is about 155°C and their M_f about 98°C. The formation of martensite in these alloys is not affected by variation in cooling rate, and therefore thick sections can be air-cooled and still be fully martensitic. Since the martensitic transformation only involves an austenitic-to-martensitic transformation of Fe-Ni, and does not involve carbon or nitrogen interstitials to any considerable extent, the martensite formed is relatively ductile and tempering reactions do not occur upon reheating.

Age Hardening

Before aging, the 18% Ni maraging steels have a yield strength in the range of 95 to 120 ksi. The hardness and strength of these alloys increases rapidly upon

Table 4-17 Nominal chemical compositions of maraging steels

Grade	% Ni	% Co	% Mo	% Al	% Ti	% C (max)
18Ni(200)	18	8	3.2	0.1	0.2	0.03
18Ni(250)	18	8	5.0	0.1	0.4	0.03
18Ni(300)	18	9	5.0	0.1	0.6	0.03
18Ni(350)	18	12	4.0	0.1	1.8	0.01

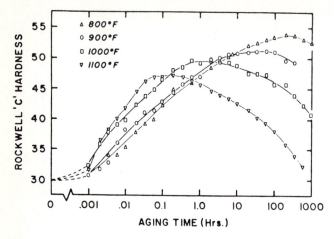

Figure 4-42 Hardness as a function of aging time for a commercial 18% Ni(250) maraging steel at four aging temperatures. [*D. T. Peters and C. R. Cupp, Trans. AIME 236(1966):1420.*]

aging, as shown in Fig. 4-42 for an 18Ni(250) maraging steel. The strength level attained depends principally on their molybdenum and titanium contents, but is also affected by the amount of cobalt and aluminum present. The highest-strength grade is 18Ni(350), which contains higher Co, Ti, and Al but slightly lower Mo.

The strengthening that is attained upon aging the 18% Ni maraging steels is believed to be caused by the formation of zones of precipitates of Ni_3Mo (see

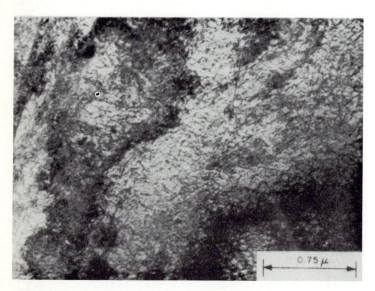

Figure 4-43 Transmission electron micrograph of the precipitate in an 18% Ni(250) maraging steel aged for 8 h at 485°C. [*After J. M. Chilton and C. J. Barton, Trans ASM 60(1967):528.*]

footnote 1 below) and Ni_3Ti.[2] Maximum hardness in the 18Ni(250) alloy occurs after 3 h at 482°C (Fig. 4-40). The precipitate formed in the 18Ni(250) alloy after aging 8 h at 482°C is shown in Fig. 4-43. The precipitate forms along dislocations and lath boundaries created by the martensitic transformation. Overaging and higher aging temperatures lead to the formation of Fe_2Mo precipitate. Cobalt is not found in any of the age-hardening precipitates. It is believed that cobalt indirectly contributes to strengthening during aging by reducing the solubility of molybdenum in the martensitic matrix.[3]

PROBLEMS

1. What are the important limitations of plain-carbon steels for engineering applications?

2. How are the standard AISI alloy steels classified?

3. List five reasons why alloying elements are added to plain-carbon steel bases to make alloy steels.

4. Which alloying elements are dissolved in the ferrite of alloy steels? Which alloying elements are partitioned between the ferrite and carbide phases in alloy steels? List them in order of increasing carbide-forming tendencies.

5. Define the hardenability of a steel. What factors determine the hardenability of a steel?

6. Describe Grossmann's method for determining hardenability.

7. (a) Calculate the *ideal critical diameter* of an AISI 1040 steel using its maximum composition limits and an ASTM grain size of 7. AISI 1040 has the following composition limits: 0.37 to 0.44% C, 0.60 to 0.90% Mn, 0.040% max P, 0.030% max S.

(b) If this steel is quenched in water with moderate agitation ($H = 1.3$), what is its *actual critical diameter*?

8. (a) Calculate the ideal critical diameter D_I of an AISI 4340 alloy steel with the maximum composition limits and an ASTM grain size of 7. AISI E4340 steel has the following composition: 0.38 to 0.43% C, 0.65 to 0.85% Mn, 0.025% max P, 0.025% max S, 0.20 to 0.35% Si, 1.65 to 2.00% Ni, 0.70 to 0.90% Cr, 0.20 to 0.30% Mo.

(b) If this alloy is quenched in oil with no agitation ($H = 0.30$), what would be its actual diameter D_o?

(c) Compare the ideal critical diameter of the AISI 1040 steel with the E4340 steel and explain the reason for the difference. (ASTM grain size = 7.)

9. (a) What is the actual critical diameter of an AISI 8640 steel quenched in oil with good agitation ($H = 0.5$) using the maximum composition limits. (ASTM grain size = 7.) The chemical composition of AISI 8640 steel is: 0.38 to 0.43% C, 0.75 to 1.00% Mn, 0.035% max P, 0.040% max S, 0.20 to 0.35% Si, 0.40 to 0.70% Ni, 0.40 to 0.60% Cr, 0.15 to 0.25% Mo.

(b) Compare the ideal critical diameter of the AISI 8640 alloy with E4340.

10. Describe the Jominy method of determining the hardenability of a steel. Why is this method preferred industrially to Grossmann's method?

11. Describe the mechanisms whereby manganese additions of 1.0 to 1.8 percent strengthen plain-carbon steels.

12. Up until about 1970, manganese alloy steels in the 2 to 5 percent range were considered too brittle to be usable. Today, a series of manganese bainitic steels with about 4% Mn and low carbon (≈ 0.04 percent) are produced. Explain the role of carbon in embrittling the 4% Mn steels.

[1]   J. M. Chilton and C. J. Barton, *Trans. ASM* 60(1967):528.
[2]   G. Thomas, I. Cheng, and J. R. Mihalisin, *Trans. ASM* 62(1969):852.
[3]   G. P. Miller and W. I Mitchell, *J. Iron Steel Inst.* 203(1965):899.

13. Quenching an alloy steel with 11 to 14% Mn and 1.0 to 1.2% C produces a useful alloy steel (Hadfield's steel) with high-impact resistance. How can the high-impact strength of this steel be explained metallurgically even though it contains such a high carbon content?

14. What is the effect of 0.5 to 0.8% Cr on the mechanical properties of plain-carbon steels?

15. Describe the microstructure of the martensite in alloy 4130.

16. Alloy 4340 has sometimes been described as a "synergistic-type" alloy. What is meant metallurgically by this description?

17. Describe one-step embrittlement in low-alloy steels.

18. What is believed to be the main cause of one-step embrittlement in low-alloy steels?

19. Describe two-step embrittlement in low-alloy steels.

20. What is believed to be the main cause of two-step embrittlement in low-alloy steels?

21. What is believed to be the mechanism of the beneficial effect of molybdenum in reducing two-step embrittlement in low-alloy steels?

22. What are maraging steels?

23. Describe the various types of maraging steels.

24. What is the heat treatment given to maraging steels to harden them and what are the mechanisms believed to be involved in strengthening maraging steels?

25. Hadfield's manganese steel (12% Mn, 1.1% C) which has been quenched from the austenitic range to room temperature has an austenitic structure and a very high work-hardening rate. Upon work hardening the austenitic structure, no martensite is formed and little, if any, ϵ carbide is precipitated. What microstructural changes could account for the abnormally high work-hardening rate observed? [Hint: see K. S. Raghavan et al., *Trans. AIME* 245(1969):1569.]

REFERENCES

1. E. C. Bain and H. W. Paxton: "Alloying Elements in Steel," 2d ed., American Society for Metals, Metals Park, OH, 1966.
2. K. Thelning: "Steel and its Heat Treatment," Butterworths, London, 1975.
3. C. A. Siebert, D. V. Doane, and D. H. Breen: "The Hardenability of Steels," American Society for Metals, Metals Park, OH, 1977.
4. D. V. Doane and J. S. Kirkaldy (eds.): "Hardenability Concepts with Applications to Steel," The Metallurgical Society of AIME, Warrendale, PA, 1978.
5. M. A. Grossmann and E. C. Bain, "Principles of Heat Treatment," American Society for Metals, Metals Park, OH, 1964.
6. R. E. Reed-Hill: "Physical Metallurgy Principles," 2d ed., Van Nostrand, New York, 1973.
7. W. W. Cias: "Austenite Transformation of Ferrous Alloys," Climax Molybdenum Co., Greenwich, CT, 1979.
8. R. M. Brick, A. W Pense, and R. B. Gordon: "Structure and Properties of Engineering Materials," McGraw-Hill, New York, 1977.
9. D. S. Clark and W. R. Varney: "Physical Metallurgy for Engineers," 2d ed., Van Nostrand, New York, 1962.
10. E. R. Petty, "Physical Metallurgy of Engineering Materials," American Elsevier, New York, 1968.
11. "Alloying Elements and their Effects. Hardenability," Republic Steel Co., Cleveland, 1976.
12. "Isothermal Transformation Diagrams," United States Steel Corporation, Pittsburgh, 1963.
13. "Alloy Steel: Semifinished; Hot Rolled and Cold Finished Bars," American Iron and Steel Institute, Washington, D. C., 1970.
14. C. L. Briant and S. K. Banerji: "Intergranular Failure in Steel: The Role of Grain-Boundary Composition," *Int. Metals Rev.* 23(4)(1978), Review 232.

15. I. Olefjord: "Temper Embrittlement," *Int. Metals Rev.* 23(4)(1978).
16. *Metals Handbook*, vol. 7: "Atlas of Microstructures of Industrial Alloys," American Society for Metals, Metals Park, OH, 1972.
17. S. Floreen: "The Physical Metallurgy of Maraging Steels," *Met. Rev.* 13(1968):115.
18. E. R. Parker: "Interrelations of Compositions, Transformation Kinetics, Morphology, and Mechanical Properties of Alloy Steels," *Met. Trans.* 8A(1977):1025.
19. R. F. Decker, "Source Book on Maraging Steels," American Society for Metals, Metals Park, OH, 1979.

ALUMINUM ALLOYS

Aluminum ranks second only to iron and steel in the metals market. In 1977 the United States shipped 13.4 billion lb,[1] with the building and construction industry taking 23.1 percent of the market and transportation another 21.7 percent (Table 5-1). Figure 5-1 shows the spectacular growth of aluminum production in the United States since 1900.

Table 5-1 Recent consumption of aluminum by major market category

Market category	1977		1976		% Change 1977–1976
	Millions of pounds	% of Market	Millions of pounds	% of Market	
Building and construction	3,078	23.1	2,939	23.0	+4.7
Transportation	2,896	21.7	2,459	19.3	+17.8
Consumer durables	1,061	7.9	1,036	8.1	+2.4
Electrical	1,336	10.0	1,312	10.3	+1.8
Machinery and equipment	921	6.9	905	7.1	+1.8
Containers and packaging	2,777	20.8	2,570	20.2	+8.1
Other	560	0.2	690	5.4	−18.8
Exports	726	5.4	836	6.6	−13.2
Total	13,355		12,747		+4.8

Source: "1977 Annual Statistical Review," The Aluminum Association, Inc., Washington, D. C., 1978.

[1] J. B. Balance, *J. Metals*, Nov. 1978, p. 37.

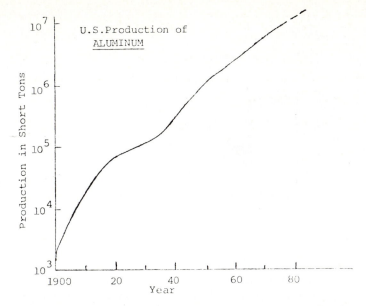

Figure 5-1 Production history of the aluminum industry in the United States and Canada. *(After W. H. C. Simmonds, J. Metals, Jan. 1976, p. 8.)*

The rapid growth of the aluminum industry is attributed to a unique combination of properties which makes it one of the most versatile of engineering and construction materials. Aluminum is light in weight, yet some of its alloys have strengths greater than that of structural steel. It has good electrical and thermal conductivities and high reflectivity to both heat and light. It is highly corrosion-resistant under a great many service conditions and is nontoxic. Aluminum can be cast and worked into almost any form and can be given a wide variety of surface finishes. With all these outstanding properties, it is not surprising that aluminum alloys have come to be of prime importance as engineering materials.

5-1 PRODUCTION OF ALUMINUM

Reduction

Aluminum is the most abundant metallic element in the earth's crust, but always occurs in the combined state with other elements such as iron, silicon, and oxygen. Bauxite, which is mainly hydrated aluminum oxides, is the chief mineral used for the production of aluminum. Pure aluminum oxide is extracted from bauxite by the *Bayer process*.

In the Bayer process, finely ground and calcined bauxite is treated with hot sodium hydroxide to convert the aluminum in the ore to sodium aluminate

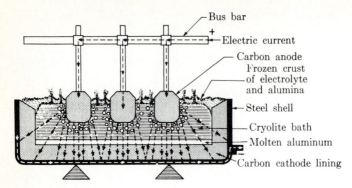

Figure 5-2 Electrolytic cell used to produce aluminum. *(Courtesy of Aluminum Company of America.)*

according to the reaction

$$Al_2O_3 + 2NaOH \rightarrow 2NaAlO_2 + H_2O \qquad (160-170°C)$$

After separation of the insoluble residue, consisting of mainly ferric oxide and silica, the aluminate solution is slowly cooled to 25 to 35°C to precipitate aluminum hydroxide $[Al(OH)_3]$ according to the reaction

$$NaAlO_2 + 2H_2O \rightarrow \underline{Al(OH)_3} + NaOH$$

The $Al(OH)_3$ is then thickened, washed, and calcined at 1100°C to produce aluminum oxide, Al_2O_3.

The aluminum oxide is dissolved in a bath of molten cryolite (Na_3AlF_6) and electrolyzed in an electrolytic cell (Fig. 5-2) using carbon anodes and cathode. In the electrolysis process (*Hall Process*), molten aluminum is deposited in the liquid state on the carbon cathode lining and sinks to the bottom of the electrolytic bath since it has a higher density. During the electrolysis, oxygen is released at the anodes where it attacks the carbon and forms CO and CO_2. The molten aluminum is periodically tapped from the cells and treated in the molten condition to remove excess oxides and gases. The cell-tapped aluminum usually contains from 99.5 to 99.9 percent aluminum, with iron and silicon being the main impurities.

Primary Fabrication

Remelting and casting The initial step in the processing of aluminum is the remelting operation. First, furnaces are charged with either liquid aluminum directly from the reduction cells or with ingots to be remelted. Alloying elements, alloying element master ingot, and scrap are also added as required. The molten metal in the remelt furnace is cleaned by skimming the surface to remove impurities and oxidized metal. The liquid metal is also "fluxed" or purged with chlorine gas to remove dissolved hydrogen gas. When chlorine gas bubbles through the liquid metal, dissolved hydrogen gas is removed by a chemical-mechanical action.

After the metal is cleaned and degassed, it is screened and cast. Ingot shapes such as sheet ingot and extrusion billet are usually cast by the direct-chill casting method. In this process, molten metal is poured into a water-cooled mold. As soon as the metal begins to solidify, the bottom of the mold is lowered so that the metal can be continuously cast into ingots about 14 ft in length. Sheet ingot cross sections can be up to about 18 in by 64 in.

Scalping In the case of sheet ingots, about $\frac{1}{2}$ in of metal is removed from ingot surfaces which will make contact with the hot-rolling-mill rolls. This is done to ensure a clean, smooth surface for the fabricated sheet or plate.

Preheating or homogenizing Most alloy ingots to be rolled are preheated about 10 to 24 h to allow atomic diffusion to homogenize the as-cast structure. Also, many constituents are taken into solid solution, such as the manganese-rich constituents in alloy 3003. The preheating temperature must be kept below the melting point of the constituent with the lowest melting point.

Hot rolling[1] Preheated ingots are reheated to hot-rolling temperature and are broken down in a four-high reversing hot-rolling mill. The slabs from this mill, which might be about 3 in thick, are then reheated and further reduced to about $\frac{3}{4}$ to 1 in in an intermediate four-high mill. Further reduction is usually carried out in tandem hot-rolling mills to produce metal about 0.1 in thick.

Cold rolling After intermediate annealing, the metal is cold-rolled to final gauge by cold-rolling mills. Intermediate annealing treatments may be required. The maximum amount of reduction that can be performed in a single pass through the mill depends on the alloy and the temper of the sheet being rolled. The percent reduction can vary from about 30 to 65 percent. Final annealing of the sheet may be required. When minimum surface oxidation is needed, special inert atmosphere furnaces are used.

5-2 CLASSIFICATION AND TEMPER DESIGNATIONS OF ALUMINUM ALLOYS[2]

Classification

Wrought aluminum and wrought aluminum alloys A system of four-digit numerical designations is used to identify wrought aluminum and wrought aluminum alloys. The first digit indicates the alloy group. The last two digits identify the aluminum alloy or indicate the aluminum purity. The second digit indicates

---

[1] Rolling mill procedures may vary in different industrial plants.
[2] See Ref. 6.

Table 5-2 Wrought aluminum alloy groups

Aluminum, 99.00 percent minimum and greater	1xxx
Aluminum alloys grouped by major alloying elements	
Copper	2xxx
Manganese	3xxx
Silicon	4xxx
Magnesium	5xxx
Magnesium and silicon	6xxx
Zinc	7xxx
Other element	8xxx
Unused series	9xxx

modifications of the original alloy or impurity limits. Table 5-2 lists the wrought aluminum alloy groups.

Casting alloys A system of four-digit numerical designations is used to identify aluminum and aluminum alloys in the form of castings and foundry ingot. The first digit indicates the alloy group. The second two digits identify the aluminum alloy or indicate the aluminum purity. The last digit, which is separated from the others by a decimal point, indicates the product form, i.e., castings or ingot. A modification of the original alloy or impurity limits is indicated by a serial letter before the numerical designation. The letter "x" is used for experimental alloys.

However, aluminum casting alloys are identified most commonly by just three digits. Table 5-3 lists the casting aluminum alloy groups.

Temper Designations

Temper designations follow the alloy designations and are separated by a hyphen (e.g., 3003-O). Subdivisions of a basic temper are in turn followed by one or more additional digits (e.g., 3003-H14).

Table 5-3 Cast aluminum alloy groups

Aluminum, 99.00 percent minimum and greater	1xx.x
Aluminum alloys grouped by major alloying elements	
Copper	2xx.x
Silicon, with added copper and/or magnesium	3xx.x
Silicon	4xx.x
Magnesium	5xx.x
Zinc	7xx.x
Tin	8xx.x
Other element	9xx.x
Unused series	6xx.x

Basic temper designations

F—As fabricated. No control over the amount of strain hardening; no mechanical property limits.

O—Annealed and recrystallized. Temper with the lowest strength and highest ductility.

H—Strain-hardened (see below for subdivisions).

T—Heat-treated to produce stable tempers other than F or O (see below for subdivisions).

Strain-hardened subdivisions

H1—Strain-hardened only. The degree of strain hardening is indicated by the second digit and varies from quarter hard (H12) to full-hard (H18), which is produced with approximately 75 percent reduction in area.

H2—Strain-hardened and partially annealed. Tempers ranging from quarter-hard to full-hard obtained by partial annealing of cold-worked materials with strengths initially greater than desired. Tempers are H22, H24, H26, and H28.

H3—Strain-hardened and stabilized. Tempers for age softening aluminum-magnesium alloys that are strain-hardened and then heated at a low temperature to increase ductility and stabilize mechanical properties. Tempers are H32, H34, H36, and H38.

Heat-treated subdivisions

T1—Naturally aged.[1] Product is cooled from an elevated-temperature shaping process and naturally aged to a substantially stable condition.

T3—Solution heat-treated, cold-worked, and naturally aged to a substantially stable condition.

T4—Solution heat-treated and naturally aged to a substantially stable condition.

T5—Cooled from an elevated-temperature shaping process and then artificially aged.[2]

T6—Solution heat-treated and then artificially aged.

T7—Solution heat-treated and stabilized.

T8—Solution heat-treated, cold-worked, and then artificially aged.

5-3 COMMERCIALLY PURE ALUMINUM

Chemical Composition and Typical Applications

Commercial purity aluminum varies from about 99.3% Al min to 99.7%. The higher-purity aluminum is selected for applications such as electrical conductor

---

[1] In *natural aging*, the aluminum alloy is aged at room temperature.

[2] In *artificial aging*, the aluminum alloy is aged at some temperature above room temperature.

Table 5-4 Chemical compositions and applications of commercially pure aluminum alloys†

Alloy	% Purity‡	% Si	% Fe	% Cu	Applications
1050	99.50	0.25	0.40	0.05	Coiled tubing, extruded
1060	99.60	0.25	0.35	0.05	Chemical equipment; railroad tank cars
1100	99.00	1.0 Si + Fe		0.12 nom.§	Sheet metal work; spun hollow ware; fin stock
1145	99.45	0.55 Si + Fe		0.05	Foil for capacitors; fin stock
1175	99.75	0.15 Si + Fe		0.10	Reflector sheet
1200	99.00	1.0 Si + Fe		0.05	Coiled tubing, extruded; sheet metal work
1230	99.30	0.7 Si + Fe		0.10	Cladding for sheets and plates
1235	99.35	0.65 Si + Fe		0.05	Foil for capacitors; tubing
1345	99.45	0.30	0.40	0.10	—
1350	99.50	0.10	0.40	0.05	Electrical conductors

† After "ASM Databook," 1979. *Metal Progress*, vol. 116, no. 1, mid-June 1979.
‡ Al min.
§ 0.05 to 0.20 range.

alloys and reflector sheet. Lower-purity metal, with iron and copper added if necessary, is used to produce the 1100 alloy, which is the "standard" commercially pure aluminum alloy. It is relatively soft and ductile, with excellent workability and weldability. Commercially pure aluminum responds well to decorative finishes and shows excellent corrosion resistance. Table 5-4 lists the chemical compositions and applications of the various types of commercially pure aluminum.

Structure

The structure of unalloyed aluminum (1xxx series) is characterized by a relatively pure aluminum matrix. The insoluble constituents in commercially pure aluminum are chiefly iron and silicon types, as shown in Figs. 5-3 and 5-4. The quantity of constituents is a function of alloy purity and their distribution functions of the type and extent of fabrication. Since every commercial aluminum alloy contains iron and silicon impurities, the insoluble iron and silicon constituents are common to them all in varying degrees.

Mechanical Properties

The mechanical properties of commercially pure aluminum are listed in Table 5-5. The tensile strength of annealed 99.99% Al is about 6.5 ksi, with a yield strength of 1.5 ksi and an elongation of 50 percent. This superpurity aluminum will not remain in the severely strain-hardened condition at ambient temperature, and will probably recrystallize. As the impurity level is increased, the

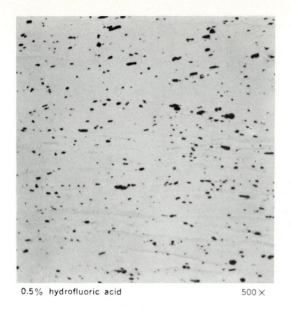

0.5% hydrofluoric acid   500 ×

Figure 5-3 Alloy 1100-H18 sheet, cold-rolled. Structure shows metal flow around insoluble particles of FeAl₃ (black). Particles are remnants of scriptlike constituents in the ingot that have been fragmented by working. *(After Metals Handbook, 8th ed., vol. 7, American Society for Metals, 1972, p. 242.)*

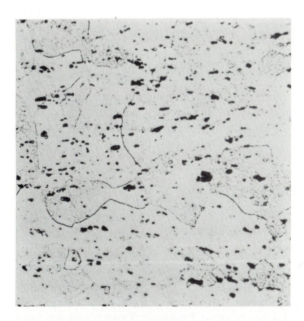

Figure 5-4 Alloy 1100-0 sheet, cold-rolled and annealed. Recrystallized, equiaxed grains, and insoluble particles of FeAl₃ (black). Size and distribution of FeAl₃ in the worked structure were unaffected by annealing. (0.5% hydrofluoric acid; 500×.) *(After Metals Handbook, 8th ed., vol. 7, American Society for Metals, 1972, p. 242.)*

Table 5-5 Typical mechanical properties of commercially pure aluminum†

Alloy	Temper	Tensile strength, psi	Tensile yield strength,‡ psi	Elongation, % in 2 in	Hardness,§ Bhn	Shear strength, psi	Fatigue limit,¶ psi
1199	O	6,500	1,500	50			
	H18	17,000	16,000	5			
1180	O	9,000	3,000	45			
	H18	18,000	17,000	5			
1060	O	10,000	4,000	43	19	7,000	3000
	H14	14,000	13,000	12	26	9,000	5000
	H18	19,000	18,000	6	35	11,000	6500
EC	O	12,000	4,000	23††		8,000	
	H14	16,000	14,000			10,000	
	H19	27,000	24,000	2.5††		15,000	
1145	O	11,000	5,000	40		8,000	
	H18	21,000	17,000	5		12,000	
1100	O	13,000	5,000	35	23	9,000	5000
	H14	18,000	17,000	9	32	11,000	7000
	H18	24,000	22,000	5	44	13,000	9000

† After Ref. 1.
‡ Yield strength, 0.2 percent offset.
§ 500-kg load, 10-mm ball, 30 s.
¶ Based on 500 million cycles using an R. R. Moore type of rotating-beam machine.
†† Elongation in 10 in.

strength of commercial purity aluminum increases reaching a maximum in the 1xxx series with 1100 alloy. Full-hard 1100 alloy has a tensile strength of about 24 ksi, with a yield strength of 22 ksi and an elongation of only 5 percent.

5-4 ALUMINUM-MANGANESE ALLOYS

Chemical Compositions and Typical Applications

The addition of about 1.2% Mn to commercial purity aluminum (0.6% Fe and 0.2% Si) produces a moderately strong non-heat-treatable aluminum alloy. The manganese addition strengthens the aluminum by solid-solution strengthening and by a fine dispersion of precipitates. Still further increases in strength are obtained by magnesium additions up to about 1 percent. These alloys are used for general purposes where moderate strength and good workability are required. Table 5-6 lists the chemical compositions and applications of aluminum-manganese and aluminum-manganese-magnesium alloys.

Table 5-6 Chemical compositions and applications of aluminum-manganese alloys†

Alloy	% Mn	% Mg	% Cu	Applications
3003	1.2		0.12	Cooking utensils, chemical equipment, pressure vessels, sheet metal work, builders' hardware
3004	1.2	1.0		Sheet metal work, storage tanks, pressure vessels
3005	1.2	0.40		Building products—siding, gutters, etc.
3105	0.5	0.50		Building products—siding, gutters, etc.

† After "ASM Databook," 1979. *Metal Progress*, vol. 116, no. 1, mid-June 1979.

Structure

The microstructure of alloy 3003 sheet (1.2% Mn) in the annealed condition is shown in Fig. 5-5. If this alloy is given a high temperature preheat (homogenization) treatment at about 600°C to dissolve many of the manganese-containing constituents, after it is cold-worked and subsequently annealed at about 340°C, a fine dispersoid of $(Mn, Fe)Al_6$ and $\alpha(Al-Fe-Mn-Si)$ constituents is formed (Fig. 5-5).

The microstructure of this alloy after preheating at 593°C, cold working 80 percent, and annealing at 343°C has been studied by Morris[1] using electron transmission microscopy. He showed that manganese-rich precipitates nucleate preferentially on the cold-worked dislocation structures during annealing (Fig. 5-6). These precipitates create a pinning action which inhibits the movement of the dislocations and subsequent formation of low-angle polygonized grain

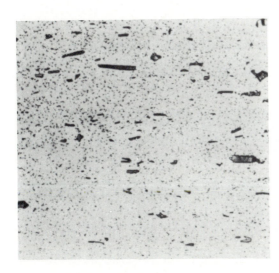

Figure 5-5 Alloy 3003 (1.2% Mn) annealed sheet; structure consists of a fine dispersion of $(Mn, Fe)Al_6$ and $\alpha(Al-Fe-Mn-Si)$ precipitates. (0.5% hydrofluoric acid; 500×.) *(After F. Keller in "Physical Metallurgy of Aluminum Alloys," American Society for Metals, 1949, p. 106.)*

[1] J. G. Morris, *Trans. ASM* 59(1966):1006.

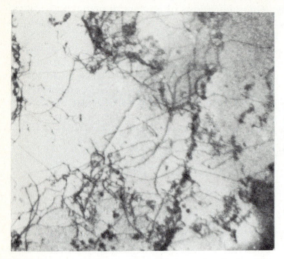

Figure 5-6 Alloy 3003 (1.2% Mn) pre-heated at 543°C, cold-rolled 80%, annealed at 343°C for 250 s. Structure shows manganese-rich constituents precipitated on dislocations during annealing. Recrystallization in the alloy is inhibited by the pinning of dislocations by the precipitates. (Electron micrograph; 25,000 × .) [*After J. G. Morris, Trans. ASM 59 (1966): 1006.*]

boundaries. The precipitates therefore inhibit recrystallization and raise the recrystallization temperature of the alloy.

Mechanical Properties

Table 5-7 lists the mechanical properties of aluminum-manganese and aluminum-manganese-magnesium alloys. The strength of 3003 alloy is about 3 to 4 ksi higher than that of 1100 alloy (e.g., 3003-O has a tensile strength of 16 ksi compared to 13 ksi for 1100-O). Alloy 3004 is further strengthened by the

Table 5-7 Typical mechanical properties of non-heat-treatable aluminum-manganese and aluminum-manganese-magnesium alloys†

Alloy	Temper	Tensile strength, psi	Tensile yield strength, psi	Elongation, % in 2 in	Hardness, Bhn	Shear strength, psi	Fatigue limit, psi
3003	O	16,000	6,000	30	28	11,000	7,000
	H14	22,000	21,000	8	40	14,000	9,000
	H18	29,000	27,000	4	55	16,000	10,000
3004	O	26,000	10,000	20	45	16,000	14,000
	H34	35,000	29,000	9	63	18,000	16,000
	H38	41,000	36,000	5	77	21,000	18,000
3005	O	19,000	8,000	25	· ·	12,000	
	H18	35,000	33,000	4	· ·	18,000	
3105	H25	26,000	24,000	8	· ·	16,000	

† After Ref. 1.

solid-solution strengthening effect of magnesium, so that in the annealed condition it has a tensile strength of 26 ksi. Several low-strength alloys such as 3005
and 3105 were introduced in 1953 and 1960, respectively. These alloys have
desirable combinations of strength, formability, and corrosion resistance for
applications in the building and specialty products areas.

5-5 ALUMINUM-MAGNESIUM ALLOYS

Chemical Compositions and Typical Applications

The binary aluminum-magnesium alloys are the basis for the 5xxx series of
non-heat-treatable aluminum alloys. Although magnesium has a substantial
solid solubility in aluminum (14.9 wt% at 451°C) and greatly decreasing solid
solubility with decreasing temperature (Fig. 5-7), aluminum-magnesium alloys
do not show appreciable precipitation hardening at concentrations less than 7%
Mg. Magnesium does, however, substantially strengthen aluminum by solid-
solution strengthening, and it causes high-work-hardening characteristics.

Table 5-8 lists the chemical compositions and applications of aluminum-
magnesium alloys. General purpose and structural Al-Mg alloys contain from 1
to slightly over 5% Mg and are in widespread industrial use. There are only a
few binary wrought aluminum-magnesium alloys such as 5005 and 5050. To
increase their strength, most aluminum-magnesium alloys contain some
manganese (0.1 to 1.0 percent) and/or chromium (0.1 to 0.25 percent). Examples
of Al-Mg alloys with additions of chromium are 5052 and 5154, while 5056 is an
example of an alloy which contains both manganese and chromium.

Many aluminum-magnesium alloys have been developed as finishing and
decorative alloys. By lowering the amount of iron, silicon and other impurities, a

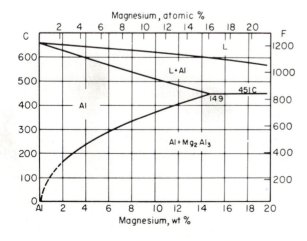

Figure 5-7 Aluminum-rich end of
aluminum–magnesium phase diagram. [*After K. R. Van Horn (ed.),
"Aluminum," vol. 1, American
Society for Metals, 1967, p. 375.*]

Table 5-8 Chemical compositions and applications of aluminum-magnesium alloys†

Alloy	% Composition	Applications
5005	0.8 Mg	Appliances; utensils; architectural trim; electrical conductors
5050	1.4 Mg	Builders' hardware; refrigerator trim; coiled tubes
5052	2.5 Mg, 0.25 Cr	Sheet metal work; hydraulic tubes; appliances; bus, truck and marine uses
5056	0.12 Mn, 5.1 Mg, 0.12 Cr	Cable sheathing; rivets for magnesium; screen wire; zippers
5083	0.7 Mn, 4.45 Mg, 0.15 Cr	⎧ Unfired, welded pressure vessels; marine, auto, and aircraft parts;
5086	0.45 Mn, 4.0 Mg, 0.15Cr	⎨ cryogenics; TV towers; drilling rigs; transportation equipment; ⎩ missile components; armor plate
5154	3.5 Mg, 0.25 Cr	Welded structures; storage tanks; pressure vessels; salt-water service
5252	2.5 Mg	Auto and appliance trim
5254	3.5 Mg, 0.25 Cr	Hydrogen peroxide and chemical storage vessels
5356	0.12 Mn, 5.0 Mg, 0.12 Cr	Welding rod, wire, and electrodes
5454	0.8 Mn, 2.7 Mg, 0.12 Cr	Welding structures; pressure vessels; marine service; tubing
5456	0.8 Mn, 5.1 Mg, 0.12 Cr	High-strength welded structures; storage tanks; pressure vessels; marine service
5457	0.3 Mn, 1.0 Mg	Anodized auto and appliance trim (good formability in annealed temper)
5652	2.5 Mg, 0.25 Cr	Hydrogen peroxide and chemical storage vessels
5657	0.8 Mg	Anodized auto and appliance trim (good brightness)

† After "ASM Databook," 1979, *Metal Progress*, vol. 116, no. 1, mid-June 1979.

series of decorative alloys[1] was created. Examples are 5053 and 5252 and the 5x57 alloys 5357, 5457, and 5657.

The aluminum-magnesium alloys have a wide range of strength, good forming and welding characteristics, and a high resistance to corrosion. An outstanding property of aluminum-magnesium alloys is the good welding response of the higher-strength alloys when argon-shielded arc-welding processes are used.

Structure

The magnesium in most aluminum-magnesium alloys is in solid solution. However, when the magnesium content in Al-Mg alloys exceeds about 3.5 percent, Mg_2Al_3 can be precipitated by low-temperature thermal treatments or by slow cooling from elevated temperatures. For example, if alloy 5086, which contains about 4% Mg, is cold-worked and heated in the 120 to 180°C range, a continuous network of Mg_2Al_3 can be precipitated at the grain boundaries (Fig. 5-8). This structure is undesirable since it can make the alloy susceptible to

[1] Iron and silicon impurities are especially detrimental to the bright finishing characteristics of aluminum alloys.

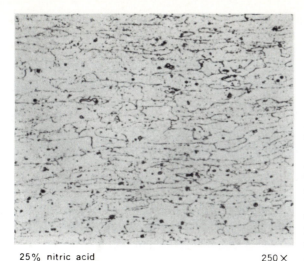

25% nitric acid   250×

Figure 5-8 Alloy 5086-H43 plate, 1/2 in thick, cold rolled and stabilized at 120 to 177°C. Undesirable continuous network of Mg_2Al_3 particles is precipitated at grain boundaries; large particles are insoluble phases. This type of structure is undesirable because under some conditions it is susceptible to stress corrosion cracking. *(After Metals Handbook, 8th ed., vol. 7, American Society for Metals, 1972, p. 244.)*

stress corrosion cracking under some conditions. It is therefore more desirable to stress-relieve this type of alloy at a higher temperature (i.e., 245°C) and with careful processing cause the Mg_2Al_3 to precipitate as a fine dispersion throughout the matrix of the alloy, as shown in Fig. 5-9.

Particles of magnesium silicide (Mg_2Si) will also be present in commercial Al-Mg alloys in proportion to the amount of silicon in the alloy because of the low solubility of Mg_2Si in the presence of excess magnesium. If the Al-Mg alloys contain chromium and manganese, other insoluble phases will also be present

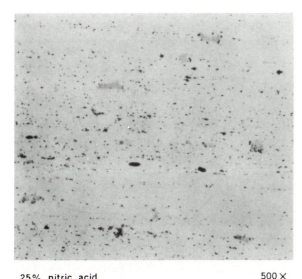

25% nitric acid   500×

Figure 5-9 Alloy 5456 plate 1/4-in thick, cold-rolled and stress-relieved at 246°C. The Mg_2Al_3 in this case is finely distributed through the matrix and there are no continuous networks of precipitates at the grain boundaries. This type of structure is more desirable and less subject to corrosive attack. Large particles are insoluble phases such as Mg_2Si (black) and (Fe, Mn)Al_6 (gray). *(After Metals Handbook, 8th ed., vol. 7, American Society for Metals, 1972, p. 244.)*

along with the ever-present iron-containing constituents found in all commercial purity aluminum alloys.

Mechanical Properties

The mechanical properties of wrought non-heat-treatable aluminum-magnesium alloys are listed in Table 5-9. The ultimate tensile strengths of the commercial aluminum-magnesium alloys in the annealed condition range from a low of 18 ksi for alloy 5005-O to a high of 45 ksi for alloy 5456-O. Alloys 5083-O and 5086-O have slightly lower strength (42 and 38 ksi, respectively) than alloy 5456-O. Wrought products of aluminum-magnesium alloys are always available in the annealed -O temper, and usually in the H3 tempers. The H3 temper is generally used for strain-hardened products since the H1 temper is not usually stable at room temperature. The H3 temper produces stable properties with higher elongation levels and improved forming characteristics.

Although aluminum-magnesium alloys are classified as non-heat-treatable, the amount of magnesium soluble at annealing temperatures for Al-Mg alloys with more than about 4% Mg (such as 5083, 5086, 5056, and 5456) is higher than that retained in solid solution at room temperature. As a result, if these alloys are severely strain-hardened and then stored for a long time at room temperature, precipitation of Mg_2Al_3 will occur along the slip bands. Also, if these alloys are exposed to elevated temperatures in the annealed condition, precipitation will occur along the grain boundaries. This precipitation makes these alloys susceptible to intergranular attack and stress corrosion in corrosive environments. For this reason, the H3xx tempers have been developed to eliminate or minimize this instability so that the higher strengths of these alloys can be utilized.

5-6 ALUMINUM-COPPER ALLOYS

Chemical Compositions and Applications

The first wrought binary aluminum-copper alloy that was developed in the United States was alloy 2025, which contains about 5.5% Cu (nominally). Although alloy 2025, introduced in about 1926, is still used to a limited extent for forgings, alloy 2219, which contains 6.3% Cu (nominally) and was developed in 1954, has in most cases replaced alloy 2025. Alloy 2219 has a much wider and higher range of strength as well as good weldability, superior resistance to stress corrosion, and higher elevated-temperature properties.

Alloy 2011 with 5.5% Cu, 0.4% Bi, and 0.4% Pb is used when good cutting and chip characteristics are necessary for high-speed production of screw-machine parts. This alloy is the basic aluminum screw-machine alloy, and is used as a reference standard for the machinability of aluminum alloys. Table

Table 5-9 Typical mechanical properties of wrought non-heat-treatable aluminum-magnesium alloys†

Alloy	Temper	Tensile strength, psi	Tensile yield strength, psi	Elon-gation, % in 2 in	Hard-ness, Bhn	Shear strength, psi	Fatigue limit, psi
5005	O	18,000	6,000	30	30	11,000	
	H14	23,000	22,000	6	41	14,000	
	H34	23,000	20,000	8	41	14,000	
	H18	29,000	28,000	4	51	16,000	
	H38	29,000	27,000	5	51	16,000	
5050	O	21,000	8,000	24	36	15,000	12,000
	H34	28,000	24,000	8	53	18,000	13,000
	H38	32,000	29,000	6	63	20,000	14,000
5052	O	28,000	13,000	25	47	18,000	16,000
	H34	38,000	31,000	10	68	21,000	18,000
	H38	42,000	37,000	7	77	24,000	20,000
5056	O	42,000	22,000	35	65	26,000	20,000
	H18	63,000	59,000	10	105	34,000	22,000
	H38	60,000	50,000	15	100	32,000	22,000
5082	H19	57,000	54,000	4			
5083	O	42,000	21,000	22	67	25,000	22,000
	H112	43,000	23,000	20	70	25,000	22,000
	H321	46,000	33,000	16	82	28,000	22,000
	H323	47,000	36,000	10	84	27,000	· · ·
	H343	52,000	41,000	8	92	30,000	· · ·
5086	O	38,000	17,000	22	60	23,000	21,000
	H32	42,000	30,000	12			
	H34	47,000	37,000	10	82	28,000	23,000
	H112	39,000	19,000	14	64	23,000	21,000
5154	O	35,000	17,000	27	58	22,000	17,000
	H34	42,000	33,000	13	73	24,000	19,000
	H38	48,000	39,000	10	80	28,000	21,000
	H112	35,000	17,000	25	63	22,000	17,000
5454	O	36,000	17,000	22	60	23,000	19,000
	H34	44,000	35,000	10	81	26,000	21,000
	H112	36,000	18,000	18	62	23,000	· · ·
	H311	38,000	26,000	14	70	23,000	· · ·
5456	O	45,000	23,000	24	70	27,000	22,000
	H24	54,000	41,000	12	· ·	31,000	
	H112	45,000	24,000	22	70	27,000	
	H311	47,000	33,000	18	75	27,000	24,000
	H321	51,000	37,000	16	90	30,000	23,000
	H323	51,000	38,000	10	90	30,000	
	H343	56,000	43,000	8	94	33,000	

† After Ref. 1.

Table 5-10 Chemical compositions and applications of aluminum-copper alloys†

Alloy	% Cu	% Mn	% Other	Applications
2011	5.5		0.4 Bi, 0.4 Pb	Screw-machine products
2025	4.5	0.8	0.8 Si	Forgings, aircraft products
2219	6.3	0.3	0.06 Ti, 0.10 V, 0.18 Zr	Structural use to 660°F, high-strength weldments for cryogenic and aircraft parts
2419‡	6.3	0.3	0.06 Ti, 0.10 V, 0.18 Zr	Same as 2219 plus high fracture toughness

† After "ASM Databook," 1979, *Metal Progress*, vol. 116, no. 1, mid-June 1977.

‡ Alloy 2419 has lower iron and silicon levels than alloy 2219.

5-10 lists the chemical compositions of wrought aluminum-copper alloys and their applications.

Binary Aluminum-Copper Alloys

Phase diagram Copper is one of the most important alloying elements for aluminum since it produces considerable solid-solution strengthening and with suitable heat treatment can provide greatly increased strength by precipitate formation. The maximum solid solubility of copper in aluminum is 5.65 percent at the eutectic temperature of 548°C (Fig. 5-10). The solubility of copper in aluminum decreases rapidly with decreasing temperature from 5.65 to less than about 0.1 percent at room temperature.

Precipitation-strengthening heat treatment for aluminum-copper alloys In order to achieve the maximum effect of precipitation strengthening (without cold deformation), the aluminum-copper alloy must be

1. Solution-heat-treated in the α-solid solution phase field (about 515°C)
2. Quenched rapidly to room temperature or below
3. Artificially aged in the 130 to 190°C range

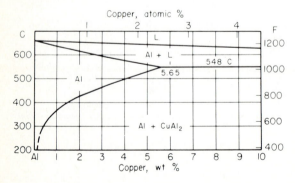

Figure 5-10 Aluminum-rich end of aluminum–copper phase diagram. [*After K. R. Van Horn (ed.), "Aluminum," vol. 1, American Society for Metals, 1967, p. 372.*]

Consider the precipitation strengthening of an Al–4% Cu alloy:

1. *Solution heat treatment.* The Al–4% Cu alloy must first be heated to about 515°C to allow the copper and aluminum atoms to diffuse randomly into a uniform solid solution. The alloy at this stage consists of solid solution α. This first step of the precipitation-strengthening heat treatment is sometimes called *solutionizing.*
2. *Quenching.* After the solution heat treatment, the alloy is quenched (rapidly cooled) to room temperature in water. This treatment produces a supersaturated solid solution of copper in aluminum. The Al–4% Cu alloy in this condition is not stable and strives to form metastable phases to lower the energy of the system. The driving force for the precipitation of metastable phases is the high-energy state of the unstable supersaturated solid solution of copper in aluminum.
3. *Aging.* If substantial precipitation of a metastable phase occurs at room temperature, the alloy is said to be *natural aging.* Although some alloys will naturally age-harden to a satisfactory strength at room temperature, most alloys must be age-hardened at an elevated temperature, or *artificially* aged. In the case of the Al–4% Cu alloy, the artificial age-hardening temperature used is generally between 130 and 190°C.

Structures formed during the aging of aluminum-copper alloys In precipitation-strengthened aluminum-copper alloys, five sequential structures can be identified: (1) supersaturated solid solution, (2) GP1 zones,[1] (3) GP2 zones (also called θ'' phase), (4) θ', and (5) θ phase, $CuAl_2$. Not all of these phases occur at all aging temperatures. GP1 and θ'' do not exist above their solvus temperatures, and θ' and θ require sufficiently high aging temperatures for their formation.

GP1 zones GP1 zones are formed at lower temperatures (i.e., below about 130°C) and are created by copper atoms segregating in the supersaturated solid solution of Al-Cu alloys. GP1 zones consist of disks of a few atoms thick (4 to 6 Å) and about 80 to 100 Å in diameter, and form on the {100} cubic planes of the matrix. We do not yet know the true structure of GP1 zones, but a recent analysis by Dahlgren (Ref. 11) indicates that GP1 zones have a *low* copper content.[2]

Since the copper atom has a diameter about 11 percent less than the aluminum atom, the cubic lattice parameter of the zone is less than that of the matrix and so is strained tetragonally. The GP1 zones can be detected in

---

[1] The letters "GP" stand for "Guinier-Preston", after two of the early scientists who explored these structures.

[2] In Dahlgren's analysis, he estimated the GP2 composition to be 17 at% Cu and, since the GP1 zones are much smaller, he estimated their copper concentration to be much less than 17 at%. Some electron and x-ray diffraction estimates are as high as 80 to 100 at% Cu for GP1 zones. Clearly, more research is needed in this area.

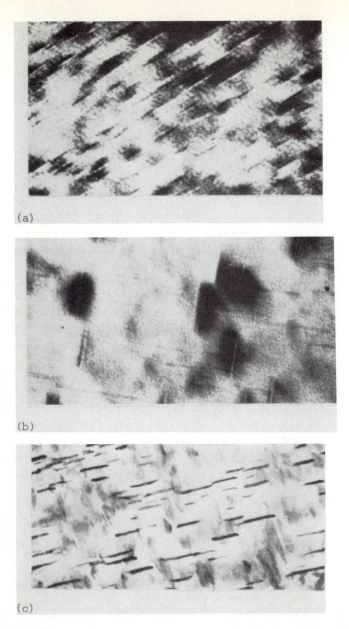

Figure 5-11 Microstructures of aged Al–4% Cu alloys. (a) Al–4% Cu, heated to 540°C, water-quenched and aged 16 h at 130°C. The Guinier-Preston zones have been formed as plates parallel to the {100} planes of the face-centered cubic matrix and at this stage are a few atoms thick and about 100 Å in diameter. Only plates lying on one crystallographic orientation are visible. (Electron micrograph; 1,000,000 × .) (b) Al–4% Cu, solution-treated at 540°C, quenched in water, and aged for 1 day at 130°C. This thin foil micrograph shows strain fields due to coherent GP (2) zones. The dark regions surrounding the zones are caused by strain fields. (Electron micrograph; 800,000 × .) (c) Al–4% Cu alloy solution heat-treated at 540°C, quenched in water, and aged for 3 days at 200°C. This thin-foil micrograph shows the incoherent and metastable phase θ' which forms by heterogeneous nucleation and growth. (Electron micrograph; 25,000 × .) *(After J. Nutting and R. G. Baker, "The Microstructure of Metals," Institute of Metals, 1965, pp. 65 and 67.)*

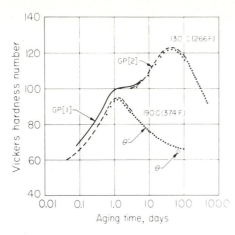

Figure 5-12 Correlation of structures and hardness of Al–4% Cu alloy aged at 130 and 190°C. [*After J. M. Silcock, T. J. Heal, and H. K. Hardy, J. Inst. Met. 82(1953–54):239, as presented in K. R. Van Horn (ed.), "Aluminum," vol. 1, American Society for Metals, 1967, p. 123.*]

the electron microscope because of strain fields associated with them, as shown in Fig. 5-11*a*. These zones impede dislocation movement, and so cause an increase in hardening and a decrease in ductility of an Al–4% Cu alloy, as indicated in Fig. 5-12.

GP2 zones (θ'') As in the case of GP1 zones, GP2 zones (θ'') have a tetragonal structure and are coherent with the {100} matrix planes of the Al–4% Cu or similar type alloys. In the early stages of their formation, GP2 zones are believed to have a low (< 17 at%) copper content (Ref. 11). As aging time is increased (e.g., at 130°C), the copper content of the zones is believed to increase, as well as their size. The size range for GP2 zones is from 10 to 40 Å thick and 100 to 1000 Å in diameter. Figure 5-11*b* shows coherent GP2 zones in an Al–4% Cu alloy. The *c* lattice parameter in the early stage of aging is 8.08 Å, and it decreases to 7.65 Å as the zones get larger at later stages of aging. Dahlgren (Ref. 11) believes this change occurs because the zones become richer in copper. GP2 zones further increase the hardness of the Al–4% Cu alloy when aged at 130 and 190°C, as shown in Fig. 5-12.

θ' phase Overaging of the Al–4% Cu alloy occurs when the completely unrelated,[1] incoherent, and metastable phase θ' forms in significant amounts. This phase nucleates heterogeneously, especially on dislocations. The size of the θ' phase depends on the time and temperature of aging, and ranges from 100 to 6000 Å or more in diameter with a thickness of 100 to 150 Å. This phase has a tetragonal structure, but with a still further reduced *c* parameter of 5.80 Å. Figure 5-11*c* shows θ' precipitates in an Al–4% Cu alloy after aging 3 days at 200°C. When this phase appears alone, the alloy is in an overaged condition, as indicated in Fig. 5-12.

---

[1] The θ' phase is not related to the GP1 or GP2 (θ'') metastable phases, but nucleates heterogeneously.

θ phase Aging at temperatures of about 190°C or above for extensive times will produce the equilibrium incoherent θ phase, $CuAl_2$. This phase has a BCT structure with $a = 6.07$ Å and $c = 4.87$ Å. θ can form from θ′ or directly from the matrix. The θ phase forms at the expense of the θ′ phase and is present when the alloy is in a highly overaged condition, as indicated by Fig. 5-12.

The general sequence of precipitation in binary aluminum-copper alloys can be represented by

Supersaturated solid solution → GP1 zones → GP2 zones (θ″ phase)

$$\rightarrow \theta' \rightarrow \theta \ (CuAl_2)$$

Commercial Wrought Aluminum-Copper Alloys

The important wrought aluminum-copper alloys in use today are alloys 2025, 2219, and 2011. The first wrought binary aluminum-copper alloy developed in the United States was alloy 2025, which contained 4.5% Cu, 0.7% Mn, and 0.8% Si. Alloy 2025 is still actively used today to a limited extent for forgings, but has been replaced for many applications by alloy 2219.

Alloy 2219, introduced in 1954, contains 6.3% Cu, 0.3% Mn, 0.25% Zr, 0.1% V, and 0.06% Ti. This alloy has a wide range of strength (25 to 69 ksi), good weldability, good stress-corrosion resistance, and excellent elevated-temperature

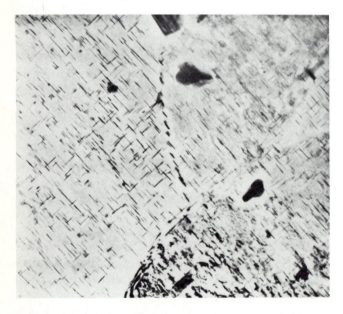

Figure 5-13 Transmission electron micrograph of alloy 2219 in the solution heat-treated and artificially aged condition. Structure shows relatively coarse θ″ precipitates. *(Courtesy of Aluminum Company of America Research Laboratories.)*

Table 5-11 Typical mechanical properties of heat-treatable aluminum-copper alloys†

Alloy	Temper	Tensile strength, psi	Tensile yield strength,‡ psi	Elongation, % in 2 in	Hardness, Bhn §	Shear strength, psi	Fatigue limit,¶ psi
2011	T3	55,000	43,000	15	95	32,000	18,000
	T6	57,000	39,000	17	97	34,000	18,000
	T8	59,000	45,000	12	100	35,000	18,000
2025	T6	58,000	37,000	19	110	35,000	18,000
2219	O	25,000	10,000	20			
	T31, T351	54,000	36,000	17	100	33,000	
	T37	57,000	46,000	11	117	37,000	
	T62	60,000	42,000	10	115	37,000	15,000
	T81, T851	66,000	51,000	10	130	41,000	15,000
	T87	69,000	57,000	10	130	41,000	15,000

† After Ref. 1.
‡ Yield strength, 0.2 percent offset.
§ 500-kg load, 10-mm ball.
¶ Based on 500 million cycles using an R. R. Moore type of rotating-beam machine.

properties for an aluminum alloy. The structure of 2219 alloy in the age-hardened condition is shown in Fig. 5-13 and consists mainly of θ'' precipitates. The excess $CuAl_2$, θ, that is not dissolved during solution heat treatment (the maximum solubility of Cu in Al is 5.65 percent) remains essentially unchanged during the heating and cooling and is expected to raise the strength of the alloy.

The mechanical properties of alloys 2025 and 2219 are listed in Table 5-11. By suitable thermomechanical treatments, the ultimate tensile strengths of alloy 2219 can be raised to 69 ksi. Increased precipitation can be produced in alloy 2219 by strain hardening after solution heat treatment and before artificial aging. The increased density of precipitation caused by strain hardening is reflected in the increased strength obtained in the T8 tempers of alloy 2219.

The presence of Mn, Zr, V, and Ti in alloy 2219 raises its recrystallization temperature so that it will retain higher strengths at elevated temperatures. Figure 5-14 shows the outstanding stress-rupture behavior of alloy 2219 after 100 and 1000 h at 200 and 315°C. A higher-purity base modification of alloy 2219, alloy 2419 was introduced in 1972. Alloy 2419, with lower iron (0.18 percent max) and silicon (0.15 percent max) levels, has higher fracture toughness for aircraft structural applications.

The wrought aluminum-copper alloy 2011 with 6.5% Cu, 0.04% Bi, and 0.04% Pb has been the basic aluminum screw-machine alloy since it was introduced in 1934. It has good cutting characteristics and produces fine, easily broken chips during machining. The lead and bismuth, however, lower the corrosion resistance of the Al-Cu alloy to some extent.

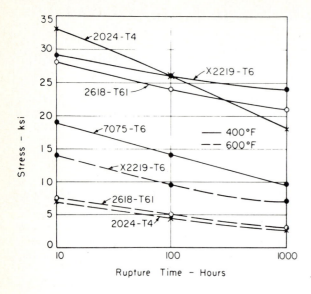

Figure 5-14 Stress rupture behavior of wrought-aluminum alloys at 400°F (204°C) and 600°F (315°C). *(After W. A. Anderson in "Precipitation from Solid Solution," American Society for Metals, 1959, p. 199.)*

5-7 ALUMINUM-COPPER-MAGNESIUM ALLOYS

Chemical Compositions and Typical Applications

Aluminum-copper-magnesium alloys were the first precipitation-hardenable alloys discovered.[1] The first alloy precipitation-hardened was a modification of alloy 2017, which now has the nominal composition 4.0% Cu, 0.60% Mg, and

Table 5-12 Chemical compositions and applications of aluminum-copper-magnesium alloys†

Alloy	% Cu	% Mg	% Mn	% Si	% Ni	% Other	Applications
2014	4.4	0.5	0.8	0.8			Truck frames, aircraft structures
2017	4.0	0.6	0.7	0.5			Screw-machine products, fittings
2018	4.0	0.7			2.0		Aircraft engine cylinder heads and pistons
2024	4.4	1.5	0.6				Truck wheels, screw-machine products, aircraft structures
2218	4.0	1.5			2.0		Jet engine impellers and compressor rings, aircraft engine cylinder heads and pistons
2618	2.3	1.6				0.18 Si, 1.0 Ni, 1.1 Fe, 0.07 Ti	Aircraft engines, temperatures to 238°C

† After "ASM Databook," 1979. *Metal Progress*, vol. 116, no. 1, mid-June 1977.

[1] A. Wilm, *Metallurgie* 8(1911):225.

0.7% Mn. Alloy 2014, with 4.4% Cu, 0.5% Mg, 0.8% Mn, and 0.8% Si, was developed later to be more responsive to artificial aging than alloy 2017, and is still one of the most popular Al-Cu-Mg alloys in use today. Alloy 2024, with 4.5% Cu, 1.5% Mg, and 0.6% Mn, was originally developed as a higher-strength naturally aging structural-aircraft alloy to replace 2017. The increased strength was attained by increasing the magnesium content from 0.5 to 1.5 percent. Table 5-12 lists the chemical compositions and typical applications for the most important Al-Cu-Mg alloys.

Structure

Additions of magnesium to aluminum-copper alloys greatly accelerate and intensify precipitation hardening in aluminum-copper alloys. In spite of their early discovery, the details of the precipitation processes in Al-Cu-Mg alloys are not completely understood. The general precipitation sequence for Al-Cu-Mg alloys is believed to be

Supersaturated solid solution $\rightarrow$ GP zones $\rightarrow$ S'(Al$_2$CuMg) $\rightarrow$ S(Al$_2$CuMg)

It is believed that GP zones are formed in the early stages of aging at low temperatures, but their form and size have not been firmly established. The zones are believed to consist of copper and magnesium atoms collected on the $\{110\}_{Al}$ planes. The acceleration of the natural aging process in Al-Cu alloys by the addition of magnesium could be due in part to an increase in diffusion rate made possible by the larger magnesium atoms compensating for the smaller-size copper atoms. The magnesium atoms would also relieve some of the stresses associated with the copper atoms in aluminum (Fig. 5-11). The overall effect of the magnesium atoms, therefore, would be to accelerate zone growth.

The mechanism of S' precipitation is firmly established since the S' metastable phase is incoherent and can be readily detected by electron microscopy. Wilson and Partridge (Ref. 12) have shown that S' is nucleated heterogeneously at dislocations and grows as laths on the $\{210\}_{Al}$ planes in the $\langle 001 \rangle$ direction.

S' precipitate, formed by solution heat treating 2024 alloy sheet at 493°C, water quenching to room temperature, and aging 12 h at 190°C, is shown in Fig. 5-15a. Since the S' phase is nucleated heterogeneously at dislocations, increasing the number of dislocations by cold working will increase the density of S' laths. By introducing 1.5 percent cold work after solution heat treatment and before aging at 190°C, the density of the S' precipitate in this sample was increased (Fig. 5-15b). Still more cold work (6 percent) between solution heat treatment and aging at 190°C further refined the S' precipitate and increased its density (Fig. 5-15c).

Mechanical Properties

The mechanical properties of the most common wrought aluminum-copper-magnesium alloys are listed in Table 5-13. The tensile strengths of alloy 2014

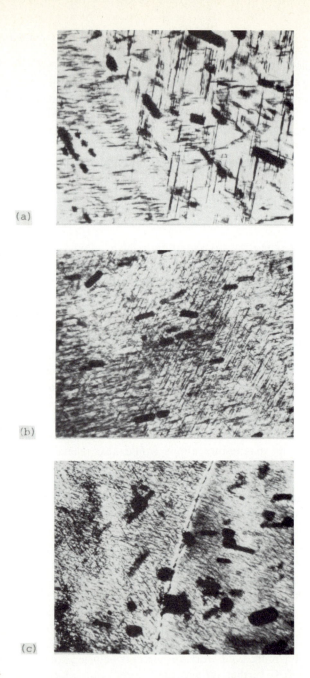

(a)

(b)

(c)

Figure 5-15 Electron transmission micrographs of 2024 alloy. (50,000 × .) (*a*) 2024-T6 alloy was solution heat-treated, quenched, and aged 12 h at 190°C. Structure consists of GP zones and coarse platelets of S'. (*b*) 2024-T81 alloy was solution heat-treated, quenched, stretched 1.5 percent, and aged 12 h at 190°C. Structure consists of GP zones and S' platelets which are smaller and more numerous than in *a*. (*c*) 2024-T86 alloy was solution heat-treated, quenched, cold-rolled 6 percent, and aged 12 h at 190°C. Structure consists of   P zones and small platelets of S'. Platelets are finer and more numerous than in *b*. [*After H. Y. Hunsicker in K. R. Van Horn (ed.), "Aluminum," vol. 1, American Society for Metals, 1967, p. 150.*]

Table 5-13 Typical mechanical properties of wrought heat-treatable aluminum-copper-magnesium alloys†

Alloy	Temper	Tensile strength, psi	Tensile yield strength,‡ psi	Elongation, % in 2 in	Hardness,§ Bhn	Shear strength, psi	Fatigue limit,¶ psi
2014	O	27,000	14,000	18	45	18,000	13,000
	T4, T451	62,000	42,000	20	105	38,000	20,000
	T6, T651	70,000	60,000	13	135	42,000	18,000
2017	O	26,000	10,000	22	45	18,000	13,000
	T4, T451	62,000	40,000	22	105	38,000	18,000
2024	O	27,000	11,000	20	47	18,000	13,000
	T3	70,000	50,000	18	120	41,000	20,000
	T36	72,000	57,000	13	130	42,000	18,000
	T4, T351	68,000	47,000	20	120	41,000	20,000
	T6	69,000	57,000	10	125	41,000	18,000
	T81, T851	70,000	65,000	6	128	43,000	18,000
	T86	75,000	71,000	6	135	45,000	18,000
2117	T4	43,000	24,000	27	70	28,000	14,000

† After Ref. 1
‡ Yield strength, 0.2 percent offset.
§ 500-kg load, 100-mm ball.
¶ Based on 500 million cycles using an R. R. Moore type of rotating-beam machine.

varies from 27 ksi in the annealed condition to 70 ksi in the T6 temper. Alloy 2024 can be age-hardened to 75 ksi if some strain hardening is introduced between solution heat treatment and aging.

The properties of wrought heat-treated Al-Cu-Mg alloys are greatly affected by the solution-heat-treatment temperature, as illustrated by the tensile properties of precipitation-hardened alloy 2014 in the T4 and T6 tempers in Fig. 5-16. If the solution-heat-treatment temperature is too low, the hardening phases are not completely dissolved prior to quenching and, therefore, lower tensile strengths will be obtained since the precipitate density will be lower. If the solution-heat-treatment temperature is too high, melting of some of the phases with low melting temperatures will occur, resulting in a decrease in strength and ductility. For the Al-Cu-Mg alloys, the normal commercial heat treatment practice is to solution-heat-treat about 5°C lower than the lowest melting eutectic.

The effect of aging in the temperature range 120 to 205°C on the tensile properties of solution-heat-treated and quenched 2014 alloy are shown in Fig. 5-17. It is noted that for each temperature precipitation strengthening is very rapid, and at temperatures over 120°C overaging also occurs rapidly. The optimum compromise for the industrial aging practice for alloy 2014 is 8 to 12 h at 170°C.

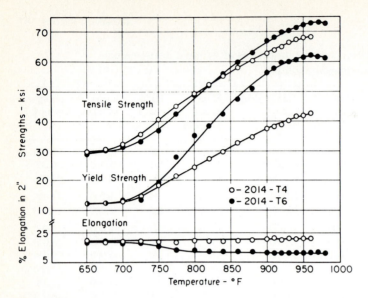

Figure 5-16 Effects of solution heat-treatment temperature on the tensile properties of 2014-T4 and 2014-T6 alloy sheet. *(After W. A. Anderson in "Precipitation from Solid Solution," American Society for Metals, 1959, p. 166.)*

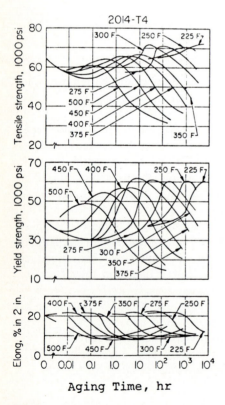

Aging Time, hr

Figure 5-17 Aging characteristics of 2014 aluminum sheet alloy. [*After H. Y. Hunsicker in K. R. Van Horn (ed.), "Aluminum," vol. 1, American Society for Metals, 1967, p. 147.*]

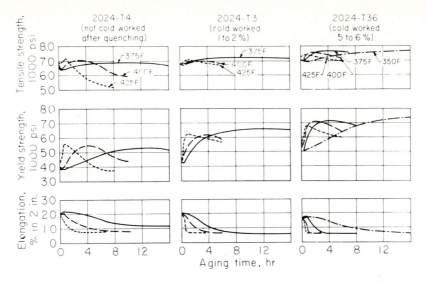

Figure 5-18 Elevated-temperature aging characteristics of 2024 alloy sheet. [*After H. Y. Hunsicker in K. R. Van Horn (ed.), "Aluminum," vol. 1, American Society for Metals, 1967, p. 149.*]

The rate and amount of precipitation strengthening can be significantly increased in some alloys by cold work after quenching, whereas in some other alloys little or no strengthening results. Alloy 2024 is particularly responsive to cold work between quenching and aging, as is shown by the increased precipitation density of S' phase in Fig. 5-15. The effect of cold work between quenching and aging on the tensile properties of alloy 2024 is shown in Fig. 5-18. Alloy 2024-T6 has a tensile yield stress of 57 ksi, but with 6 percent cold work introduced between quenching and aging the yield stress is raised to 71 ksi, which is an increase of 14 ksi.

5-8 ALUMINUM-MAGNESIUM-SILICON ALLOYS

Chemical Compositions and Typical Applications

The combination of magnesium (0.6 to 1.2 percent) and silicon (0.4 to 1.3 percent) in aluminum forms the basis for the 6xxx series of wrought precipitation-hardenable aluminum-magnesium-silicon alloys. In most cases, the magnesium and silicon are present in the alloy in amounts to nominally combine to form metastable phases of the intermetallic compound Mg_2Si, but silicon in excess of that required for Mg_2Si may also be used. Manganese or chromium are added to most 6xxx series alloys for increased strength and grain-size control. Copper also increases the strength of these alloys, but if present in amounts over 0.5 percent reduces their corrosion resistance. Table 5-14 lists the chemical

Table 5-14 Chemical compositions and applications of aluminum-magnesium-silicon alloys†

Alloy	% Mg	% Si	% Mn	% Cr	% Cu	% Other	Applications
6003	1.2	0.7					Cladding for sheets and plates
6005	0.5	0.8					Trucks and marine structures; railroad cars; furniture
6009	0.6	0.8	0.5		0.38		Auto body sheet
6010	0.8	1.0	0.5		0.38		Auto body sheet
6053	1.3	0.7		0.25			Wire and rods for rivets
6061	1.0	0.6		0.2	0.27		Heavy-duty structures where corrosion resistance is needed; truck and marine structures; railroad cars; furniture; bridge railing; hydraulic tubing
6063	0.7	0.4					Pipe; railings; furniture; architectural extrusions; truck flooring
6066	1.1	1.3	0.8		0.9		Forging and extrusions for welded structures
6070	0.8	1.4	0.7		0.3		Heavy-duty welded structures; pipelines
6101	0.6	0.5					High-strength bus conductors
6151	0.6	0.9		0.25			Moderate-strength intricate forgings for machine and auto parts
6162	0.9	0.6					Structures requiring moderate strength; busbars
6201	0.8	0.7					Electrical conductor wire (high strength)
6253	1.2	0.7		0.25		2.0 Zn	Component of clad rod and wire
6262	1.0	0.6		0.09	0.27	0.55 Pb; 0.55 Bi	Screw-machine products (better corrosion resistance than 2011)
6463	0.7	0.4				Low iron (0.15 max)	Architectural and trim extrusions

† After "ASM Databook," 1979, *Metal Progress*, vol. 116, no. 1, mid-June 1977.

composition and applications of some of the more important wrought aluminum-magnesium-silicon alloys.

The first aluminum alloy with balanced Mg_2Si content was 6053, which was developed in the 1930s and contains 2% Mg_2Si and 0.25% Cr. This alloy was followed by 6061, which is also a balanced alloy containing 1.5% Mg_2Si, 0.25% Cr, and 0.27% Cu. Alloy 6061 is an intermediate-strength general-purpose structural alloy. Used to a great extent today, it is one of the most important aluminum alloys. Higher-strength Al-Mg-Si alloys such as 6066 and 6070 with higher silicon contents were introduced in the 1960s.

For ease of extrudability for shapes, a slightly lower-strength alloy, 6063, was developed which contains about 1.0% Mg_2Si. This alloy can be quenched during the extrusion operation as it comes out of the press, thus avoiding the expense of solution heat treatment. Variations of alloy 6063 such as 6463 have been developed for better finishing characteristics. In alloy 6463, the iron level is kept low so that the brightness of the aluminum will be improved after anodizing.

Structure

Precipitation hardening in the Al-Mg-Si system is made possible by the decrease in solid solubility of the intermetallic compound Mg_2Si as the temperature is decreased. Figure 5-19 shows a vertical binary section of the Al-Mg-Si ternary system at the Mg_2Si composition. As seen in Fig. 5-17, a quasibinary eutectic is formed between the aluminum solid solution and Mg_2Si. The solubility of Mg_2Si in aluminum decreases from 1.85 percent at the eutectic temperature to about 0.1 percent at room temperature. Alloys which contain about 0.6 percent or more Mg_2Si show marked precipitation hardening.

If an Al-Mg-Si alloy containing 1.3 wt% Mg_2Si is solution-heat-treated at 565°C, water-quenched, and aged at 160°C, GP zones form which are believed to have a needlelike shape and which are oriented in the $\langle 001 \rangle$ directions of the matrix. When the maximum strength is reached during aging at 160°C for 24 h, a high density of β' precipitate is formed, with some short needles being observed (Fig. 5-20). Reheating the fully hardened Al-Mg-Si alloy 15 min at 275°C causes a coarsening of the β' needles, as is observed in Fig. 5-21.

The general sequence of precipitation in the Al-Mg-Si system is thus represented by

Supersaturated solid solution $\rightarrow$ GP zones (needles?) $\rightarrow$ $\beta'(Mg_2Si) \rightarrow \beta(Mg_2Si)$

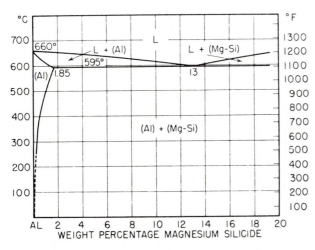

Figure 5-19 Binary section, aluminum–magnesium silicide phase diagram. *(After "Physical Metallurgy of Aluminum Alloys," American Society for Metals, 1949, p. 78.)*

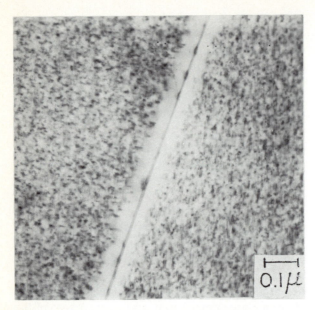

Figure 5-20 Al–1.3% Mg$_2$Si alloy solution heat-treated at 565°C, quenched, and aged 24 h at 160°C to produce the fully precipitation strengthened condition; structure consists of GP zones and β' precipitates. [*After W. F. Smith, Met. Trans. 4(1973):2435.*]

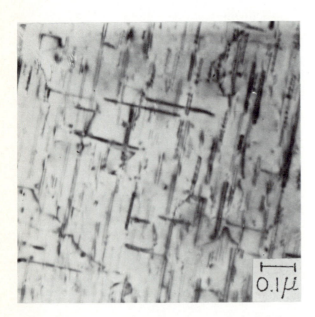

Figure 5-21 Al–1.3% Mg$_2$Si alloy precipitation strengthened by aging 24 h at 160°C. Reheated 15 min at 275°C; structure shows coarse needles of β' precipitates. [*After W. F. Smith, Met. Trans. 4(1973):2435.*]

Since coherency strains are not observed in the GP zones or β' transition stages of precipitation, it has been suggested that the increase in strength of the Al-Mg$_2$Si alloy is due to the increased energy required for the dislocations to break the magnesium-silicon bonds as they pass through the precipitates.

Mechanical Properties

The mechanical properties of selected wrought heat-treatable Al-Mg-Si alloys are listed in Table 5-15. The wrought Al-Mg-Si alloys are only of intermediate strength (45 to 57 ksi in the T6 temper) since only relatively small amounts of Mg$_2$Si (1 to 2 wt%) can be alloyed for precipitation hardening in these alloys. The highest-strength alloys of this class are 6066 and 6070, which have an excess of silicon above that necessary to provide for about 1 to 2 wt% Mg$_2$Si.

Table 5-15 Typical mechanical properties of wrought heat-treatable aluminum-magnesium-silicon alloys†

Alloy	Temper	Tensile strength, psi	Tensile yield strength,‡ psi	Elon-gation, % in 2 in	Hard-ness, Bhn§	Shear strength, psi	Fatigue limit psi ¶
6053	O	16,000	8,000	35	26	11,000	8,000
	T6	37,000	32,000	13	80	23,000	13,000
6061	O	18,000	8,000	25	30	12,000	9,000
	T4, T451	35,000	21,000	22	65	24,000	13,000
	T6, T651	45,000	40,000	12	95	30,000	14,000
	T81	55,000	52,000	15	..	32,000	
	T91	59,000	57,000	12	..	33,000	14,000
	T913	67,000	66,000	10	..	35,000	
6066	O	22,000	12,000	18	43	14,000	
	T4, T451	52,000	30,000	18	90	29,000	
	T6, T651	57,000	52,000	12	120	34,000	16,000
6070	O	21,000	10,000	20	35	14,000	9,000
	T6	57,000	52,000	12	120	34,000	14,000
6101	T6	32,000	28,000	15	71	20,000	· · ·
6151	T6	48,000	43,000	17	100	32,000	12,000
6201	T81	48,000	· · ·	6	..	· · ·	15,000
6262	T9	58,000	55,000	10	120	35,000	13,000
6351	T4, T451	42,000	27,000	20	60	22,000	13,000
	T6, T651	49,000	43,000	13	95	29,000	13,000
6951	O	16,000	6,000	30	28	11,000	
	T6	39,000	33,000	13	82	26,000	

† After Ref. 1.
‡ Yield strength, 0.2 percent offset.
§ 500-kg load, 100-mm ball.
¶ Based on 500 million cycles using an R. R. Moore type of rotating-beam machine.

Alloy 6061 has a tensile strength of 45 ksi in the T6 temper and contains 1.6 wt% Mg_2Si. By reducing the amount of Mg_2Si to 1.1 wt%, the strength of alloy 6063 is reduced to 35 ksi in the T6 temper. The lower strength of alloy 6063 is necessary for ease of extrudability.

Al-Mg-Si alloys are usually solution-heat-treated at about 520°C. Since this temperature is well below the eutectic melting temperature for these alloys, there is little chance of melting by slightly overheating. Alloy 6061 can be solution-heat-treated at higher temperatures than 520°C with some increase in strength being obtained since not all the Mg_2Si present is soluble at this temperature. As in the case of the Al-Cu-Mg alloys, rapid quenching is required to obtain maximum strength.

The artificial aging characteristics of 6061 alloy are shown in Fig. 5-22. It should be noted that the highest strengths are obtained at the lower temperatures for long times (135°C for 500 h). Industrially, for economical purposes this alloy is aged 16 to 20 h at 160°C.

The highest strengths in Al-Mg-Si alloys are obtained when artificial aging is started immediately after quenching. Losses of 3 to 4 ksi in strength occur if these alloys are room-temperature-aged for 1 to 7 days. Although there is some recovery of strength with a month or more of room temperature aging, the

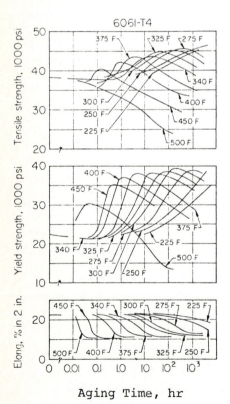

Aging Time, hr

Figure 5-22 Aging characteristics of 6061 aluminum sheet alloy. [*After H. Y. Hunsicker in K. R. Van Horn (ed.), "Aluminum," vol. 1, American Society for Metals, 1967, p. 147.*]

maximum strength never reaches that obtained by aging immediately after quenching.

Corrosion Resistance

The Al-Mg-Si alloys have excellent corrosion resistance in all natural atmospheric environments and in many artificial ones. The corrosion resistance of these alloys is best in materials which are quenched rapidly and aged artificially to the desired temper.

5-9 ALUMINUM-ZINC-MAGNESIUM AND ALUMINUM-ZINC-MAGNESIUM-COPPER ALLOYS

Chemical Compositions and Typical Applications

Combinations of 4 to 8 wt% Zn and 1 to 3 wt% Mg in aluminum are used to produce the 7xxx series of wrought heat-treatable aluminum alloys. Some of these alloys develop the highest-strength properties of any commercial aluminum-base alloys. Zinc and magnesium both have high solid solubilities in aluminum and develop unusually high-precipitation-hardening characteristics. Copper additions of 1 to 2 wt% increase the strength properties of Al-Zn-Mg alloys to make the high-strength aircraft aluminum alloys.

After extensive research, alloy 7075 was introduced in 1943. The successful development of this outstanding member of the 7xxx series was made possible by the beneficial effect of chromium, which greatly improved the stress-corrosion cracking resistance of sheet made from this alloy. Alloy 7075 contains 5.6% Zn, 2.5% Mg, 1.6% Cu, and 0.30% Cr. A higher-strength modification of 7075, alloy 7178, was developed in 1951 and contains higher levels of Zn, Mg, and Cu. The highest-strength alloy in commercial production, 7001, was introduced in 1960 and contains 7.4% Zn, 3.0% Mg, and 2.1% Cu.

Aluminum-zinc-magnesium alloys without copper (less than 0.1 percent) have been developed which have intermediate strength and are weldable. Alloys such as 7004 and 7005 are used for truck bodies, trailer parts, portable bridges, and railroad cars. Table 5-16 lists the chemical compositions and typical applications for Al-Zn-Mg and Al-Zn-Mg-Cu alloys.

Structure

Al-Zn-Mg alloys Aluminum-zinc-magnesium wrought alloys are strengthened by precipitation reactions during aging after solution heat treatment and quenching. The precipitation sequence upon aging the supersaturated solid solution is generally recognized to be

Supersaturated solid solution $\rightarrow$ GP zones $\rightarrow \eta'(MgZn_2) \rightarrow \eta(MgZn_2)$

The GP zones are coherent with the matrix and have a spherical shape. The

Table 5-16 Chemical compositions and applications of aluminum-zinc-magnesium and aluminum-zinc-magnesium-copper alloys†

					Aluminum-zinc-magnesium alloys	
Alloy	% Zn	% Mg	% Cr	% Mn	% Zr	Applications
7004	4.2	1.5		0.45	0.15	Truck bodies and trailer parts; portable bridges; railroad cars; extruded products
7005	4.5	1.4	0.13	0.40	0.14	

				Aluminum-zinc-magnesium-copper alloys	
	% Zn	% Mg	% Cu	% Cr	Applications
7001	7.4	3.0	2.1	0.30	Missile structurals
7049	7.7	2.5	1.6	0.15	Aircraft and other structures; hydraulic fittings
7075	5.6	2.5	1.6	0.30	Aircraft and other structures; hydraulic fittings
7475	Lower impurity limits than 7075				Aircraft and other structures (good fracture toughness)
7178	6.8	2.7	2.0	0.30	Aircraft and other structures

† After "ASM Databook," 1978, *Metal Progress*, vol. 114, no. 1, mid-June 1978.

interfacial energy for GP zones in the Al-Zn-Mg system is so low that a high density of very-small-size zones (~ 30 Å) can be produced at low temperatures (e.g., 20 to 120°C). The semicoherent intermediate metastable phase η' has been described as having a monoclinic unit cell, while the incoherent equilibrium phase, $MgZn_2$, η, is hexagonal.

The highest strength obtained for an Al–5% Zn–2% Mg alloy is found to be associated with a high density of small GP zones, which is produced by duplex aging first for 5 days at 20°C and then for 48 h at the higher temperature of 120°C. The matrix structure formed by this treatment consists of a high density of small GP zones and shows no evidence of semicoherent intermediate-phase precipitates (Fig. 5-23a). The first stage of the duplex aging creates a high density of small stable GP zones with a narrow size distribution. Aging at the higher temperature of the second stage dissolves some of the small zones, but many others grow larger at the expense of the smaller ones (Ostwald ripening). In this way, a high density of relatively small GP zones is formed at the higher temperature.

By duplex aging the Al–5% Zn–2% Mg alloy at higher temperatures (16 h at 80°C plus 24 h at 150°C), a slightly coarser precipitate structure is produced, as can be seen by the size of the grain boundary precipitates in Fig. 5-23b. Single-stage aging this alloy for 24 h at 150°C produces a fine dispersion of intermediate η' precipitates with wide precipitate-free zones (Fig. 5-23c). The alloy in this condition has a lower strength of 40 ksi as compared to 51 ksi for

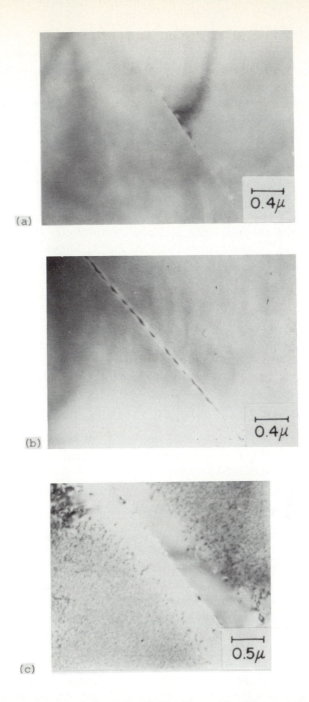

Figure 5-23 Microstructures of an Al–5% Zn–2% Mg alloy aged by different treatments to produce different precipitate structures. (*a*) Alloy was aged 5 days at 20°C ksi plus 48 h at 120°C (UTS = 51 ksi). Structure consists of GP zones only. (*b*) Alloy was aged 16 h at 80°C plus 24 h at 150°C (UTS = 49 ksi). Structure consists of GP zones and possibly some η'. (*c*) Alloy was aged 24 h at 150°C. (UTS = 40 ksi). Structure consists of η'. (Electron transmission micrographs.) [*After W. F. Smith and N. J. Grant, Met. Trans. 1(1970):979.*]

the 20°C plus 120°C duplex-aged material. The increased strength of the alloy with the high density of GP zones is attributed to the increased resistance to dislocation movement arising from the strong atomic bonds existing in the zones. Dislocation movement is easier in the coarser further-spaced-apart semi-coherent intermediate η' precipitates.

Al-Zn-Mg-Cu alloys The addition of up to about 2% Cu to Al-Zn-Mg alloys does not appear to change their precipitation mechanisms. During zone formation, the copper in Al-Zn-Mg-Cu alloys appears to be uniformly distributed (Ref. 1, p. 125). Copper in the GP zones does, however, increase their stability, as it enables the zones to exist at higher temperatures than in comparable Al-Zn-Mg alloys.[1] Copper strengthens Al-Zn-Mg alloys primarily by solid-solution strengthening, but also makes some contribution to precipitation strengthening.

Microstructures of 7075 alloy (one of the most important of the 7xxx series) in the fully hardened and overaged conditions are shown in Fig. 5-24. In the fully age-hardened T651 condition, the GP zones are ≤ 75 Å with some η' (~ 150 Å) also present (Fig. 5-24a). The larger, darker particles are chromium-rich precipitates which are found in many Al-Zn-Mg-Cu alloys. After overaging the T651 material at 170°C for 9 h to produce the T7351 temper, the microstructure consists of η' (100 to 300 Å) and η (400 to 800 Å) (Fig. 5-24b).

As in the case of Al-Zn-Mg alloys, overaging and coarsening of the precipitates results in lower strengths. For example, the 7075-T651 material has an ultimate tensile strength of 76.7 ksi and a yield (0.2 percent) of 66.4 ksi, while the 7075-T7351 with $\eta + \eta'$ precipitates has an ultimate tensile strength of 63.7 ksi and a yield strength of 54.3 ksi.

Mechanical Properties

The mechanical properties of selected wrought heat-treatable Al-Zn-Mg and Al-Zn-Mg-Cu alloys are listed in Table 5-17. The highest room-temperature strengths of all aluminum alloys are developed in Al-Zn-Cu-Mg alloys.

Alloy 7001, with 7.4% Zn, 3.0% Mg, and 2.1% Cu, has an ultimate tensile strength of 98 ksi with an elongation of 9 percent when it is heat-treated to the T651 temper. This is one of the highest-strength 7xxx series alloys. Alloy 7075, which is one of the most commonly used in the 7xxx series, has lower zinc, magnesium, and copper levels (5.6% Zn, 2.5% Mg, and 1.6% Cu) and has an ultimate tensile strength of 83 ksi with 11 percent elongation when heat-treated to the T651 temper. These high strengths are attributed to the high density of GP zones and η' precipitates that can be developed in these alloys by duplex-aging treatments.

---

[1] I. J. Polmear, *J. Inst. Metals* 87(1958–59):65.

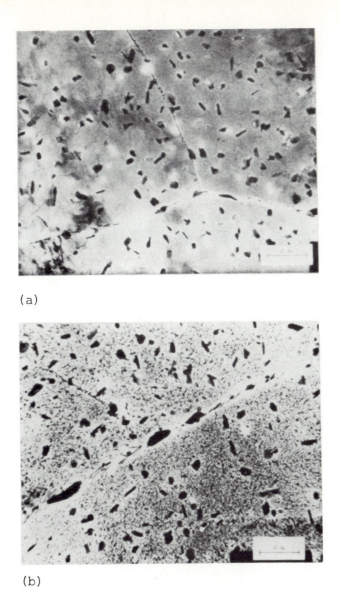

(a)

(b)

Figure 5-24 Microstructures of 7075 alloy in (*a*) the T651 fully hardened condition and (*b*) the T7351 overaged condition. (*a*) 7075-T651 in fully hardened condition structure shows GP zones (< 75 Å) and η' ($\approx$ 150 Å) in matrix and 700 Å precipitate-free zone at grain boundary; very large particles are chromium-rich precipitates. (*b*) 7075-T651 aged at 175°C for 9 h to overaged T7351 temper, with η' (100 to 300 Å) and η (400 to 800 Å) in matrix and 900 Å precipitate-free zone at grain boundary. (Electron transmission micrographs.) [*After P. N. Adler et al., Met. Trans. 3(1972):319.*]

Table 5-17 Typical mechanical properties of wrought heat-treatable aluminum-zinc-magnesium and aluminum-zinc-magnesium-copper alloys†

Alloy	Temper	Tensile strength, psi	Tensile yield strength,‡ psi	Elongation, % in 2 in	Hardness,§ Bhn	Shear strength, psi	Fatigue limit,¶ psi
7001	O	37,000	22,000	14	60		
	T6	98,000	91,000	9	160	· ·	22,000
	T651	98,000	91,000	9	160	· · ·	22,000
	T75	84,000	72,000	12			
7005	O	28,000	12,000	20			
	W	50,000	30,000	20			
	T6	51,000	42,000	13	· · ·	31,000	22,000
7075	O	33,000	15,000	17	60	22,000	17,000
	T6	83,000	73,000	11	150	48,000	23,000
	T651	83,000	73,000	11	150	48,000	23,000
	T73	73,000	63,000	13			
7178	O	33,000	15,000	15	60	22,000	
	T6	88,000	78,000	10	160	52,000	22,000
	T651	88,000	78,000	10	160	52,000	22,000

† After Ref. 1
‡ Yield strength, 0.2 percent offset.
§ 500-kg load, 10-mm ball.
¶ Based on 500 million cycles using an R. R. Moore type of rotating-beam machine.

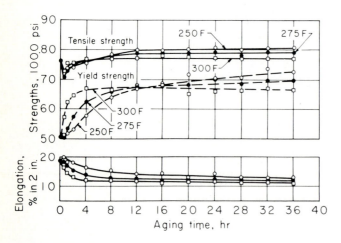

Figure 5-25 Aging of 7075 aluminum alloy sheet at 120 to 150°C. [*After J. A. Nock, Jr. in K. R. Van Horn (ed.), "Aluminum," vol. 1, American Society for Metals, 1967, p. 153.*]

The artificial aging characteristics of 7075 sheet are shown in Fig. 5-25. In order to reduce furnace time for artificial aging, short-time duplex-aging treatments have been developed. In one aging practice, 7075 sheet is aged 4 h at 100°C plus 8 h at 157°C, while in another 3 h at 120°C plus 3 h at 175°C is used. In these aging treatments, a high density of small GP zones are nucleated and grown so that at the higher aging temperature a high density of slightly larger zones will be retained.

In contrast to the Al-Cu-Mg alloys, cold working the Al-Zn-Mg and Al-Zn-Mg-Cu alloys between quenching and aging does not strengthen them significantly. The 7xxx series alloys do not respond favorably to cold-work treatments between quenching and aging since they are strengthened almost exclusively by zone formation and precipitates which nucleate from the zones. Thus, introducing many new dislocations by cold work after solution heat treatment and quenching does not greatly accelerate the precipitation of an intermediate metastable phase as is the case in the Al-Cu-Mg alloys.

5-10 ALUMINUM CASTING ALLOYS

Chemical Compositions and Typical Applications

Aluminum casting alloys have been developed for casting qualities such as fluidity[1] and feeding ability,[2] as well as for properties such as strength, ductility, and corrosion resistance. Thus their chemical compositions differ widely from those of the wrought aluminum alloys. Table 5-18 lists the chemical compositions and typical applications of sand-, permanent-mold-, and die-cast aluminum alloys. They are classified according to the Aluminum Association numbering system. As listed in Table 5-2, the major alloying elements for aluminum casting alloys are

2xx	Copper
3xx	Silicon with copper and/or magnesium
4xx	Silicon
5xx	Magnesium
7xx	Zinc
8xx	Tin

Aluminum-Silicon Casting Alloys

Aluminum casting alloys with silicon as the major alloying element are the most important commercial casting alloys because of their superior casting characteristics. Aluminum-silicon alloys have comparatively high fluidity in the molten state, excellent feeding during solidification, and comparative freedom from hot shortness. Silicon does not reduce the good corrosion resistance of pure aluminum, and in some cases increases its corrosion resistance in mildly acidic environments.

Binary Al-Si alloys are not considered heat-treatable since only a small amount of silicon is soluble in aluminum (1.65 percent maximum) and since the silicon that does reprecipitate from solid solution causes very little hardening. The Al-Si system is a simple eutectic type with the eutectic composition at 12.6% Si (Fig. 5-26).

The most important commercial binary aluminum-silicon alloys are 443, which contains 5.3% Si (nominal), and 413, which contains 12% Si (nominal). Alloy 443 is used mainly for sand and permanent-mold casting while alloy 413 is

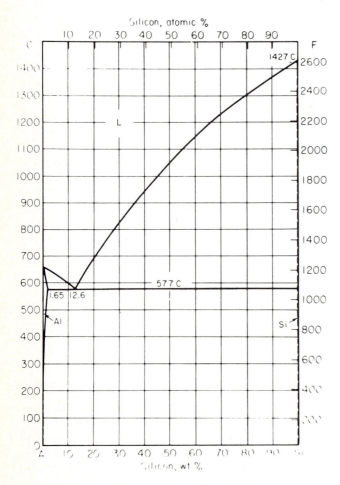

Figure 5-26 The aluminum–silicon phase diagram [*After K. R. Van Horn (ed.), "Aluminum," vol. 1, American Society for Metals, 1967, p. 378.*]

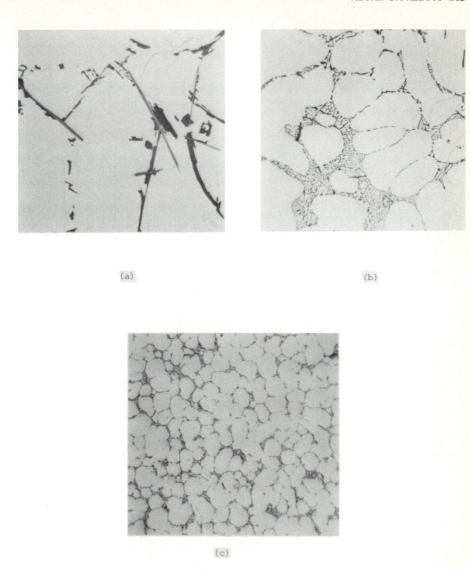

(a)

(b)

(c)

Figure 5-27 Aluminum casting alloy 443 (Al–5% Si) cast at different rates. Note the decreased dendrite cell size as the solidification rate is increased. (*a*) Alloy 443-F, as sand cast. Large dendrite cells resulted from slow cooling in sand mold. Interdendritic structure: silicon (dark gray), Fe_3SiAl_{12} (medium-gray script) and $Fe_2Si_2Al_9$ (light-gray needles) (0.5% hydrofluoric acid; 500 × .) (*b*) Alloy B443-F, as permanent mold cast. Constituents are same as in Fig. 5-25*a* but dendrite cells are smaller, because of faster cooling in the metal permanent mold. (0.5% hydrofluoric acid; 500 × .) (*c*) Alloy C443-F, as die cast. Same constituents as in Fig. 5-25*a*, but dendrite cells are much smaller because of the very rapid cooling of the water-cooled die-casting die. (0.5% hydrofluoric acid; 500 × .) *(Courtesy of F. Krill, Kaiser Aluminum Co., and as in the Metals Handbook, 8th ed., vol. 7, American Society for Metals, 1972, p. 259.)*

Table 5-18 Chemical compositions and typical applications for aluminum casting alloys†

				Sand- and permanent-mold-casting alloys	

Aluminum-copper casting alloys

Alloy designation‡	% Cu	% Si	% Mg		Typical applications
208	4.0	3.0			General-purpose sand castings; manifold and valve bodies
213	7.0	2.0			Washing machine agitators; automotive cylinder heads and timing gears
222	10.0	—	0.25		Primarily a piston alloy; also used for air-cooled cylinder heads
242	4.0	—		2% Ni	Air-cooled cylinder heads; pistons in high-performance gasoline engines
295	4.5	1.1			General structural castings requiring high strength and shock resistance
B295	4.5	2.5			Permanent-mold version of 295; aircraft fittings; fuel-pump bodies

Aluminum-silicon-copper alloys

	% Si	% Cu	% Mg	Typical applications
308	5.5	4.5		General-purpose permanent mold castings and ornamental grilles
319	6.3	3.5		General-purpose alloy; engine parts; automotive cylinder heads
333	9.0	3.5	0.25	General-purpose alloy used for engine parts, meter housings,
354	9.0	1.8	0.5	aircraft, missile, and other applications requiring high-strength castings

Aluminum-silicon-magnesium alloys

	% Si	% Cu	% Mg	% Other	Typical applications
F332	9.5	3.0	1.0		Automotive pistons; parts requiring elevated-temperature strength
355	5.0	1.2	0.5		General use where high strength and pressure tightness are required such as pump bodies and liquid-cooled cylinder heads; crankcases, accessory housings, and aircraft fixtures; stressed castings such as blower housings, snow removal equipment, and scaffold pedestals
C355	5.0	1.2	0.5	0.20% max Fe	Similar to 355, but stronger and more ductile; aircraft, missile, and other structural uses requiring high strength; parts requiring high strength-to-weight ratios, such as crankcases and wheels in aerospace applications

Alloy	%Si	%Mg	Typical applications
356	7.0	0.3	Intricate castings requiring good strength and ductility; transmission cases, truck axle housings, truck wheels, cylinder blocks, railway tank car fittings, marine hardware, valve bodies, and bridge railing parts; outboard motor parts, cylinder heads, fan blades, pneumatic tools, storage tank fittings, gray anodized architectural components
A356	7.0	0.3 (0.20% max Fe)	Similar to 356, but stronger and more ductile; aircraft and missile components requiring strength, ductility, and corrosion resistance
357	7.0	0.5	Highly stressed castings requiring a high strength-to-weight ratio and excellent corrosion resistance; aircraft and missile components, machine parts, high-velocity fan blades
A357	7.0	0.5 (0.05 Be)	Aircraft and missile parts
359	9.0	0.6	High-strength aircraft, missile, and other structural applications

Aluminum-copper-magnesium-nickel alloys

Alloy	%Si	%Cu	%Mg	%Ni	Typical applications
A332	12	1.0	1.0	2.5	Automotive pistons; diesel engine pistons; pulley sheaves and engine parts operating at elevated temperatures

Die-casting alloys§

Alloy	%Si	%Fe*	%Mg	%Cu	Typical applications
413	12.0	2.0			Large, intricate castings with thin sections—instrument cases, typewriter frames
A413	12.0	1.3			
C443	5.3	2.0			Castings requiring high resistance to corrosion and shock
360	9.5	2.0	0.5		General-purpose castings—instrument cases and cover plates
A360	9.5	1.3	0.5		
380	8.5	2.0		3.5	General-purpose castings
A380	8.5	1.3		3.5	
383	10.5	1.3		2.5	
384	11.3	1.3		3.8	

† After "ASM Databook," 1979, *Metal Progress*, vol. 116, no. 1, mid-June 1979.
‡ Aluminum Association numbering system.
§ After "ASM Databook," 1978, *Metal Progress*, vol. 114, no. 1, mid-June 1978.
* Maximum values.

used mainly for die casting. During solidification of 443 (Al–5% Si), dendrites of almost pure aluminum solidify first. The spaces between these dendrites are then filled with aluminum-silicon eutectic. When the eutectic freezes, it decomposes into almost pure aluminum and silicon. As the solidification rate is increased, the dendrite cells become smaller. This relationship is shown in Fig. 5-27 for as-cast 443 alloy solidified at increasingly faster rates. In Fig. 5-27a, the 443-F alloy when sand-cast produces large dendrite cells resulting from slow cooling during solidification. Faster-cooling alloy B443-F in a permanent mold produces a smaller dendrite cell size (Fig. 5-27b). Die casting the alloy, which produces an even faster rate of cooling, produces an even smaller dendrite cell size (Fig. 5-27c).

The eutectic structure of sand-cast Al-Si alloys can be greatly refined by the addition of small amounts of sodium (0.025 percent), either as metallic sodium or as sodium salts just before casting. Figure 5-28 shows the effect of sodium modification on the eutectic structure in an Al–7% Si alloy. Sodium modification of aluminum sand castings leads to higher tensile strengths, as indicated in Fig. 5-29a. Faster solidification rates, which produce crystallization with high undercooling, can also lead to similar refinement of the eutectic structure in Al-Si alloys and also to higher strengths, as shown in Fig. 5-29b. The structure of the modified Al-Si eutectic has been found to be of an irregular fibrous form, as shown by the scanning electron micrographs in Fig. 5-30a and b.

Aluminum-Silicon-Magnesium Casting Alloys

The strength properties of cast aluminum-silicon binary alloys can be improved by the addition of small amounts of magnesium (about 0.35 percent). The most

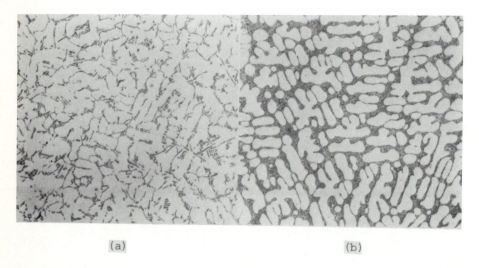

(a)   (b)

Figure 5-28 Microstructure of sand cast Al–7% Si alloy (a) without sodium modification and (b) with sodium modification. Note the refinement of the eutectic in the sodium modified alloy. [*After B. Chamberlain and V. J. Zabek, AFS Trans. 81(1973):322.*]

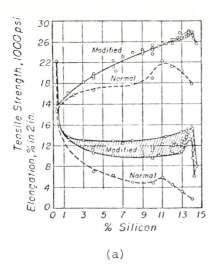
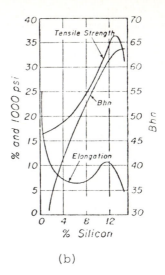

(a)   (b)

Figure 5-29 Tensile properties of Al–Si alloys. (*a*) Sand-cast 0.5-in-diameter test bars in as-cast condition with and without sodium modification. (*b*) Chill-cast test bars in as-cast condition. *(After Metals Handbook, 1948 edition, American Society for Metals, 1948, p. 805.)*

important aluminum casting alloy of this type is 356, which contains 7% Si for castability and 0.35% Mg to make the alloy heat-treatable. The magnesium silicide (Mg_2Si) content of the alloy is in the range 0.5 to 0.6 percent and the precipitation strengthening is attributed to a metastable phase of Mg_2Si (see Structure, in Sec. 5-8).

The microstructure of alloy 356 in several cast and heat-treated conditions is shown in Fig. 5-31. The slow solidification rate of sand casting leads to silicon

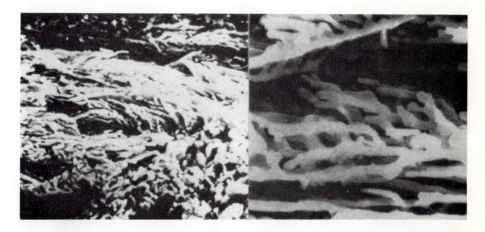

Figure 5-30 Structure of modified aluminum–silicon eutectic alloy. Heavy etching shows undissolved silicon fibers. [Electron scanning micrographs. (*a*) 2500 × ; (*b*) 12,000 × .] *[After M. G. Day and A. Hellawell, J. Inst. Metals 95(1967):377.]*

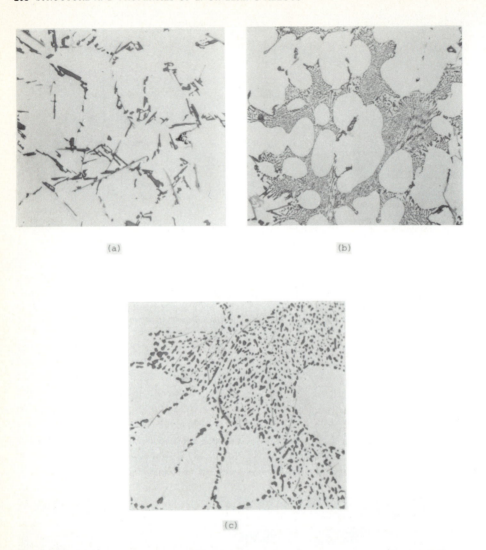

Figure 5-31 Microstructures of alloy 356 (Al–7% Si–0.3% Mg) cast and heat-treated in different conditions. (*a*) Alloy 356-T51, sand-cast, artificially aged. The angular dark gray constituent is silicon; black script is Mg_2Si; blades are $Fe_2Si_2Al_9$; light script is $FeMg_3Si_6Al_8$. (0.5% hydrofluoric acid; 250 × .) (*b*) Alloy 356-F, modified by the addition of 0.025% sodium; as sand-cast. Structure consists of interdendritic network of silicon particles. (0.5% hydrofluoric acid; 250 × .) (*c*) Alloy 356-T7 modified by sodium addition sand-cast, solution heat-treated and stabilized. Structure: rounded particles of silicon and blades of $Fe_2Si_2Al_9$. (0.5% hydrofluoric acid; 250 × .) *(Courtesy of F. Krill, Kaiser Aluminum Co., and as in the Metals Handbook, 8th ed., vol. 7, American Society for Metals, 1972, p. 258.)*

**Table 5-19 Effect of iron impurity on mechanical properties
of chill-cast Al–10% Si alloys†**

% Si	% Fe	Tensile strength, psi	% Elongation in 2 in	Bhn
10.8	0.29	31,100	14.0	62
10.8	0.79	30,900	9.8	65
10.3	0.90	30,000	6.0	65
10.1	1.13	24,500	2.5	66
10.4	1.60	18,000	1.5	68
10.2	2.08	11,200	1.0	70

† After D. Stockdale and I. Wilkinson, *J. Inst. Metals* 36(1926):313.

particles in the interdendritic Al-Si eutectic (Fig. 5-31a). Artificially aging this alloy in the as-cast condition does not change the optical microstructure, but does produce a fine dispersion of metastable precipitates which strengthen the alloy.

If the 356 alloy is modified by the addition of 0.025% Na to the melt, the sand-cast eutectic structure is refined and the particles of silicon in the eutectic are smaller and less angular. This refinement provides some improvement in the mechanical properties of slowly solidified sand castings, but the principal benefit is in the improvement in the feeding characteristics of both sand- and permanent-mold castings. It appears that the smaller silicon particles produce less interference with the flow of liquid metal during solidification. As a result, sodium-modified alloys produce a superior finish and less microshrinkage between the dendrites than unmodified metals.

Figure 5-31b shows the sand-cast alloy 356 structure after modification. When the modified structure is solution-heat-treated, quenched, and overaged to the T7 temper, the silicon particles agglomerate to produce larger rounded particles (Fig. 5-31c).

It has been known since the 1920s that plates and needles of Al-Fe-Si constituent reduce the strength of aluminum-silicon casting alloys (Table 5-19). By reducing the iron level of 356 alloy to about 0.10 percent, considerable improvement in the strength of this alloy can be attained.

Aluminum-Copper Casting Alloys

Aluminum-copper casting alloys have been almost completely replaced by aluminum-silicon-magnesium alloys. The main reasons for the replacement of the aluminum-copper alloys is that they have poorer casting characteristics, are not as corrosion-resistant, and have higher specific gravities than the aluminum-silicon-magnesium alloys.

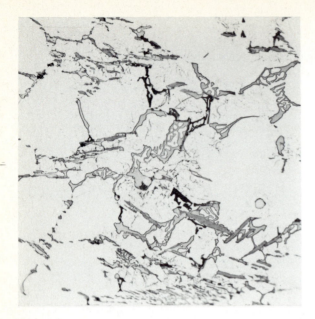

Figure 5-32 Alloy 242-T571, permanent-mold cast and artificially aged. Structure shows blades of NiAl$_3$ (dark gray) in the medium gray Cu$_3$NiAl$_6$ script. CuAl$_2$ particles (light) and Mg$_2$Si (black) are also present. *(Courtesy of F. Krill, Kaiser Aluminum Co., and as in the Metals Handbook, 8th ed., vol. 7, American Society for Metals, 1972, p. 256.)*

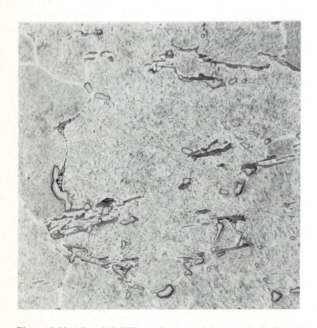

Figure 5-33 Alloy 242-T77, sand-cast and heat-treated. Constituents are same as in Fig. 5-32 but particles of NiAl$_3$ and Cu$_3$NiAl$_5$ have been rounded by solution heat treatment. Precipitation is caused by overaging treatment. *(Courtesy of F. Krill, Kaiser Aluminum Co., and as in the Metals Handbook, 8th ed., vol. 7, American Society for Metals, 1972, p. 256.)*

Alloy 222, which contains 10% Cu and 0.2% Mg, was originally developed for internal combustion engine pistons, but has been replaced for this application by alloys 242 and F332. Alloy 242 contains 4% Cu, 2% Ni, and 1.5% Mg, and has higher strength at elevated temperatures than the older alloys. The nickel content of the alloy is largely responsible for the higher temperature strengths. The microstructure of alloy 242 is shown in the permanent-mold cast and artificially aged T571 temper in Fig. 5-32, and in the as-cast, solution-heat-treated, quenched, and overaged condition in Fig. 5-33.

Mechanical Properties

The mechanical properties of selected aluminum casting alloys are listed in Table 5-20. The tensile strengths of aluminum casting alloys usually range from

Table 5-20 Typical mechanical properties of sand-, permanent-mold-, and die-cast aluminum alloys†

Alloy and temper	Type of casting‡	Tensile strength, ksi	Yield strength, ksi	Elongation, %
208	SC	21	14	2.5
213	SC	24	15	1.5
213	PM	28	19	2.0
222-T551	PM	37	35	<0.5
242-T571	PM	40	34	1.0
295-T6	SC	36	24	5.0
B295-T6	PM	40	26	5.0
308	PM	28	16	2
319-F	PM	34	19	2.5
319-T6	PM	40	27	3.0
F332-T5	PM	36	28	1.0
355-T6	PM	42	27	4.0
C355-T61	PM	44	34	3.0
356-T51	SC	25	20	2.0
356-T6	SC	33	24	3.5
356-T6	PM	37	27	5.0
357-T6	PM	52	43	5.0
359-T61	PM	47	37	7.0
A332-T551	PM	36	28	0.5
413	DC	43	21	2.5
443	DC	33	16	9.0
360	DC	47	25	3.0
A360	DC	46	24	5.0
380	DC	48	24	3.0
A380	DC	47	23	4.0

† After Ref. 1; "ASM Databook," 1977, *Metal Progress*, vol. 112, no. 1, mid-June 1977; and "ASM Databook," 1978, *Metal Progress*, vol. 114, no. 1, mid-June 1978.

‡ SC: sand-cast; PM: permanent-mold-cast; DC: die-cast.

about 18 to 48 ksi. Sand-cast aluminum alloys, because of their relatively large dendritic cell size due to slow solidification rates, have lower tensile strengths than permanent-mold- or die-cast aluminum alloys. Higher strengths are obtained in the two latter methods by higher solidification rates and lesser gas porosity due to the use of metal molds. In die casting, pressure feeding reduces gas porosity also. As an example, alloy 356 has a minimum tensile strength of 30 ksi when sand-cast and aged to peak strength (T6 temper), but when it is permanent-mold-cast it has a peak strength of 33 ksi.

PROBLEMS

1. What is the chief mineral from which aluminum is extracted commercially? How is pure aluminum oxide extracted chemically from this mineral?

2 Describe the Hall process for electrolytically producing aluminum metal from aluminum oxide. What are the main impurities contained in aluminum produced by this method?

3 How are wrought aluminum alloys designated? How are aluminum casting alloys designated?

4 What are the basic temper designations for aluminum alloys? The strain-hardened subdivisions? The heat-treated subdivisions?

5 What constituents are observed in the microstructure of annealed 1100 alloy at 500× ?

6 What constituents are observed in the microstructure of alloy 3003 at 500× ?

7 How does manganese strengthen alloy 3003?

8 By what mechanism is manganese believed to raise the recrystallization temperature of 3003 alloy?

9 How does magnesium strengthen aluminum?

10 Why is a continuous network of Mg_2Al_3 in the grain boundaries of Al-Mg alloys undesirable? How can such a network be avoided in Al-Mg alloys with 3.5 to 5% Mg?

11 By what two methods can copper strengthen aluminum?

12 Outline the three principal steps necessary for precipitation strengthening an Al–4% Cu alloy.

13 What is the difference between natural and artificial aging heat treatments?

14 What are the five sequential structures that can be identified during the precipitation hardening of an Al–4% Cu alloy?

15 At lower temperatures why do metastable phases form instead of the equilibrium $CuAl_2$ phase during the aging of an Al–4% Cu alloy?

16 What is believed to be the structure of GP1 zones, GP2 zones, and θ' phase?

17 How do the hardness data of an Al–4% Cu alloy correlate with aging time at 130 and 190°C for an Al–4% Cu alloy?

18 Why are the elements Mn, Ti, V, and Zr added to Al–6% Cu to make 2219 alloy? What special properties does alloy 2219 have?

19 What is the general sequence of precipitation in Al-Cu-Mg alloys?

20 Describe the GP zones and S′ and S phases found in aged Al-Cu-Mg alloys.

21 Why does cold working alloy 2024 after solution heat treatment and quenching but before aging increase the strength of this alloy after aging?

22 What is the general sequence of precipitation in Al-Mg-Si alloys? What is the effective hardening compound in these alloys?

23 What advantages do the 6xxx alloys offer for engineering applications? How are these applications related to their microstructures?

24 What is the general sequence of aging in the Al-Zn-Mg alloys?

25 Explain the effects of duplex aging on the GP zones and intermediate precipitates formed in solution-heat-treated and quenched Al-Zn-Mg alloys?

26 What engineering advantages and disadvantages do the Al-Zn-Mg and Al-Zn-Mg-Cu alloys have?

27 How are the highest strengths in Al-Zn-Mg-Cu alloys developed by (*a*) composition and (*b*) heat treatment?

28 Why does not cold working after solution heat treatment and quenching and before aging lead to increased strengths in Al-Zn-Mg-Cu alloys?

29 Define fluidity and feeding ability as pertains to casting metals.

30 By what system are aluminum casting alloys designated by the Aluminum Association?

31 How do the chemical compositions of aluminum casting alloys differ from wrought alloys?

32 Why is silicon so important an alloying element for aluminum casting alloys?

33 How does the solidification rate affect the dendritic cell size of aluminum casting alloys?

34 What is sodium modification of aluminum-silicon alloys? How is the eutectic structure of sand-cast Al-Si alloys affected by this process?

35 Why are small additions of magnesium added to Al-Si alloys?

36 Why does an increase in iron content of Al-Si alloys decrease their ductility and in some cases their tensile strength?

37 Why have aluminum-silicon-magnesium alloys almost completely replaced aluminum-copper-type casting alloys?

38 Why is nickel added to the aluminum casting alloy 242?

39 The kinetics of recrystallization in alloy 3003 (Al–1.2% Mn) are greatly affected by the distribution of the manganese in the alloy. When this alloy is preheated at 510°C for 4 h, cold-rolled 80 percent, and annealed at 343°C for 64 s, a polygonized structure develops. However, when this alloy is preheated at 600°C, cold-rolled 80 percent, and annealed at 343°C for 64 s, no polygonization takes place. What has happened to prevent the development of the polygonized structure in the high-temperature (600°C) preheated 3003 alloy? [See J. G. Morris, *Trans. ASM* 59(1966):1007.]

REFERENCES

1. K. R. Van Horn: "Aluminum," vol. 1, American Society for Metals, Metals Park, OH, 1967.
2. L. F. Mondolfo: "Aluminum Alloys: Structure and Properties," Butterworths, London, 1976.
3. M. Van Lancker: "Metallurgy of Aluminum Alloys," John Wiley, New York, 1967.
4. "Precipitation From Solid Solution," American Society for Metals, Metals Park, OH, 1959.
5. "Physical Metallurgy of Aluminum Alloys," American Society for Metals, Metals Park, OH, 1949.
6. "Aluminum Standards and Data," Aluminum Association, Inc., Washington, D. C., 1976.
7. *Metals Handbook*, vol. 7: "Atlas of Microstructures of Industrial Alloys," American Society for Metals, Metals Park, OH, 1972.
8. "Symposium on the Advances in the Physical Metallurgy of Aluminum Alloys," *Met. Trans.* 6A(1975):625.
9. M. V. Hyatt et al.: "Improved Aluminum Alloys for Aircraft Applications," *Metal Progress*, March 1977, p. 56.
10. J. M. Van Orden and D. E. Petit: "A Close Look at 7475 and 2024 Aluminum for Aircraft Structures," *Metal Progress*, Dec. 1977, p. 28.
11. S. D. Dahlgren: "Coherency Stresses, Composition and Dislocation Interactions for θ'' Precipitates in Age-Hardened Aluminum-Copper," *Met. Trans.* 7A(1976):1401.
12. R. N. Wilson and P. G. Partridge: "The Nucleation and Growth of S′ Precipitates in an Aluminum-2.5% Copper-1.2% Magnesium Alloy," *Acta Met.* 13(1965):1321.

13. J. M. Silcock, T. J. Heal, and H. K. Hardy: "Structural Aging Characteristics of Binary Aluminum-Copper Alloys," *J. Inst. Metals* 82(1953–54):239.

14. W. F. Smith: "Effect of Reversion Treatments on Precipitation Mechanisms in an Al-1.35 at.% Mg₂Si Alloy," *Met. Trans.* 4(1973):2435.

15. J. T. Staley: "Aging Kinetics of Aluminum Alloy 7050," *Met. Trans.* 5(1974):929.

16. W. F. Smith and N. J. Grant: "Mechanism of Formation of Precipitate-Free Zones in an Al-4.7% Zn-3.9% Mg Alloy," *Trans. ASM* 62(1969):724.

17. W. F. Smith and N. J. Grant: "The Effect of Multiple-Step Aging on the Strength Properties and Precipitate-Free Zone Widths in Al-Zn-Mg Alloys," *Met. Trans.* 1(1970):979.

18. A. J. De Ardo and C. J. Simensen: "A Structural Investigation of Multiple Aging of Al-7% Zn-2.3% Mg Alloy," *Met. Trans.* 4(1973):2413.

19. W. F. Smith and N. J. Grant: "The Effect of Two-Step Aging on the Quench Sensitivity of an Al-5% Zn-2% Mg Alloy with and without 0.1% Cr," *Met. Trans.* 1(1970):1735.

20. P. N. Adler et al.: "Influence of Microstructure on the Mechanical Properties and Stress Corrosion Susceptibility of 7075 Alloy," *Met. Trans.* 3(1972):3191.

21. W. F. Smith and N. J. Grant: "Effects of Chromium and Copper Additions on Precipitation in Al-Zn-Mg Alloys," *Met. Trans.* 2(1971):1333.

22. P. E. Marth, H. I. Aaronson, G. W. Lorimer, T. L. Bartel, and K. C. Russell: "Application of Heterogeneous Nucleation Theory to Precipitation Nucleation at GP Zones," *Met. Trans.* 7A(1976):1519.

23. M. G. Day and A. Hellawell: "The Structure of Modified Aluminum-Silicon Eutectic Alloy," *J. Inst. Metals* 95(1967):377.

24. B. Chamberlain and V. J. Zabek: "Reappraisal of the Tensile Properties of Al-Si-Mg Casting Alloys," *AFS Trans.* 81(1973):322.

25. K. C. Russell and H. I. Aaronson: "Precipitation Processes in Solids," *Met. Soc. AIME* 1978.

COPPERS AND COPPER ALLOYS

Copper is an important engineering metal since it is widely used in its *unalloyed* condition as well as in alloys with other metals. In the unalloyed form, it has an extraordinary combination of properties which make it the basic material in the electrical industry, some of those properties being its high electrical conductivity and corrosion resistance, ease of fabrication, reasonable tensile strength, controllable annealing properties, and general soldering and joining characteristics. The wide variety of brasses and bronzes it forms with other metals, however, also have associated useful properties that make *alloyed* copper indispensable for many additional engineering applications.

6-1 PRODUCTION OF COPPER

Copper comes from two principal sources: ores and copper scrap. Most copper is derived from copper sulfide ore deposits from which the copper sulfide is concentrated by various ore-dressing procedures to yield a product that can be smelted at a profit. Figure 6-1 outlines the processing steps necessary to produce high-purity copper from sulfide ores.

The copper sulfide concentrates are smelted in a reverberatory furnace (similar to the steelmaking open-hearth furnace) to produce a matte which is a mixture of copper and iron sulfides and a slag which is separated from the matte. The copper sulfide in the matte is then chemically converted to blister copper by blowing air through the matte. In this process, the iron sulfide is oxidized before the copper sulfide and is slagged off. Further blowing then converts the copper sulfide into blister copper, which is elemental copper with

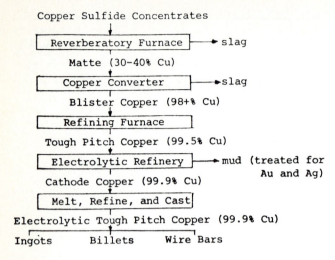

Copper Sulfide Concentrates

Reverberatory Furnace ──► slag

Matte (30–40% Cu)

Copper Converter ──►slag

Blister Copper (98+% Cu)

Refining Furnace

Tough Pitch Copper (99.5% Cu)

Electrolytic Refinery ──► mud (treated for Au and Ag)

Cathode Copper (99.9% Cu)

Melt, Refine, and Cast

Electrolytic Tough Pitch Copper (99.9% Cu)

Ingots   Billets   Wire Bars

Figure 6-1 Processing steps in the production of high-purity copper from copper sulfide concentrates.

impurities. The reaction for the oxidation of the copper sulfide is

$$2Cu_2S + 2O_2 \rightarrow 4Cu + 2SO_2$$

Near the end of the converter blow some of the copper is oxidized to Cu_2O.

The blister copper, which contains about 2 percent impurities, is fire-refined in a casting furnace by a process called *poling*. In this operation, most of the impurities in the copper are oxidized and removed as a slag, while at the same

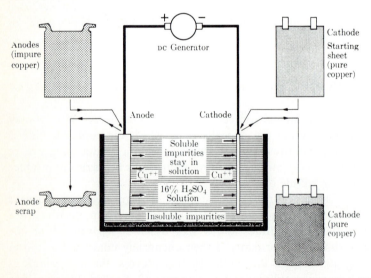

Figure 6-2 Electrolytic refining of impure anode copper. *(After A. G. Guy, "Elements of Physical Metallurgy," 2d Ed., © 1959, Addison-Wesley, Reading, Massachusetts, Fig. 2-10, p. 26.)*

time some of the copper is also oxidized. The oxidized copper is reduced back to elemental copper by a blanket of coke or charcoal and green tree trunks or poles which are put into the metal. The action of the carbon and greenwood in the copper results in the reduction of the Cu_2O to copper. The reduction is stopped when the Cu_2O content is about 0.5 percent so that sound castings will be produced.

Although this copper, called *tough-pitch* copper, can be used for some applications, most fire-refined copper is further refined electrolytically to produce 99.95% electrolytic tough-pitch (ETP) copper. Figure 6-2 illustrates the process whereby fire-refined anodes of copper are converted to cathodes of high-purity copper. Finally, the copper is melted and cast into shapes such as ingots for remelting and wire bar and billets for further fabrication. Oxygen-free high-conductivity copper is produced by casting cathode copper under a reducing atmosphere to prevent oxidation.

6-2 CLASSIFICATION OF COPPERS AND COPPER ALLOYS

Copper and copper alloys are classified according to a designation system administered by the Copper Development Association (CDA). In this system, numbers from C100 through C799 designate wrought alloys and numbers from C800 to C999 cast alloys. Within these two main classes, the system is divided into groups and subgroups:

Wrought alloys	
C1xx	Coppers[1] and high-copper alloys[2]
C2xx	Copper-zinc alloys (brasses)
C3xx	Copper-zinc-lead alloys (leaded brasses)
C4xx	Copper-zinc-tin alloys (tin brasses)
C5xx	Copper-tin alloys (phosphor bronzes)
C6xx	Copper-aluminum alloys (aluminum bronzes), copper-silicon alloys (silicon bronzes) and miscellaneous copper-zinc alloys
C7xx	Copper-nickel and copper-nickel-zinc alloys (nickel silvers)
Cast alloys	
C8xx	Cast coppers, cast high-copper alloys, the cast brasses of various types, cast manganese-bronze alloys, and cast copper-zinc-silicon alloys
C9xx	Cast copper-tin alloys, copper-tin-lead alloys, copper-tin-nickel alloys, copper-aluminum-iron alloys, and copper-nickel-iron and copper-nickel-zinc alloys

[1] "Coppers" have a minimum copper content of 99.3 percent or higher.

[2] High-copper alloys have less than 99.3% Cu, but more than 96 percent, and do not fit into the other copper alloy groups.

6-3 THE WROUGHT COPPERS

Unalloyed copper is an important engineering metal. Because of its high electrical conductivity (Table 6-1), it is used to a large extent in the electrical industry. As noted previously, other properties which make unalloyed copper attractive as an engineering material are its high corrosion resistance, ease of fabrication, reasonable tensile strength, controllable annealing properties, and good soldering and joining characteristics.

The wrought coppers are classified according to their oxygen and impurity contents. Table 6-2 lists the chemical compositions, mechanical properties, and typical applications of most of the important wrought coppers. The following three types of coppers will be discussed in this chapter: (1) electrolytic tough-pitch, (2) oxygen-free, and (3) phosphorus deoxidized.

Electrolytic Tough-Pitch Copper (Type ETP, CDA 110)

This copper has a minimum of 99.9%[1] Cu and a nominal 0.04% O content. The normal limits of oxygen in ETP copper are between 0.02 and 0.05 percent. It is the least expensive of the industrial coppers and is used extensively for the production of wire, rod, plate, and strip. The oxygen converts some impurity

Table 6-1 Relative electrical and thermal conductivities of commercially pure metals (At 20°C)

Metal	Relative electrical conductivity (copper = 100)	Relative thermal conductivity (copper = 100)
Silver	106	108
Copper	100	100
Gold	72	76
Aluminum	62	56
Magnesium	39	41
Zinc	29	29
Nickel	25	15
Cadmium	23	24
Cobalt	18	17
Iron	17	17
Steel	13–17	13–17
Platinum	16	18
Tin	15	17
Lead	8	9
Antimony	4.5	5

[1] Unless otherwise indicated all percentages are weight percents.

elements to their oxides; on casting the oxygen forms an even dispersion of blowholes that prevents pipe cavities from forming. Upon hot working, these small blowholes are welded together. This action is similar to that which occurs in rimmed steel (see Fig. 3-10).

Oxygen is almost insoluble in copper, as shown in the Cu-O phase diagram of Fig. 6-3, and forms Cu_2O interdendritic eutectic upon solidification. Figure 6-4 shows the interdendritic eutectic Cu_2O in as-cast electrolytic tough-pitch copper. Hot working breaks up the interdendritic network of Cu_2O and causes it to be strung out as particles in the direction of working (Fig. 6-5). Oxygen, if its concentration is about 0.04 percent, has the beneficial effect of increasing the electrical conductivity of ETP copper since it removes some of the impurities which lower the conductivity.

At temperatures above about 400°C, solid tough-pitch copper is deoxidized by reducing gases, especially hydrogen-containing gases. Since the hydrogen atoms are so small, they are able to diffuse into the solid copper and react with the internally dispersed Cu_2O to form steam according to the reaction

$$Cu_2O + H_2 \text{ (dissolved in Cu)} \rightarrow 2Cu + H_2O \text{ (steam)}$$

Since the steam formed by the reaction is insoluble in copper, high pressures build up so that the grain boundaries of the copper rupture. Thus, electrolytic

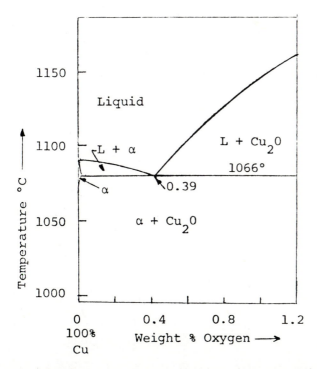

Figure 6-3 The copper-rich end of the copper–oxygen phase diagram.

Table 6-2 Chemical compositions, mechanical properties, and typical applications of selected wrought coppers†

Name and number	Nominal composition, %	Commercial forms‡	Mechanical properties			Corrosion resistance§	Machinability rating¶	Fabricating characteristics and typical applications
			Tensile strength, 1000 psi	Yield strength, 1000 psi	Elongation in 2 in, %			
101 Oxygen-free electronic	99.99 Cu	F, R, W, T, P, S	32–66	10–53	55–4	G–E	20	Excellent hot and cold workability; good forgeability. Fabricated by coining, coppersmithing, drawing and upsetting, hot forging and pressing, spinning, swaging, stamping. Uses: busbars, bus conductors, waveguides, hollow conductors, lead-in wires and anodes for vacuum tubes, vacuum seals, transistor components, glass-to-metal seals, coaxial cables and tubes, klystrons, microwave tubes, rectifiers.
102 oxygen-free copper	99.95 Cu	F, R, W T, P, S	32–66	10–53	55–4	G–E	20	Fabricating characteristics same as copper No. 101. Uses: busbars, waveguides.
103 Oxygen-free, extra low phosphorus	99.95 Cu, 0.003 P	F, R, T, P, S	32–55	10–50	50–6	G–E	20	Fabricating characteristics same as copper No. 101. Uses: busbars, electrical conductors, tubular bus and applications requiring good conductivity and welding or brazing properties.
104, 105, 107 oxygen-free, silver-bearing	99.95 Cu	F, R, W, S	32–66	10–53	55–4	G–E	20	Fabricating characteristics same as copper No. 101. Uses: auto gaskets, radiators; busbars, conductivity wire, contacts, radio parts, windings, switches, terminals, commutator segments; chemical process equipment, printing rolls, clad metals, printed circuit foil.
108 Oxygen-free, low phosphorus	99.95 Cu, 0.009 P	F, R, T, P	32–55	10–50	50–4	G–E	20	Fabricating characteristics same as copper No. 101. Uses: refrigerators, air conditioners, gas and heater lines, oil burner tubes, plumbing pipe and tube, brewery tubes, condenser and heat exchanger tubes, dairy and distiller tubes, pulp and paper lines, tanks; air, gasoline, hydraulic, and oil lines.
110 Electrolytic tough-pitch copper	99.90 Cu, 0.04 O	F, R, W, T, P, S	32–66	10–53	55–4	G–E	20	Fabricating characteristics same as copper No. 101. Uses: downspouts, gutters, roofing, gaskets, auto radiators, busbars, nails, printing rolls, rivets, radio parts.
111 Electrolytic tough-pitch, anneal-resistant	99.90 Cu, 0.04 O, 0.01 Cd	W	66	—	1–5 in 60 in	G–E	20	Fabricating characteristics same as copper No. 101. Uses: electrical power transmission where resistance to softening under overloads is desired.

113, 114, 115, 116 Silver-bearing tough-pitch copper	99.90 Cu, 0.04 O, Ag	F, R, W, T, S	32–66	10–53	55–4	G–E	20	Fabricating characteristics same as copper No. 101. Uses: gaskets, radiators, busbars, windings, switches, chemical process equipment, clad metals, printed-circuit foil.
120, 121	99.9 Cu	F, T, P	32–57	10–53	55–4	G–E	20	Fabricating characteristics same as copper No. 101. Uses: busbars, electrical conductors, tubular bus, and applications requiring welding or brazing.
122 Phosphorus deoxidized copper, high residual phosphorus	99.90 Cu, 0.02 P	F, R, T, P	32–55	10–50	45–8	G–E	20	Fabricating characteristics same as copper No. 101. Uses: gas and heater lines; oil burner tubing; plumbing pipe and tubing; condenser evaporator, heat exchanger, dairy, and distiller tubing; steam and water lines; air, gasoline, and hydraulic lines.
125, 127, 128, 129, 130 Fire-refined tough-pitch with silver	99.88 Cu	F, R, W, S	32–67	10–53	55–4	G–E	20	Fabricating characteristics same as copper No. 101. Uses: same as copper No. 110.
142 Phosphorus deoxidized, arsenical	99.63 Cu, 0.3 As, 0.02 P	F, R, T	32–55	10–50	45–8	G–E	20	Fabricating characteristics same as copper No. 101. Uses: plates for locomotive fireboxes, staybolts, heat exchanger and condenser tubes.
143 Deoxidized cadmium copper	99.90 Cu, 0.07 Cd	F	32–65	10–62	45–2	G–E	20	Fabricating characteristics same as copper No. 101. Uses: cooling fins for automotive and heavy-duty radiators, air conditioners, motor commutators, electrical terminals and connectors. Anneal-resistant electrical applications.
145 Phosphorus deoxidized, tellurium-bearing	99.5 Cu, 0.50 Te, 0.008 P	F, R, W, T	32–56	10–51	50–3	G–E	85	Fabricating characteristics same as copper No. 101. Uses: forgings and screw-machine products, and parts requiring high conductivity, extensive machining, corrosion resistance, copper color, or a combination of these; electrical connectors, motor and switch parts, plumbing fittings, soldering coppers, welding torch tips, transistor bases, and furnace brazed articles.

† After Ref. 6.
‡ F: flat products; R: rod; W: wire; T: tube; P: pipe; S: shapes.
§ G: good; E: excellent.
¶ Based on 100% for copper alloy 360.

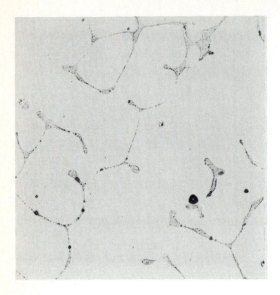

Figure 6-4 As-cast electrolytic tough-pitch copper (99.95% Cu, 0.03% O); structure shows Cu_2O interdendritic eutectic; the dark spots are gas pores. (Etch: potassium dichromate; 150 × .) *(Courtesy of Amax Base Metals Research Inc.)*

tough-pitch copper cannot be used where joining processes involve temperatures above 400°C. Figure 6-6 shows internal holes developed in ETP copper exposed to H_2 at 850°C for $\frac{1}{2}$ h.

Oxygen-Free Copper

Oxygen-free copper can be produced from electrorefined cathode copper by melting and casting under a reducing atmosphere of carbon monoxide and

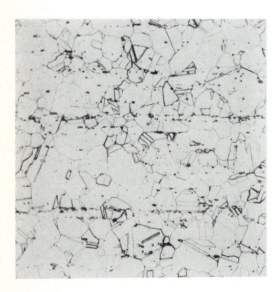

Figure 6-5 Hot-worked electrolytic tough-pitch copper; structure shows stringers of Cu_2O and complete recrystallization. (Etch: NH_4OH + H_2O_2; 150 × .) *(Courtesy of Amax Base Metals Research, Inc.)*

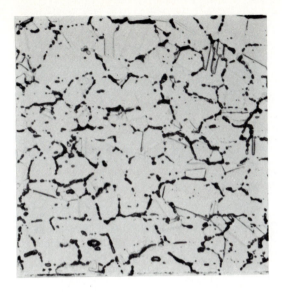

Figure 6-6 Electrolytic tough-pitch copper exposed to hydrogen at 850°C for 1/2 h; structure shows internal holes developed by steam, which makes the copper brittle. (Etch: potassium dichromate; 150 × .) *(Courtesy of Amax Base Metals Research, Inc.)*

nitrogen so that oxygen is prevented from entering the copper. The as-cast structure of oxygen-free copper (99.95% Cu) (Fig. 6-7) does not contain the interdendritic eutectic Cu_2O or the gas porosity found in ETP coppers (Fig. 6-4). After hot working, the oxygen-free copper has a clean wrought structure (Fig. 6-8) free of the Cu_2O stringers found in the ETP copper (Fig. 6-5). Also, the oxygen-free copper is not susceptible to hydrogen embrittlement at elevated temperatures since it does not contain Cu_2O (Fig. 6-9).

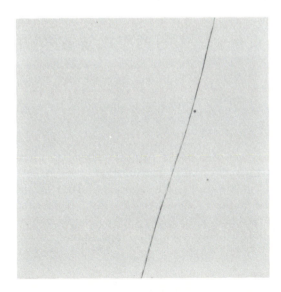

Figure 6-7 As-cast oxygen-free copper; structure shows a grain boundary and traces of microporosity. Note the absence of Cu_2O interdendritic eutectic. (Etch: potassium dichromate; 150 × .) *(Courtesy of Amax Base Metals Research, Inc.)*

Figure 6-8 Hot-worked oxygen-free copper; structure is clean and free of Cu_2O stringers. (Etch: potassium dichromate; 150 × .) *(Courtesy of Amax Base Metals Research, Inc.)*

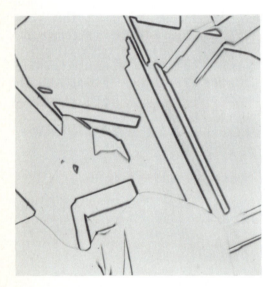

Figure 6-9 Oxygen-free copper exposed to hydrogen at 850°C for 1/2 h. Note clear structure and absence of hydrogen embrittlement. (Etch: potassium dichromate; 150 × .) *(Courtesy of Amax Base Metals Research, Inc.)*

The electrical conductivity of oxygen-free (99.95 percent) copper is about the same as ETP copper which is 101 percent IACS.[1] The increased conductivity due to the elimination of the Cu_2O in the oxygen-free copper is offset by the increased solubility of some of the impurity elements, i.e., iron. If selected cathodes of high-purity copper are remelted, 99.99% oxygen-free copper can be produced (CDA 101) and is preferred for many electronics applications. Due to

[1] International Annealed Copper Standard.

the special processing of the oxygen-free coppers, they are more expensive than the ETP coppers.

Deoxidized Coppers

With the addition of sufficient phosphorus, all the available oxygen in the copper will be converted to phosphorus pentoxide (P_2O_5). Since phosphorized high-conductivity coppers contain very little retained phosphorus (normally less than 0.009 percent), the high-conductivity of copper is maintained. Higher levels of phosphorus are used in the deoxidized high-phosphorus copper (CDA 122), so that these alloys may contain as high as 0.040% residual P, resulting in a lower electrical conductivity of about 85 percent IACS. The excess phosphorus in the copper prevents the adsorption of oxygen during hot working and annealing and allows this material to be welded. A disadvantage of the deoxidized coppers is that pipe cavities are formed during solidification. Since the piped region must be cut off, the yield of usable metal is lowered.

6-4 COPPER-ZINC ALLOYS (BRASSES)

Chemical Compositions and Typical Applications

The copper-zinc brasses consist of a series of alloys of copper with up to about 40% Zn. As the percentage of zinc changes, the properties of the Cu-Zn alloys change also. Copper-zinc brasses containing additional elements such as tin, aluminum, silicon, manganese, nickel, and lead are referred to as "alloy brasses." The alloying additions, which rarely exceed about 4 percent, improve some of the properties of the straight Cu-Zn brasses so they can be used for other applications. Table 6-3 lists the chemical compositions and typical applications of selected copper-zinc brasses.

   The uses of the solid-solution α brasses (solid solutions of zinc in copper) depend mainly on the property of high ductility coupled with sufficient strength, good corrosion resistance, pleasing colors, and solderability. Brasses are also able to be nickel- and chromium-plated and have sufficient thermal conductivity to be used for heat-transfer media. The best combination of ductility and strength occurs at 70% Cu and 30% Zn, and hence this alloy can be used for its excellent deep-drawing ability. The 70% Cu–30% Zn alloy is descriptively called "cartridge brass," but is used for other applications such as radiator cores and tanks, and lamp fixtures.

Structure

Phase diagram for the copper-zinc system The phase diagram for the copper-zinc system is shown in Fig. 6-10. Zinc has extensive solid solubility in copper and forms α-solid solutions with up to 39% Zn at 456°C. With increasing zinc

Table 6-3 Chemical compositions and typical applications of selected copper-zinc alloys (brasses)†

Name and number	Nominal composition, %	Fabricating characteristics and typical applications
	Unalloyed brasses	
210 Gilding, 95%	95.0 Cu, 5.0 Zn	Excellent cold workability; good hot workability for blanking, coining, drawing, piercing and punching, shearing, spinning, squeezing and swaging, stamping. Uses: coins, medals, bullet jackets, fuse caps, primers, plaques, jewelry base for gold plate.
220 Commercial bronze, 90%	90.0 Cu, 10.0 Zn	Fabricating characteristics same as copper alloy No. 210, plus heading and upsetting, roll threading and knurling, hot forging and pressing. Uses: etching bronze, grillwork, screen cloth, weatherstripping, lipstick cases, compacts, marine hardware, screws, rivets.
226 Jewelry bronze, 87.5%	87.5 Cu, 12.5 Zn	Fabricating characteristics same as copper alloy No. 210, plus heading and upsetting, roll threading and knurling. Uses: angles, channels, chain, fasteners, costume jewelry, lipstick cases, compacts, base for gold plate.
230 Red brass, 85%	85.0 Cu, 15.0 Zn	Excellent cold workability; good hot formability. Uses: weatherstripping, conduit, sockets, fasteners, fire extinguishers, condenser and heat exchanger tubing, plumbing pipe, radiator cores.
240 Low brass, 80%	80.0 Cu, 20.0 Zn	Excellent cold workability. Fabricating characteristics same as copper alloy No. 230. Uses: battery caps, bellows, musical instruments, clock dials, pump lines, flexible hose.
260 Cartridge brass, 70%	70.0 Cu, 30.0 Zn	Excellent cold workability. Fabricating characteristics same as copper alloy No. 230, except for coining, roll threading, and knurling. Uses: radiator cores and tanks, flashlight shells, lamp fixtures, fasteners, locks, hinges, ammunition components, plumbing accessories, pins, rivets.
268, 270 Yellow brass	65.0 Cu, 35.0 Zn	Excellent cold workability. Fabricating characteristics same as copper alloy No. 230. Uses: same as copper alloy No. 260, except not used for ammunition.
280 Muntz metal	60.0 Cu, 40.0 Zn	Excellent hot formability and forgeability for blanking, forming and bending, hot forging and pressing, hot heading and upsetting, shearing. Uses: architectural, large nuts and bolts, brazing rod, condenser plates, heat exchanger and condenser tubing, hot forgings.
	Alloy brasses	
443, 444, 445 Inhibited admiralty	71.0 Cu, 28.0 Zn, 1.0 Sn	Excellent cold workability for forming and bending. Uses: condenser, evaporator and heat exchanger tubing, condenser tubing plates, distiller tubing, ferrules.

Table 6-3, Continued

Name and number	Nominal composition, %	Fabricating characteristics and typical applications
464 to 467 Naval brass	60.0 Cu, 39.25 Zn, 0.75 Sn	Excellent hot workability and hot forgeability. Fabricated by blanking, drawing, bending, heading and upsetting, hot forging, pressing. Uses: aircraft turnbuckle barrels, balls, bolts, marine hardware, nuts, propeller shafts, rivets, valve stems, condenser plates, welding rod.
667 Manganese brass	70.0 Cu, 28.8 Zn, 1.2 Mn	Excellent cold formability. Fabricated by blanking, bending, forming, stamping, welding. Uses: brass products resistance-welded by spot, seam, and butt welding.
674	58.5 Cu, 36.5 Zn, 1.2 Al, 2.8 Mn, 1.0 Sn	Excellent hot formability. Fabricated by hot forging and pressing, machining. Uses: bushings, gears, connecting rods, shafts, wear plates.
675 Manganese bronze, A	58.5 Cu, 1.4 Fe, 39.0 Zn, 1.0 Sn, 0.1 Mn	Excellent hot workability. Fabricated by hot forging and pressing, hot heading and upsetting. Uses: clutch disks, pump rods, shafting, balls, valve stems, and bodies.
687 Aluminum brass, arsenical	77.5 Cu, 20.5 Zn, 2.0 Al, 0.1 As	Excellent cold workability for forming and bending. Uses: condenser, evaporator and heat exchanger tubing, condenser tubing plates, distiller tubing, ferrules.
688	73.5 Cu, 22.7 Zn, 3.4 Al, 0.40 Co	Excellent hot and cold formability. Fabricated by blanking, drawing, forming and bending, shearing and stamping. Uses: springs, switches, contacts, relays, drawn parts.
694 Silicon red brass	81.5 Cu, 14.5 Zn, 4.0 Si	Excellent hot formability for fabrication by forging, screw-machine operations. Uses: valve stems where corrosion resistance and high strength are critical.

† After Ref. 6.

content, a second solid solution of zinc in copper is formed which is designated the β phase. The α-solid solution has the FCC structure. The β phase has the BCC crystal structure and transforms upon cooling through the 468 to 456°C temperature range from a disordered β-phase structure to an ordered β' structure. Figure 6-11 illustrates the difference between the ordered and disordered unit cells of β brass at 50% Cu–50% Zn (atomic percent). With more than about 50% Zn, the γ-phase solid solution forms, which has a complex structure and which is very brittle. Copper-zinc alloys containing the brittle γ phase are of little engineering use.

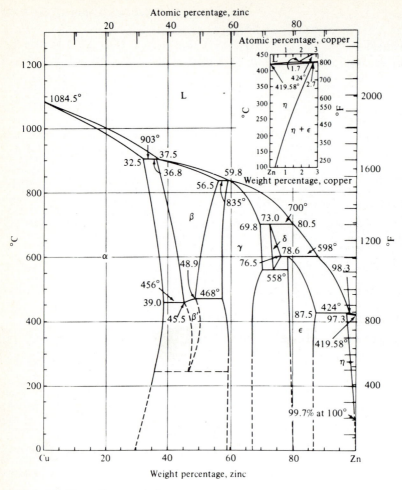

Figure 6-10 Phase diagram of the copper–zinc system. *(After Metals Handbook, 8th ed., vol. 8, American Society for Metals, 1973, p. 301.)*

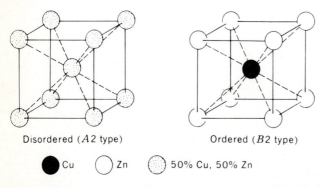

Figure 6-11 The disordered and ordered unit cells of 50% Cu–50% Zb β-brasses. *(After P. G. Shewmon, "Transformations in Metals," McGraw-Hill, New York, 1969, p. 261.)*

On the basis of the Cu-Zn phase diagram, commercial brasses can be divided into two important groups:

1. α brasses with the α structure and containing up to about 35% Zn
2. $\alpha + \beta$ brasses with the $\alpha + \beta$ two-phase structure, which are mainly based on a 60 : 40 ratio of copper to zinc

Microstructure of α brasses

The microstructures of the single-phase α brasses consist of α solid solutions. This structure is illustrated by the annealed wrought commercial bronze alloy (90% Cu–10% Zn) in Fig. 6-12a and by cartridge brass (70% Cu–30% Zn) in Fig.

(a)

(b)

Figure 6-12 Microstructures of (a) commercial bronze (90% Cu–10% Zn) and (b) cartridge brass (70% Cu–30% Zn) in the annealed condition. (Etchant: $NH_4OH + H_2O_2$; 75 × .)

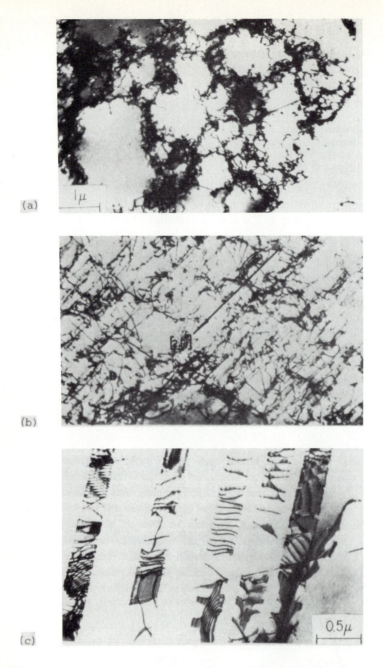

(a)

(b)

(c)

Figure 6-13 Effect of small amounts (5 to 10 percent) plastic deformation on the dislocation distributions in pure copper and brasses. (*a*) Pure copper deformed 5 percent: structure shows cellular distribution of dislocation tangles. (*b*) Red brass (85% Cu–15% Zn) deformed 10 percent: structure shows planar arrays of dislocations developing. (*c*) High brass (63% Cu–37% Zn) deformed 10 percent: structure shows well-defined planar arrays of dislocations. [*a and c are from P. R. Swann and J. D. Embury, "High-Strength Materials," ed. by V. F. Zackay, John Wiley & Sons, Inc., 1965, pp. 333–334 and are reprinted by permission of John Wiley & Sons, Inc. b is after J. Hedworth and G. Pollard, Met. Sc. J. 5(1971):41.*]

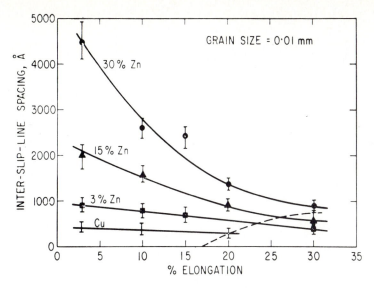

Figure 6-14 Effect of zinc content on the interslip-line spacing of deformed copper–zinc brasses. [*After P. R. Swann and J. Nutting, J. Inst. Metals 90(1961–62):133.*]

6-12*b*. With the higher zinc contents, more annealing twins are observed in the alpha grains.

The dislocation substructures for the α brasses, given the same amount of cold deformation, change as the amount of zinc is increased. This change is shown in Fig. 6-13 for deformations of 5 to 10 percent for (1) pure copper, (2) red brass (85% Cu–15% Zn), and (3) a 63% Cu–37% Zn brass. Pure copper shows a cellular distribution of dislocation tangles characteristic of deformed pure metals (Fig. 6-13*a*). The distance between slip lines in pure copper is relatively small (Fig. 6-14). As the amount of zinc is increased to 15 percent, the interslip line spacing is increased and arrays of dislocations begin to form (Fig. 6-13*b*). When the zinc content is increased to 37 percent (about the maximum soluble in copper), the interslip spacing is increased greatly and the dislocations are arranged in well-defined planar arrays (Fig. 6-13*c*).

The reason for this change in dislocation arrangements with increasing zinc content of Cu-Zn alloys is attributed to the lowering of the stacking-fault energy (SFE) of copper by zinc (Fig. 6-15). In pure copper, the SFE is relatively high and dislocations can cross slip easily and thereby produce fine slip during deformation. With the addition of zinc, the SFE of the copper is lowered, making cross slip more difficult so that dislocations tend to remain in their slip planes either as pileups or short stacking-fault ribbons.

Microstructure of $\alpha + \beta$ brasses

When the zinc content of copper-zinc alloys is about 40 percent, these alloys have a duplex structure containing both α and β phases. The most commonly

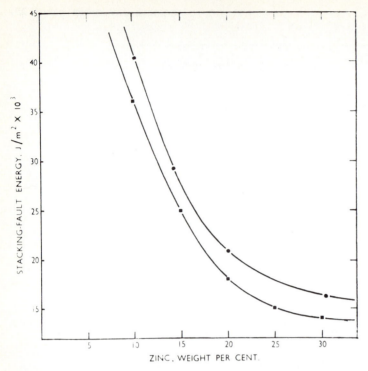

Figure 6-15 Effect of zinc content on the stacking-fault energies of copper–zinc brasses. [*After J. Hedworth and G. Pollard, Met. Sc. J. 5(1971):42.*]

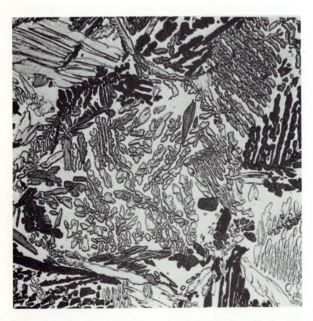

Figure 6-16 Cast structure of muntz metal (60% Cu–40% Zn, CDA Alloy 280). Structure consists of dendrites of alpha in a matrix of beta. (Etch: NH_4OH + H_2O_2; 100 × .) *(Courtesy of Chase Copper and Brass Co.)*

Figure 6-17 Hot-rolled muntz metal sheet (60% Cu–40% Zn). Structure consists of beta phase (dark) and alpha phase (light). Note the twinning in the alpha crystals, which is a result of the strain accompanying the transformation of the beta phase to alpha. (Etchant: NH₄OH + H₂O₂; 75 × .) *(Courtesy of Anaconda American Brass Co.)*

used α + β brass is the 60% Cu–40% Zn alloy called "Muntz metal" (Table 6-3). Muntz metal is difficult to cold work since it contains the β phase, and so is essentially a hot-working alloy having excellent hot-working properties. The presence of the β phase makes this alloy heat-treatable, but lowers its ductility. The cast structure of 60% Cu–40% Zn shows dendrites of α phase in a matrix of β (Fig. 6-16). The grain structure of the α + β brasses can easily be refined by hot working, as is indicated in the microstructure of the hot-rolled 60% Cu–40% Zn alloy sheet of Fig. 6-17.

Decomposition of β' ***in*** α + β ***Cu-Zn alloys*** If Cu-Zn brasses containing 40 to 45% Zn are heated to about 830°C and then are hot-quenched to temperatures in the 700 to 710°C range, unstable β or β' will transform isothermally, producing some α phase (Fig. 6-18).

Flewitt and Towner (Ref. 23) have investigated the isothermal decomposition of β for a 58.4% Cu–41.6% Zn alloy and found that two distinct types of α phase were formed: a rodlike type and a platelike type. The rod-type α precipitate formed at higher temperatures (500 to 700°C) above a temperature limit designated the B_s (bainitic start) temperature and precipitated in a Widmanstätten pattern. Below the B_s temperature, bainitic plates of α were nucleated uniformly throughout any β grain and grew rapidly in the lengthwise direction.

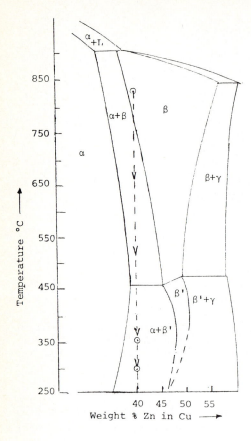

Figure 6-18 Section of the copper–zinc phase diagram showing the cooling path at 60% Cu–40% Zn giving a beta to alpha + beta transformation.

Figure 6-19 Copper–41.6% zinc alloy heated to 830°C, quenched to 250°C and held for 20 h; structure shows alpha plates transformed from beta matrix. (400 × .) [*After P. E. Flewitt and J. M. Towner, J. Inst. Met. 95(1967):273.*]

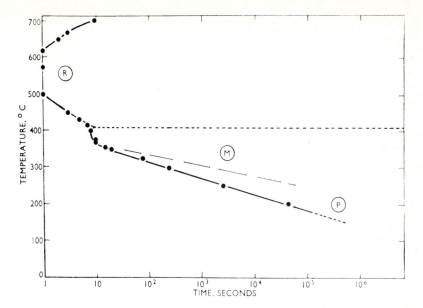

Figure 6-20 Isothermal transformation diagram for the start of visible precipitation from the metastable β' phase of a 58.4% Cu–41.6% Zn alloy. The R stands for rods, the M for mixed rods and plates, and the P for plates. The dotted horizontal line represents the bainite start temperature. That is, at temperatures below this line, the bainitic reaction is found. [*After P. E. Flewitt and J. M. Towner, J. Inst. Met. 95(1967):273.*]

Figure 6-19 shows the α plates after isothermally transforming the β' for 20 h at 250°C. Figure 6-20 shows the IT diagram for the decomposition of a 58.4% Cu–41.6% Zn alloy. The α-phase rods which form in a Widmanstätten pattern can clearly be seen protruding from a grain boundary in the scanning electron micrograph of Fig. 6-21. Both platelike and rodlike decomposition products are associated with stress-relief effects similar to those observed in the bainitic reaction of plain-carbon steels.

Microstructure of alloy brasses

Leaded brasses Lead is soluble in liquid copper at high temperatures, as is indicated in the copper-lead phase diagram of Fig. 6-22. However, at room temperature copper and lead are essentially insoluble in each other.

Consider the slow cooling of a 97% Cu–3% Pb alloy from the liquid state at about 1080°C to below 326°C. From about 1080 to 955°C, pure copper crystals will nucleate and grow and will result in the progressive enrichment of the remaining liquid in lead until, at 955°C, the liquid will contain 36% Pb. Then, the remaining liquid undergoes a monotectic reaction as follows:

$$\text{Liquid}_1(36\% \text{ Pb}) \underset{955°C}{\rightleftharpoons} \alpha(100\% \text{ Cu}) + \text{liquid}_2(87\% \text{ Pb})$$

Figure 6-21 Widmanstätten alpha phase rods protruding from grain boundary precipitates formed by the isothermal decomposition of a 55.9 at %Cu–44.1 at %Zn brass at 400° C (1500 × .) [*After G. R. Purdy, Met. Sc. J. 5(1971):81.*]

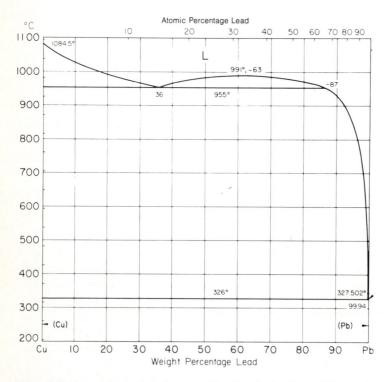

Figure 6-22 Copper–lead phase diagram. (*After Metals Handbook, 8th ed., vol. 8, American Society for Metals, 1973, p. 296.*)

Figure 6-23 Free-cutting brass extruded rod showing elongated lead globules. Remainder of structure is α phase. (Etch: $NH_4OH + H_2O_2$; 75 × .) *(Courtesy of Anaconda American Brass Co.)*

During slow cooling from 955°C to 326°C, the remaining liquid will increase in lead content until it reaches 99.94% Pb at 326°C. At that temperature, the liquid will undergo a eutectic reaction as follows:

$$\text{Liquid}_2(99.94\% \text{ Pb}) \underset{326°C}{\rightleftharpoons} \alpha(100\% \text{ Cu}) + \beta(99.993\% \text{ Pb})$$

The essentially pure lead (99.99% Pb) produced by the eutectic reaction will be distributed interdendritically in the copper as small globules. During deformation, these globules will be strung out as indicated in the micrograph of a cold-drawn rod of free-cutting brass shown in Fig. 6-23. Small amounts of lead (0.5 to 3.0 percent) are added to many types of brasses to improve their machinability and are called "leaded brasses."

Tin and aluminum brasses The addition of 1% Sn to cartridge brass (70% Cu–30% Zn) improves its corrosion resistance in sea water. Since this alloy was adopted by the British Admiralty in the 1920s, it became known as "*admiralty brass*." It was later found that small additions of arsenic (≈ 0.04 percent) could almost eliminate a common corrosion condition called "dezincification" (discussed later in this chapter), and hence arsenical admiralty brass was used for a long time for marine condensers. Still later it was discovered that replacing the tin with aluminum gave the brass a "self-healing" protective oxide on its surface. The hard aluminum-type oxide film makes the alloy more resistant than is admiralty brass to the impingement of high-velocity water. Today, a 77.5%

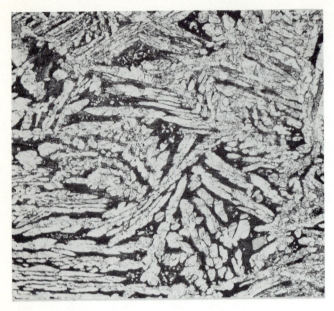

Figure 6-24 Microstructure of extruded naval brass rod; (60% Cu–39.25% Zn–0.75% Sn); structure consists of α phase in β phase matrix (black). (Etch: NH_4OH + H_2O_2; 75 × .) *(Courtesy of Anaconda American Brass Co.)*

Cu–20.5% Zn–2.0% Al alloy (aluminum brass) with an arsenic addition to inhibit dezincification has replaced admiralty brass for marine condensers.

The addition of 1% Sn to Muntz metal (60% Cu–40% Zn) improves its corrosion resistance and forms an alloy called *"naval brass."* Figure 6-24 shows the microstructure of an extruded naval-brass rod.

Mechanical Properties

The tensile properties of selected copper-zinc brasses are listed in Table 6-4. In general, the mechanical properties of Cu-Zn alloys are closely related to the phases present in the alloy.

Low brasses (80 to 95% Cu, 20 to 5% Zn) Increasing the zinc content of these brasses increases their strength, hardness, and ductility (Fig. 6-25). Their color changes from red through gold to the green yellows. Their hot-working properties are comparable to those of commercial copper, and they may be hot-worked in the 730 to 900°C range. However, their lead content should be kept below 0.01 percent to avoid hot-working difficulties. Low brasses in the annealed condition are extremely ductile and malleable at room temperature, having elongations in the 45 to 50 percent range (Table 6-4), and therefore can be cold-worked by any conventional method.

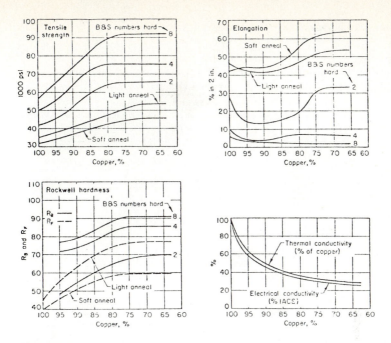

Figure 6-25 The effect of zinc content on the mechanical and electrical properties of alpha brass: (*a*) tensile strength; (*b*) elongation; (*c*) Rockwell hardness; (*d*) electrical conductivity. (*After Metals Handbook, 8th ed., vol. 1, American Society for Metals, 1961, p. 1015.*)

Table 6-4 Typical mechanical properties and corrosion ratings for copper-zinc alloys (brasses)†

Name and number	Nominal Composition, %	Mechanical properties			Corrosion resistance‡	Machinability rating§
		Tensile strength, 1000 psi	Yield strength, 1000 psi	Elongation in 2 in, %		
		Unalloyed brasses				
210 Gilding, 95%	95.0 Cu, 5.0 Zn	34–64	10–58	45–4	G–E	20
220 Commercial bronze, 90%	90.0 Cu, 10.0 Zn	37–72	10–62	50–3	G–E	20
226 Jewelry bronze, 87.5%	87.5 Cu, 12.5 Zn	39–97	11–62	46–3	G–E	30
230 Red brass, 85%	85.0 Cu, 15.0 Zn	39–105	10–63	55–3	G–E	30
240 Low brass, 80%	80.0 Cu, 20.0 Zn	42–125	12–65	55–3	F–E	30
260 Cartridge brass, 70%	70.0 Cu, 30.0 Zn	44–130	11–65	66–3	F–E	30
268, 270 Yellow brass	65.0 Cu, 35.0 Zn	46–128	14–62	65–3	F–E	30
280 Muntz metal	60.0 Cu, 40.0 Zn	54–74	21–55	52–10	F–E	40

Table 6-4, Continued

Name and number	Nominal Composition, %	Mechanical properties			Corrosion resistance‡	Machinability rating§
		Tensile strength, 1000 psi	Yield strength, 1000 psi	Elongation in 2 in, %		
Alloy brasses						
443, 444, 445 Inhibited admiralty	71.0 Cu, 28.0 Zn, 1.0 Sn	48–55	18–22	65–60	G–E	30
464 to 467 Naval brass	60.0 Cu, 39.25 Zn, 0.75 Sn	55–88	25–66	50–17	F–E	30
667 Manganese brass	70.0 Cu, 28.8 Zn, 1.2 Mn	45.8–100	12–92.5	60–2	G–E	30
674	58.5 Cu, 36.5 Zn, 1.2 Al, 2.8 Mn, 1.0 Sn	70–92	34–55	28–20	F–E	25
675 Manganese bronze, A	58.5 Cu, 1.4 Fe, 39.0 Zn, 1.0 Sn, 0.1 Mn	65–84	30–60	33–19	F–E	30
687 Aluminum brass, arsenical	77.5 Cu, 20.5 Zn, 2.0 Al, 0.1 As	60	27	55	G–E	30
688	73.5 Cu, 22.7 Zn, 3.4 Al, 0.40 Co	82–129	55–114	36–2	G–E	—
694 Silicon red brass	81.5 Cu, 14.5 Zn, 4.0 Si	80–100	40–57	25–20	G–E	30

† After Ref. 6.
‡ G: good; E: excellent; F: fair.
§ Based on 100% for copper alloy 360.

High brasses (60 to 80% Cu, 40 to 20% Zn) These brasses, because of their high zinc contents, have increased strengths. Their ductility also increases with increasing zinc content and reaches a maximum at about 30% Zn. When the zinc content exceeds 36 percent, the ductility of these alloys decreases rapidly due to the presence of the β phase, but strength and hardness continue to increase to about 45% Zn.

The α brasses with copper contents between 80 and 64 precent (20 and 36% Zn) are relatively poor for hot working and their lead content must be kept down to a trace. The $\alpha + \beta$ brasses, due to the presence of β brass, can be hot-worked (usually in the 650 to 769°C range) much more easily than the high-α brasses. However, the $\alpha + \beta$ brasses are difficult to cold-work, with this difficulty increasing as the β-phase content increases.

Alloy brasses Small additions of alloying elements, such as 1% Sn, to the brasses do not greatly affect their mechanical properties. However, multiple additions of manganese, iron, and tin, for example, to convert Muntz metal to manganese bronze[1] significantly increase the strength of Muntz metal. Because of its increased strength, manganese bronze is best worked in the hot condition. The addition of up to about 3% Pb to improve the machinability of brasses has practically no effect on the tensile strength and hardness of the leaded brasses. However, ductility, and hence cold-working ability, of the brasses is reduced by the Pb additions.

Corrosion of Brasses

Stress-corrosion cracking (season cracking) α brasses in the cold-worked condition and containing more than about 15% Zn are susceptible to *stress-corrosion cracking* if in contact with a trace of ammonia in the presence of oxygen and moisture. The stress-corrosion cracking which occurs in α brasses usually is along the grain boundaries (intergranular cracking). Cracks through the grains or transgranular cracking may occur if the alloy is severely plastically deformed. Figure 6-26 shows intergranular cracking in cartridge brass which was exposed to the corrosive action of the atmosphere. This type of stress-corrosion cracking is sometimes called *season cracking.*[2] Stress-corrosion cracking can be alleviated in cold-worked brasses by a low-temperature stress relief (recovery treatment) which reduces residual and internal stresses.

Dezincification Another type of corrosion attack to which some brasses are susceptible is known as *dezincification*. In the course of it, the zinc corrodes preferentially and leaves a porous residue of copper and corrosion products (Fig. 6-27). Although the exact mechanism of dezincification is still not fully understood, it is believed that the zinc diffuses to the brass surface and reacts there preferentially, leaving a copper-rich alloy residue. As a result, a porous plug of dezincified metal is created as can be seen in Fig. 6-27.

---

[1] The term "manganese bronze" is a misnomer since it is essentially a brass.

[2] The origin of the term "season cracking" is uncertain. It is sometimes ascribed to the fact that long ago brass cartridge cases stored in India were observed to be prone to cracking during the monsoon season.

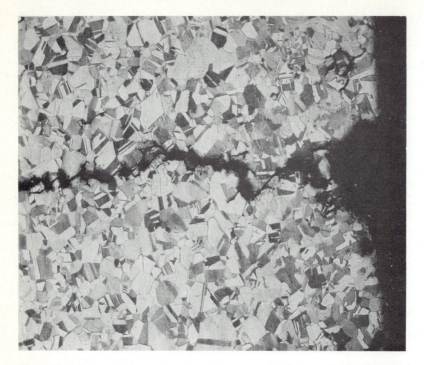

Figure 6-26 Intergranular stress-corrosion crack in cartridge brass (70% Cu–30% Zn) resulting from release of internal stresses and by the corrosive action of the atmosphere. (Etch: $NH_4OH + H_2O_2$; 75 × .) *(Courtesy of Chase Brass Co.)*

Figure 6-27 Dezincification of cartridge brass (70% Cu–30% Zn) tube. Note the porous plug of copper-rich alloy residue. (Etch: $NH_4OH + H_2O_2$; 75 × .) *(Courtesy of Chase Brass Co.)*

6-5 COPPER-TIN ALLOYS

Alloys consisting of principally copper and tin are properly called *tin bronzes*. Since phosphorus is usually added to these alloys as a deoxidizing agent during casting, the tin bronzes are commercially known as "phosphor bronzes." These alloys possess desirable properties such as high strength, wear resistance, and good sea-water corrosion resistance.

Phase Diagram of the Copper-Tin System

The phase diagram of the copper-tin system is shown in Fig. 6-28. The solid solubility of tin in copper reaches a maximum of 15.8 percent between 520 and 586°C, which is much less than the solubility of zinc in copper. From this phase diagram, it would appear that Cu-Sn alloys with up to about 11% Sn should precipitate the ϵ phase when cooled to room temperature from above about 350°C. This transformation must be sluggish since the ϵ phase is not observed in the optical microscope in a Cu–5% Sn alloy. However, GP zones and metastable ϵ' have been found in a Cu–5% Sn alloy which was cold-rolled to 97 percent reduction and subsequently solution-heat-treated, quenched, and aged (Ref. 16).

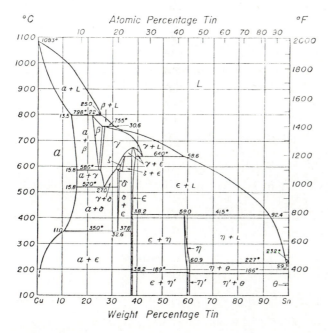

Figure 6-28 Phase diagram for the copper–tin system. *(After Metals Handbook, 1948 ed., American Society for Metals, 1948, p. 1204.)*

Wrought Copper-Tin Bronzes

Wrought copper-tin bronzes containing from 1.25 to 10% Sn are termed *phosphor bronzes* since they usually contain up to about 0.1% P, which is added to improve castability and act as a deoxidizer. If any phosphorus is retained after deoxidizing, it forms the hard compound Cu_3P, which increases the strength and hardness of the tin bronze. The wrought tin bronzes are stronger than the brasses, especially in the cold-worked condition, and have better corrosion resistance. Table 6-5 lists the chemical compositions, mechanical properties, and typical applications of selected wrought tin bronzes. The microstructure of the 92% Cu–8% Sn phosphor bronze in the annealed condition is shown in Fig. 6-29 and consists of a recrystallized equiaxed grains of α solid solution.

Cast Copper-Tin Bronzes

High tin contents over about 10 percent make copper-tin alloys unworkable, but castings containing up to 16% Sn are used for high-strength bearings and gear blanks. Gear-blank castings are often made by centrifugally casting to ensure sound castings. Tin levels of about 10 percent are common for bearings, with variable quantities of lead being added to improve plasticity and adaptability for bearing surfaces.

6-6 COPPER-ALUMINUM ALLOYS

Chemical Compositions and Typical Applications

Copper-aluminum alloys are called *aluminum bronzes*, although a better name would be aluminum brasses. These alloys are quite hard, have high tensile

Figure 6-29 Microstructure of phosphor bronze (CDA 521) 92% Cu–8% Sn–trace P. Structure consists of recrystallized alpha grains with annealing twins. (Etch: NH_4OH + H_2O_2; 75 × .) *(Courtesy of Anaconda American Brass Co.)*

Table 6-5 Chemical compositions, mechanical properties, and typical applications of selected phosphor bronzes (tin bronzes)†

Name and number	Nominal composition, %	Commercial forms‡	Mechanical properties				Machinability rating¶	Fabricating characteristics and typical applications
			Tensile strength, 1000 psi	Yield strength, 1000 psi	Elongation in 2 in, %	Corrosion resistance§		
505 Phosphor bronze, 1.25% E	98.75 Cu, 1.25 Sn, trace P	F, W	40–79	14–50	48–4	G–E	20	Excellent cold workability; good hot formability. Fabricated by blanking, bending, heading and upsetting, shearing and swaging. Uses: electrical contacts, flexible hose, pole-line hardware.
510 Phosphor bronze, 5% A	95.0 Cu, 5.0 Sn, trace P	F, R, W, T	47–140	19–80	64–2	G–E	20	Excellent cold workability. Fabricated by blanking, drawing, bending, heading and upsetting, roll threading and knurling, shearing, stamping. Uses: bellows, bourdon tubing, clutch disks, cotter pins, diaphragms, fasteners, lock washers, wire brushes, chemical hardware, textile machinery, welding rod.
511	95.6 Cu, 4.2 Sn, 0.2 P	F	46–103	50–80	48–2	G–E	20	Excellent cold workability. Uses: bridge bearing plates, locator bars, fuse clips, sleeve bushings, springs, switch parts, truss wire, wire brushes, chemical hardware, perforated sheets, textile machinery, welding rod.
521 Phosphor bronze, 8% C	92.0 Cu, 8.0 Sn trace P	F, R, W	55–140	24–80	70–2	G–E	20	Good cold workability for blanking, drawing, forming and bending, shearing, stamping. Uses: generally for more severe service conditions than copper alloy No. 510.
524 Phosphor bronze, 10% D	90.0 Cu, 10.0 Sn, trace P	F, R, W	66–147	28 (annealed)	70–3	G–E	20	Good cold workability for blanking, forming and bending, shearing. Uses: heavy bars and plates for severe compression, bridge and expansion plates and fittings, articles requiring good spring qualities, resiliency, fatigue resistance, good wear and corrosion resistance.

† After Ref. 6.
‡ F, flat products; R, rod; W, wire; T, tube.
§ G: good; E: excellent.
¶ Based on 100% for copper alloy 360.

245

Table 6-6 Chemical compositions, mechanical properties, and typical applications of selected aluminum bronzes†

Name and number	Nominal composition, %	Commercial forms‡	Mechanical properties			Corrosion resistance§	Machinability rating¶	Fabricating characteristics and typical applications
			Tensile strength, 1000 psi	Yield strength, 1000 psi	Elongation in 2 in, %			
608 Aluminum bronze, 5%	95.0 Cu, 5.0 Al	T	60	27	55	G–E	20	Good cold workability; fair hot formability. Uses: condenser, evaporator and heat exchanger tubes, distiller tubes, ferrules.
610	92.0 Cu, 8.0 Al	R, W	70–80	30–55	65–25	G–E	20	Good hot and cold workability. Uses: bolts, pump parts, shafts, tie rods, overlay on steel for wearing surface.
613	92.65 Cu, 0.35Sn, 7.0 Al	F, R, T, P, S	70–85	30–58	42–35	G–E	30	Good hot and cold formability. Uses: nuts, bolts, stringers and threaded members, corrosion-resistant vessels and tanks, structural components, machine parts, condenser tube and piping systems, marine protective sheathing and fastening, munitions, mixing troughs and blending chambers.
614 Aluminum bronze, D	91.0 Cu, 7.0 Al, 2.0 Fe	F, R, W, T, P, S	76–89	33–60	45–32	G–E	20	Similar to copper alloy No. 613.
618	89.0 Cu, 1.0 Fe, 10.0 Al	R	80–85	39–42.5	28–23	G–E	40	Fabricated by hot forging and hot pressing. Uses: bushings, bearings, corrosion-resistant applications, welding rods.
619	86.5 Cu, 4.0 Fe, 9.5 Al	F	92–152	49–145	30–1	G–E	—	Excellent hot formability for fabricating by blanking, forming, bending, shearing, and stamping. Uses: springs, contacts, and switch components.
623	87.0 Cu, 3.0 Fe, 10.0 Al	F, R	75–98	35–52	35–22	G–E	50	Good hot and cold formability. Fabricated by bending, hot forging, hot pressing, forming, and welding. Uses: bearings, bushings, valve guides, gears, valve seats, nuts, bolts, pump rods, worm gears, and cams.

Table 6-6, Continued

Name and number	Nominal composition, %	Commercial forms‡	Mechanical properties			Corrosion resistance§	Machinability rating¶	Fabricating characteristics and typical applications
			Tensile strength, 1000 psi	Yield strength, 1000 psi	Elongation in 2 in, %			
624	86.0 Cu, 3.0 Fe, 11.0 Al	F, R	90–105	40–52	18–14	G–E	50	Excellent hot formability for fabrication by hot forging and hot bending. Uses: bushings, gears, cams, wear strips, nuts drift pins, tie rods.
625	82.7 Cu, 4.3 Fe, 13.0 Al	F, R	100	55	1	G–E	20	Excellent hot formability for fabrication by hot forging and machining. Uses: guide bushings, wear strips, cams, dies, forming rolls.
630	82.0 Cu, 3.0 Fe, 10.0 Al, 5.0 Ni	F, R	90–118	50–75	20–15	G–E	30	Good hot formability. Fabricated by hot forming and forging. Uses: nuts, bolts, valve seats, plunger tips, marine shafts, valve guides, aircraft parts, pump shafts, structural members.
632	82.0 Cu, 4.0 Fe, 9.0 Al, 5.0 Ni	F, R	90–105	45–53	25–20	G–E	30	Good hot formability. Fabricated by hot forming and welding. Uses: nuts, bolts, structural pump parts, shafting requiring corrosion resistance.
638	95.0 Cu, 2.8 Al, 1.8 Si, 0.40 Co	F	82–130	54–114	36–4	G–E	—	Excellent cold workability and hot formability. Uses: springs, switch parts, contacts, relay springs, glass sealing and porcelain enameling.
642	91.2 Cu, 7.0 Al, 1.8 Si	F, R	75–102	35–68	32–22	G–E	60	Excellent hot formability. Fabricated by hot forming, forging, machining. Uses: valve stems, gears, marine hardware, pole-line hardware, bolts, nuts, valve bodies and components.

† After Ref. 6.
‡ F, flat products; R, rod; W, wire; T, tube; P, pipe; S, shapes.
§ G: good; E: excellent.
¶ Based on 100% for copper alloy 360.

strengths, and are tough. They resist wear and fatigue and have excellent corrosion resistance due to the "self-healing" surface film of aluminum oxide. Table 6-6 lists the chemical compositions, mechanical properties, and typical applications for some selected aluminum bronzes.

Structure

Phase diagram The copper-rich end of the copper-aluminum phase diagram is shown in Fig. 6-30. The solid solubility of aluminum in copper extends to about 9.4 percent at 565°C. The solubility of aluminum increases considerably with a decrease in temperature along the $\alpha/(\alpha + \beta)$ boundary. Also the β phase decomposes by a eutectoid reaction into the $\alpha + \gamma_2$ phases at 565°C and 11.8% Al.

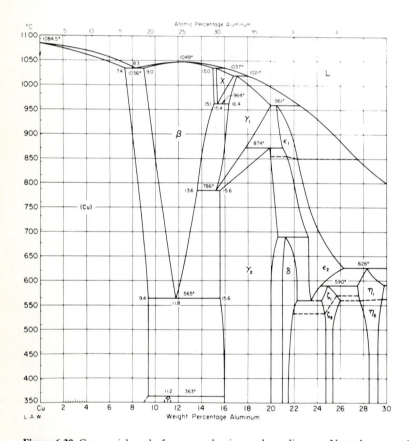

Figure 6-30 Copper-rich end of copper–aluminum phase diagram. Note the eutectoid reaction at 11.8% Al and 565°C whereby β reacts to form the $\alpha + \gamma_2$ phases under equilibrium cooling conditions. If the β phase (11.8% Al) is quenched rapidly from above 565°C, a martensite transformation occurs, resulting in the formation of a tetragonal β' structure. (*After Metals Handbook, 8th ed., vol. 8, American Society for Metals, 1973, p. 259.*)

Figure 6-31 Microstructure of a copper–5% aluminum bronze in the annealed condition; structure shows alpha grains with twin bands inside. (Etch: potassium dichromate; 75 × .) *(Courtesy of American Anaconda Brass Co.)*

Microstructure of the α aluminum bronzes α aluminum bronzes contain from 5 to 8% Al and consist of single-phase α solid solutions. The microstructure of Cu–5% Al bronze is shown in Fig. 6-31, and is similar to that of the α brasses. The α aluminum bronzes are strong and tough, and have good cold-working properties and corrosion resistance.

Microstructure and heat treatment of the complex aluminum bronzes When the aluminum content is above 8 percent in Cu-Al alloys and the temperature above 900°C, the β phase is introduced into the structure and produces alloys with duplex structures. Since there is an increase in solubility of α as the temperature decreases, the rate of cooling will markedly affect the structure obtained at room temperature. Also, when the aluminum composition increases above about 9.5 percent, the possibility of eutectoid decomposition is incurred (Fig. 6-30). If this type of alloy is rapidly quenched to room temperature, a martensitic transformation occurs (similar to the martensitic reaction in plain-carbon steels) whereby a metastable β' tetragonal structure is produced.

For example, consider a Cu–9.8% Al alloy cooled under different conditions. If the alloy is first heated to 900°C and held there 1 h, and then quenched to room temperature, the structure will be almost all β' martensite (Fig. 6-32a) and will have high strength and low ductility (Table 6-7). If the alloy is slowly cooled to 800°C or 650°C and then quenched to room temperature, less β' martensite will form (Fig. 6-32b and c) and the strength of the alloy will decrease and its ductility increase. However, if the alloy is slowly cooled to 500°C, which is below the eutectoid temperature, and then quenched to room temperature, the β phase will decompose by the eutectoid reaction into $\alpha + \gamma_2$ phases (Fig. 6-32d) as

$$\beta \rightleftharpoons \alpha + \gamma_2 \quad \text{(aluminum bronze pearlite)}$$

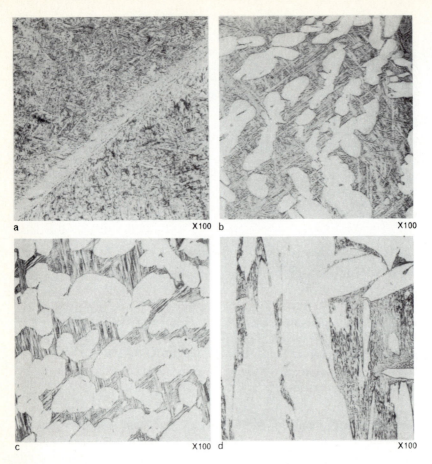

Figure 6-32 Binary alloy of Cu–9.8% Al. (*a*) Soaked 1 h at 900°C and quenched in water. (*b*) Slowly cooled to 800°C and quenched. (*c*) Slowly cooled to 650°C and quenched. (*d*) Slowly cooled to 500°C and quenched. (See Table 6-7 for associated properties.) *(After P. J. Macken and A. A. Smith, "The Aluminum Bronzes," United Kingdom Copper Development Association, 1966.)*

Table 6-7 Influence of variation in quenching temperature of Cu–9.8% Al alloy†

Heat treatment	0.1% Proof stress, ksi (kg/mm²)	Tensile strength, ksi (kg/mm²)	Elonga-tion, % on 2 in	Hardness, Bhn
Heated at 900°C and quenched	46.6 (32.8)	97.3 (68.4)	4	255
Heated at 900°C, slowly cooled to 800°C, and quenched	42.9 (30.2)	85.7 (60.3)	9	216
Heated at 900°C, slowly cooled to 650°C, and quenched	21.5 (15.1)	61.6 (43.3)	17	138
Heated at 900°C, slowly cooled to 500°C, and quenched	19.8 (13.9)	42.9 (30.2)	5	136

† After Ref. 8, p. 195.

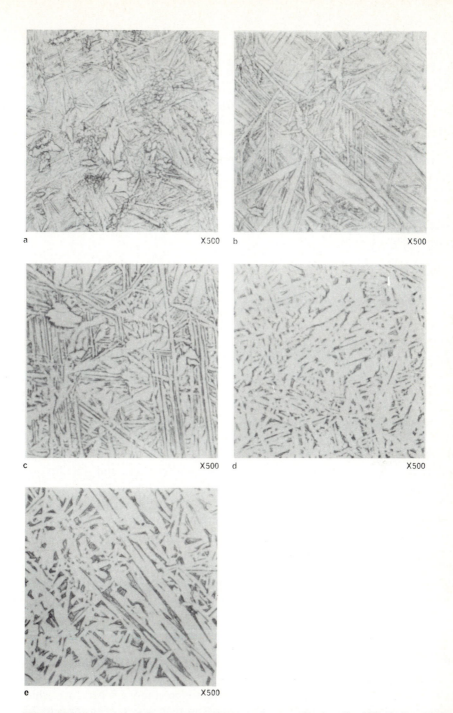

Figure 6-33 Formation of β martensite and tempered martensites in a Cu–10% Al alloy. (*a*) Soaked 1 h at 900°C and quenched. (*b*) Tempered 1 h at 400° C. (*c*) Tempered 1 h at 500°C. (*d*) Tempered 1 h at 600°C. (*e*) Tempered 1 h at 650°C. *(After P. J. Macken and A. A. Smith, "The Aluminum Bronzes," United Kingdom Copper Development Association, 1966.)*

Table 6-8 Mechanical properties of rod after quenching and tempering (Cu–9.4% Al)†

Heat treatment	0.1% Proof stress, ksi (kg/mm²)	Tensile strength, ksi (kg/mm²)	Elonga-tion, %	Hardness, HV‡
Heated 1 h at 900°C and quenched	28.1 (19.8)	109 (76.5)	29	187
Quenched from 900°C and tempered at 400°C for 1 h	30.7 (21.6)	107 (76.4)	29	185
Quenched from 900°C and tempered at 600°C for 1 h	34.5 (24.3)	102 (71.2)	34	168
Quenched from 900°C and tempered at 650°C for 1 h	32.3 (22.7)	93.6 (65.8)	48	150

† After Ref. 8, p. 198.
‡ Vickers hardness number.

The structure now will have low strength and ductility due to the presence of the brittle γ_2 phase. The γ_2 phase is thus avoided in commercial alloys because of its detrimental effect on ductility.

The most commonly used heat treatment of the 90 Cu–10% Al bronzes consists of quenching the alloy from 900°C or above, which results in an all β'-martensitic structure (Fig. 6-33a). The alloy is then tempered in the 400° to 650°C range to obtain the desired properties (Fig. 6-33b to e). Since the α phase is precipitated along crystallographic planes and results in a much finer precipitate than that obtained by continuous cooling, good strength and ductility are obtained (Table 6-8).

Aluminum bronzes containing up to about 10% Al and with additions of about 5% Fe and 5% Ni are exceptionally strong and tough, and have excellent corrosion and oxidation resistance at elevated temperatures. These alloys are used for high-strength bearings and gear wheels and can be used for dies for deep drawing some types of stainless steels. Table 6-6 lists the compositions, mechanical properties, and applications of these complex aluminum bronzes.

Mechanical Properties

Typical mechanical properties of selected aluminum bronzes are listed in Table 6-6. In general, the tensile strength of the α-aluminum bronzes increases linearly up to the maximum solubility of aluminum in copper (about 8 percent), while the elongation increases up to about 5 percent and then levels off in the 5 to 7.5 percent range (Fig. 6-34). As the amount of β phase increases, the tensile strength increases but the elongation drops off rapidly (Fig. 6-34).

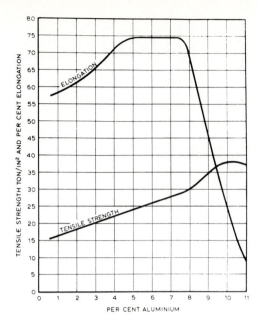

Figure 6-34 Effect of aluminum content on the mechanical properties of copper–aluminum bronzes. *(After P. J. Macken and A. A. Smith, "The Aluminum Bronzes," United Kingdom Copper Development Association, 1966.)*

6-7 COPPER-SILICON ALLOYS

Chemical Compositions and Typical Applications

Copper-silicon alloys are usually referred to as *silicon bronzes* or by their trade names such as Everdur or Herculoy. Most silicon bronzes contain between 1 to 3% Si. Small additions of manganese and iron are sometimes added to improve their properties. Table 6-9 lists the chemical compositions, mechanical properties, and typical applications for two of the most commonly used silicon bronzes.

Silicon bronzes find engineering application because of their resistance to corrosion and relatively high strength and toughness compared to low-carbon steels. For many uses they are low-cost substitutes for the tin bronzes since, except in regard to impingement attack, they have good sea-water corrosion resistance. These alloys can be cast or hot- or cold-worked.

Structure

Phase diagram The copper-rich end of the copper-silicon diagram is shown in Fig. 6-35. Silicon has a maximum solid solubility in copper of 5.3 percent at 843°C. Since this solubility decreases to only about 4 percent at room temperature, these alloys are not precipitation-hardenable.

Microstructure The microstructure of the silicon bronze alloy Everdur (96% Cu, 3% Si, 1% Mn) in the annealed condition is shown in Fig. 6-36. As in the case of

Table 6-9 Chemical compositions, mechanical properties, and typical applications of selected silicon bronzes†

Name and number	Nominal composition, %	Commercial forms‡	Mechanical properties			Corrosion resistance§	Machinability rating¶	Fabricating characteristics and typical applications
			Tensile strength, 1000 psi	Yield strength, 1000 psi	Elongation in 2 in, %			
651 Low-silicon bronze, B	98.5 Cu, 1.5 Si	R, W, T	40–95	15–69	55–11	G–E	30	Excellent hot and cold workability. Fabricated by forming and bending, heading and upsetting, hot forging and pressing, roll threading and knurling, squeezing and swaging. Uses: hydraulic pressure lines, anchor screws, bolts, cable clamps, cap screws, machine screws, marine hardware, nuts, pole-line hardware, rivets, U bolts, electrical conduits, heat exchanger tubing, welding rod.
655 High-silicon bronze, A	97.0 Cu, 3.0 Si	F, R, W, T	56–145	21–70	63–3	G–E	30	Excellent hot and cold workability. Fabricated by blanking, drawing, forming and bending, heading and upsetting, hot forging and pressing, roll threading and knurling, shearing, squeezing and swaging. Uses: similar to copper alloy No. 651, including propeller shafts.

† After Ref. 6.
‡ F, flat products; T, tube; W, wire; R, rod.
§ G: good; E: excellent.
¶ Based 0n 100% for copper alloy 360.

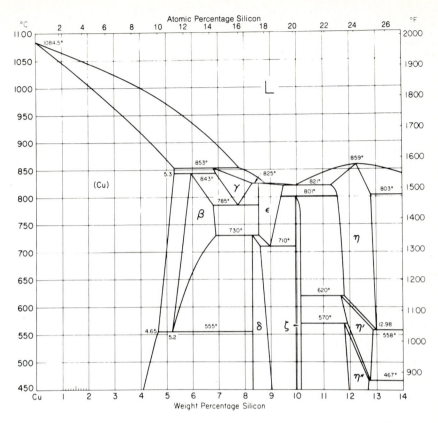

Figure 6-35 Copper-rich end of the copper–silicon phase diagram. *(After Metals Handbook, 8th ed., vol. 8, American Society for Metals, 1973, p. 298.)*

Figure 6-36 Microstructure of silicon bronze, Everdur (96% Cu, 3% Si, 1% Mn) in the annealed condition; structure shows alpha grains with twin bands inside. (75 × .) *(Courtesy of American Anaconda Brass Co.)*

the other α bronzes, the annealed microstructure consists of α grains with twin bands inside the grains.

Mechanical Properties

The mechanical properties of low- and high-silicon bronzes are listed in Table 6-9. The tensile strengths of these alloys vary from 40 to 56 ksi in the annealed condition. By severe cold working to produce the spring temper, their strength can be raised as high as 145 ksi. Since the solid solubility of silicon in copper does not decrease substantially, these alloys cannot be precipitation-hardened.

6-8 COPPER-BERYLLIUM ALLOYS

Chemical Compositions and Applications

Commercial copper-beryllium alloys contain between 0.6 to 2% Be with additions of cobalt from 0.2 to 2.5 percent. These alloys are precipitation-hardenable and can be heat-treated to produce tensile strengths as high as 212 ksi, which is the highest strength developed in commercial copper alloys. Table 6-10 lists the chemical compositions, mechanical properties, and typical applications of selected copper-beryllium alloys.

Copper-beryllium alloys are used for tools which require high-hardness and nonsparking characteristics such as may be needed in the chemical industry. The corrosion and fatigue resistance and strength of these alloys have made them useful for springs, gears, diaphragms, and valves. They are also used for electrical contacts and molds for forming plastics. Even though these alloys contain only a small amount of beryllium, their cost is relatively high and thus their use is only justified when other lower-cost alloys will not meet the engineering requirement of an application.

Structure

Phase diagram for Cu-Be system The maximum solid solubility of beryllium in copper is 2.7 percent, which occurs at 866°C as indicated in the Cu-Be phase diagram of Fig. 6-37. Cu-Be alloys with up to about 2% Be are precipitation-hardenable since there is a rapid decrease in solid solubility from 2.7 percent at 866°C to less than 0.5 percent at room temperature and since a coherent metastable precipitate forms during aging at lower temperatures.

Precipitation sequence and microstructure The general precipitation sequence in the Cu–2% Be system has been studied by x-ray and electron microscopic

Table 6-10 Chemical compositions, mechanical properties, and typical applications of selected copper-beryllium alloys†

Name and number	Nominal composition, %	Commercial forms‡	Tensile strength, 1000 psi	Yield strength, 1000 psi	Elongation in 2 in, %	Corrosion resistance§	Machinability rating¶	Fabricating characteristics and typical applications
170 Beryllium copper	98.1 Cu, 1.7 Be, 0.20 Co	F, R	70–190	32–170	45–3	G–E	20	Fabricating characteristics same as copper alloy No. 162. Commonly fabricated by blanking, forming and bending, turning, drilling, tapping. Uses: bellows, bourdon tubing, diaphragms, fuse clips, fasteners, lockwashers, spring, switch parts, roll pins, valves, welding equipment.
172 Beryllium copper	97.9 Cu, 1.9 Be, 0.20 Co	F, R, W, T, P, S	68–212	25–195	48–1	G–E	20	Similar to copper alloy No. 170, particularly for its nonsparking characteristics.
173 Beryllium copper	97.7 Cu, 1.9 Be, 0.40 Pb	R	68–200	25–182	48–3	G–E	50	Combines superior machinability with good fabricating characteristics of copper alloy No. 172.
175 Copper-cobalt-beryllium alloy	96.9 Cu, 2.5 Co, 0.6 Be	F, R	45–115	25–110	28–5	G–E	—	Fabricating characteristics same as copper alloy No. 162. Uses: fuse clips, fasteners, springs, switch and relay parts, electrical conductors, welding equipment.
182, (184, 185) Chromium copper	99.1 Cu, 0.9 Cr	F, W, R, S, T	34–86	14–77	40–5	G–E	20	Excellent cold workability; good hot workability. Uses: resistance welding electrodes, seam welding steels, switch gear, electrode holder jaws, cable connectors, current-carrying arms and shafts, circuit breaker parts, molds, spot welding tips, flash welding electrodes, electrical and thermal conductors requiring strength, switch contacts.

† After Ref. 6.
‡ F, flat products; R, rod; W, wire; T, tube; P, pipe; S, shapes.
§ G: good; E: excellent.
¶ Based on 100% for copper alloy 360.

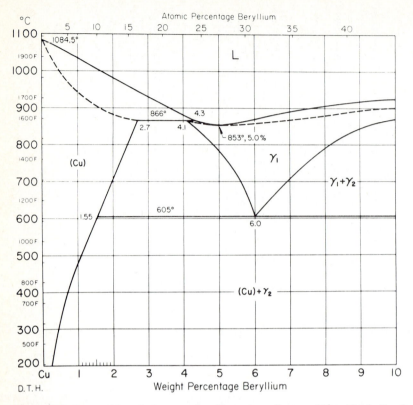

Figure 6-37 Copper-rich end of copper–beryllium phase diagram. *(After Metals Handbook, 8th ed., vol. 8, American Society for Metals, 1973, p. 271.)*

methods and shown to be

$$\underset{\text{solid solution}}{\text{Supersaturated}} \rightarrow \underset{\text{(flat plates)}}{\text{GP zones}} \rightarrow \underset{\substack{\text{(rods, plates)} \\ \text{BCT with} \\ a = b = 2.79 \text{ Å,} \\ c = 2.54 \text{ Å}}}{\gamma'} \rightarrow \underset{\substack{\text{CuBe (ordered),} \\ \text{BCC,} \\ a = 2.70 \text{ Å}}}{\gamma}$$

The GP zones in Cu–2% Be alloys are monolayer plates that form coherently on the {100} matrix planes. These zones can vary in size depending on the time and temperature of aging. Zones of 10 to 30 Å in diameter and 2 to 3 Å thick are produced after 100 h of aging at 100°C, and up to 70 Å in diameter and 1 to 3 atom planes thick after 1 h at 198°C. The GP zones formed in a Cu–2% Be alloy by aging 1 h at 198°C are shown in Fig. 6-38.

Further aging produces the γ' intermediate partially coherent precipitate, which nucleates on the GP zones when they are present. Above the GP zone solvus, which is about 320°C, γ' nucleates heterogeneously. The γ' phase, which is BCT ($a = b = 2.79$ Å, and $c = 2.54$ Å), is shown in Fig. 6-39. Both γ' and the

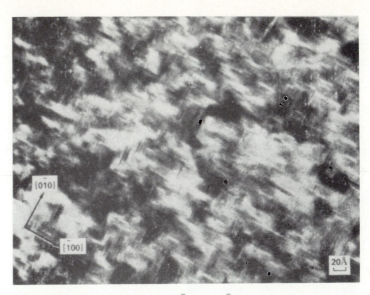

Figure 6-38 GP zones parallel to (0$\bar{1}$0) and ($\bar{1}$00) matrix planes in a Cu–2% Be alloy solution heat-treated at 800°C, quenched and aged 1 h at 198°C. [*After V. A. Phillips and L. E. Tanner, Acta Met. 21(1973):441.*]

Figure 6-39 Cu–1.87% Be alloy solution heat-treated at 800°C, quenched, and aged 4 h at 350°C. Structure shows the intermediate ordered (γ') CuBe phase. *(After W. K. Armitage et al., 5th Int. Congr. Electron Microscopy, vol. 1. p. K4, Academic Press, New York, 1962.)*

Figure 6-40 Cu–1.87% Be alloy solution heat-treated at 800°C, quenched, and aged 16 h at 400°C. Structure shows an eutectoid-type precipitation of ordered CuBe γ phase in a disordered α-matrix. *(After W. K. Armitage et al., "5th Int. Congr. Electron Microscopy," vol. 1, p. K-4, Academic Press, New York, 1962.)*

GP zones are present when the γ' is first formed. The formation of the γ' is associated with softening of the alloy during aging.

Increasing the aging temperature to above 380°C produced the equilibrium ordered BCC phase CuBe, which grows by a discontinuous phase transformation to give an eutectoid-type structure (Fig. 6-40). The discontinuous γ precipitate is nucleated at grain boundaries and gradually spreads throughout the grains so that, after 16 h at 400°C, the whole microstructure is of the eutectoid type. The γ phase is associated with further overaging and decrease in hardness as its amount increases.

Mechanical Properties

The mechanical properties of selected copper-beryllium alloys are listed in Table 6-10. These alloys are usually solution-heat-treated at about 800°C, quenched in water, and precipitation-hardened between 250 and 330°C. Cold working these alloys before aging greatly increases their strength, as can be seen in Fig. 6-41. The explanation given for the effect of cold work is that it provides an increased defect concentration for the formation of GP zones, and hence leads to a higher density of GP zones formed (Ref. 19, p. 415). By combining cold work and precipitation hardening, tensile strengths of over 200 ksi can be attained.

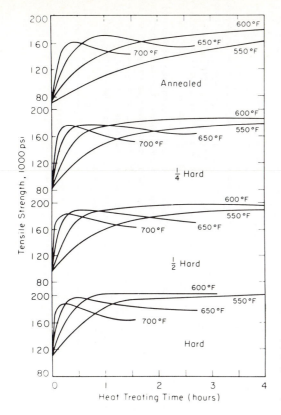

Figure 6-41 Aging curves for a Cu–2% Be–0.2% Co alloy aged at different temperatures after 0, 11, 21, and 27 percent reduction in thickness. (650°F = 343°C; 600°F = 315°C; 550°F = 288°C.) *(After "Precipitation from Solid Solution," American Society for Metals, 1959, p. 348.)*

6-9 COPPER-NICKEL ALLOYS

Chemical Compositions and Applications

Nickel is added to copper to form a series of solid-solution alloys of approximately 10, 20, and 30% Ni called *cupronickels*. Table 6-11 lists the chemical compositions, mechanical properties, and typical applications of some of these alloys. The nickel additions increase the strength, oxidation, and corrosion resistance of copper. The cupronickels are used for marine condensers and tubing for conducting sea water because of their moderately high to high strength and resistance to the corrosive and erosive effects of high-velocity sea water. Since the cupronickels do not work-harden rapidly, they are used for condenser tubes and plates, heat exchangers, and a wide variety of chemical process equipment.

Structure

Phase diagram Copper and nickel are completely soluble in all proportions in the solid state, as shown in the copper-nickel phase diagram of Fig. 6-42.

Table 6-11 Chemical compositions, mechanical properties, and typical applications of selected copper-nickel alloys†

Name and number	Nominal composition, %	Commercial forms‡	Mechanical properties			Corrosion resistance§	Machinability rating¶	Fabricating characteristics and typical applications
			Tensile strength, 1000 psi	Yield strength, 1000 psi	Elongation in 2 in, %			
706 Copper nickel, 10%	88.7 Cu, 1.3 Fe, 10.0 Ni	F, T	44–60	16–57	42–10	E	20	Good hot and cold workability. Fabricated by forming and bending, welding. Uses: condensers, condenser plates, distiller tubing, evaporator and heat exchanger tubing, ferrules, salt-water piping.
710 Copper nickel, 20%	79.0 Cu, 21.0 Ni	F, W, T	49–95	13–85	40–3	E	20	Good hot and cold formability. Fabricated by blanking, forming and bending, welding. Uses: communication relays, condensers, condenser plates, electrical springs, evaporator and heat exchanger tubes, ferrules, resistors.
715 Copper nickel, 30%	70.0 Cu, 30.0 Ni	F, R, T	54–75	20–70	45–15	E	20	Similar to copper alloy No. 706.
717	67.8 Cu, 0.7 Fe, 31.0 Ni, 0.5 Be	F, R, W	70–200	30–180	40–4	G–E	20	Good hot and cold formability. Uses: high-strength constructional parts for sea-water corrosion resistance, hydrophone cases, mooring cable wire, springs, retainer rings, bolts, screws, pins for ocean telephone cable applications.
725	88.2 Cu, 9.5 Ni, 2.3 Sn	F, R, W, T	55–120	22–108	35–1	E	20	Excellent cold and hot formability. Fabricated by blanking, brazing, coining, drawing, etching, forming and bending, heading and upsetting, roll threading and knurling, shearing, spinning, squeezing, stamping and swaging. Uses: relay and switch springs, connectors, brazing alloy, lead frames, control and sensing bellows.

† After Ref. 6.
‡ R, rod; W, wire; T, tube; F, flat products.
§ E: excellent; G: good.
¶ Based on 100% for copper alloy 360.

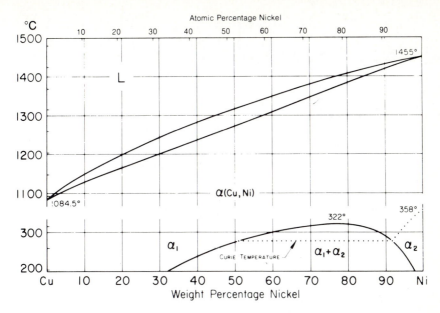

Figure 6-42 Copper–nickel phase diagram. *(After Metals Handbook, 8th ed., vol. 7, American Society for Metals, 1972, p. 294.)*

Microstructure The microstructure of the cupronickels consists of α-phase solid solutions. This is exemplified by the recrystallized α-phase grains in the 70% Cu–30% Ni cupronickel shown in Fig. 6-43.

Mechanical and Electrical Properties

The mechanical properties of the cupronickels are listed in Table 6-11. Nickel and copper both have the FCC structure, and so Cu-Ni alloys are ductile

Figure 6-43 The microstructure of the cupronickel 70% Cu–30% Ni. Structure shows recrystallized alpha grains with twin bands inside. 150 × .) *(Courtesy of American Anaconda Brass Co.)*

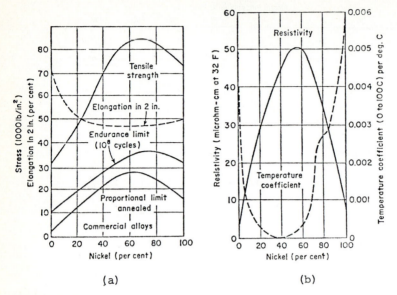

Figure 6-44 Effect of nickel on (*a*) the mechanical properties and (*b*) the electrical properties of copper–nickel alloys. *(After F. T. Sisco, "Modern Metallurgy for Engineers," 2d ed., Pitman Publishing Co., 1948, as presented in Physical Metallurgy for Engineers, 2d ed. by D. S. Clark and W. R. Varney © 1962 by Litton Educational Publishing, Inc. Reprinted by permission of D. Van Nostrand Co.)*

through their whole composition range. Figure 6-44*a* shows how additions of nickel increase the tensile strength of copper by solid-solution strengthening while maintaining high ductility.

Nickel greatly increases the electrical resistivity of copper, as shown in Fig. 6-44*b*. An alloy of 55% Cu–45% Ni has very high resistance but an extremely low temperature coefficient of resistivity. That is, the electrical resistivity changes very little with temperature. Thus, this alloy is useful for wire-wound resistances for electrical instruments.

6-10 COPPER-NICKEL-ZINC ALLOYS (NICKEL SILVERS)

Chemical Compositions and Applications

The nickel-silver alloys are essentially ternary copper-nickel-zinc alloys and do not contain silver. The misnomer arises from the silver color of these alloys rather than from any silver content. The zinc content of the nickel silvers ranges from about 17 to 27% Zn, while their nickel content varies from about 8 to 18 percent. As the nickel content is increased, the color of the nickel silvers varies from soft ivory to silvery white. Table 6-12 lists the chemical compositions, mechanical properties, and typical applications of some selected nickel silvers.

Table 6-12 Chemical compositions, mechanical properties, and typical applications of selected copper-nickel-zinc alloys (nickel-silvers)†

Name and number	Nominal composition, %	Commercial forms‡	Mechanical properties			Corrosion resistance§	Machinability rating¶	Fabricating characteristics and typical applications
			Tensile strength, 1000 psi	Yield strength, 1000 psi	Elongation in 2 in, %			
745 Nickel silver, 65-10	65.0 Cu, 25.0 Zn, 10.0 Ni	F, W	49–130	18–76	50–1	E	20	Excellent cold workability. Fabricated by blanking, drawing, etching, forming and bending, heading and upsetting, roll threading and knurling, shearing, spinning, squeezing and swaging. Uses: rivets, screws, slide fasteners, optical parts, etching stock, hollow ware, nameplates, platers' bars.
752 Nickel silver, 65-18	65.0 Cu, 17.0 Zn, 18.0 Ni	F, R, W	56–103	25–90	45–3	E	20	Fabricating characteristics similar to copper alloy No. 745. Uses: rivets, screws, table flatware, truss wire, zippers, bows, camera parts, core bars, temples, base for silver plate, costume jewelry, etching stock, hollow ware, nameplates, radio dials.
754 Nickel silver, 65-15	65.0 Cu, 20.0 Zn, 15.0 Ni	F	53–92	18–79	43–2	E	20	Fabricating characteristics similar to copper alloy No. 745. Uses: camera parts, optical equipment, etching stock, jewelry.
757 Nickel silver, 65-12	65.0 Cu, 23.0 Zn, 12.0 Ni	F, W	52–93	18–79	48–2	E	20	Fabricating characteristics similar to copper alloy No. 745. Uses: slide fasteners, camera parts, optical parts, etching stock, nameplates.
770 Nickel silver, 55-18	55.0 Cu, 27.0 Zn, 18.0 Ni	F, R, W	60–145	27–90	40–2	E	30	Good cold workability. Fabricated by blanking, forming and bending, and shearing. Uses: optical goods, springs and resistance wire.
782 Leaded nickel silver, 65-8-2	65.0 Cu, 2.0 Pb, 25.0 Zn, 8.0 Ni	F	53–91	23–76	40–3	E	60	Good cold formability. Fabricated by blanking, milling and drilling. Uses: key blanks, watch plates, watch parts.

† After Ref. 6.
‡ F, flat products; W, wire; R, rod.
§ E: excellent.
¶ Based on 100% for copper alloy 360.

60 g FeCl₃ + 20 g Fe(NO₃)₃ + 2000 ml H₂O   100 ×

Figure 6-45 Nickel silver alloy CDA 745 (65% Cu–10% Ni–25% Zn) cold-rolled sheet, 0.100-in thick, annealed at 650 to 700°C.; longitudinal section. The structure consists of equiaxed recrystallized grains of solid solution alpha which contain twin bands. *(After Metals Handbook, 8th ed., vol. 7, American Society for Metals, 1972, p. 283.)*

Structure

Most nickel-silver alloys are single-phase solid solutions and have structures similar to the α brasses. Figure 6-45 shows the recrystallized microstructure of a 65% Cu–10% Ni–25% Zn nickel silver which consists of solid-solution α. The structure closely resembles that of cartridge brass (70% Cu–30% Zn) because of the large number of internal twins.

Mechanical and Corrosion Properties

The mechanical properties of some selected nickel-silver alloys are listed in Table 6-12. The single-phase nickel silvers have medium to high strengths and good cold workability, but only fair hot workability. Because of their good cold formability, they are used widely in cold-forming operations and for articles which require a smooth surface for plating. Lead is sometimes added to these alloys to improve their machinability, but it reduces their ductility.

   The corrosion resistance of these alloys is in general considered excellent since the presence of nickel improves their corrosion properties (Table 6-12). However, their high zinc content makes a number of them subject to dezincification under some conditions. If the nickel silvers are highly stressed during fabrication, they should be stress-relieved to prevent stress-corrosion cracking.

PROBLEMS

1. Describe the processing steps necessary to produce electrolytic tough-pitch (ETP) copper from copper-sulfide concentrates.

2. How are wrought and cast coppers and copper alloys classified by the Copper Development Association?

3. What properties make unalloyed copper a useful engineering material?

4. What is the chemical composition of ETP copper?

5. What do cast ingots of ETP copper and rimmed low-carbon steel have in common?

6. Where is most of the oxygen located in ETP copper cast ingots? What happens to the Cu_2O constituents when ingots of ETP copper are rolled into sheet?

7. Why cannot ETP copper be used for joining processes which involve temperatures above 400°C? Write the chemical equation for the reaction that takes place when hydrogen enters ETP copper which is heated above 400°C.

8. How is oxygen-free high-purity copper produced from ETP copper?

9. What are the properties that make the α brasses attractive engineering materials?

10. Describe the crystal structure unit cells for (a) α brass, (b) disordered β brass, β, and (c) ordered β brass, β'.

11. Why are copper-zinc alloys containing the γ phase of little engineering use?

12. Describe the microstructure of annealed α brasses as observed in the optical microscope at about 100 × .

13. How does the dislocation distribution differ in (a) copper and (b) a 63% Cu–37% Zn brass after 10 percent cold deformation? What causes this dislocation distribution to change?

14. Why is it difficult to cold-work Muntz metal (60% Cu–40% Zn)?

15. What types of transformations occur when a 58% Cu–42% Zn alloy is heated to about 830°C, hot-quenched in the 600 to 250°C range, and isothermally transformed?

16. Why are lead additions of about 0.5 to 3 percent made to brasses? How is the lead distributed in the brass?

17. What alloying elements are used to make the alloy brasses?

18. Under what conditions are the α brasses with high zinc contents susceptible to stress-corrosion cracking?

19. How can the tendency to stress-corrosion cracking be reduced?

20. What is the dezincification of brasses? What is believed to be the mechanism for this type of corrosion behavior?

21. Why is phosphorus added to the tin bronzes? What desirable engineering properties do the tin bronzes have? What is the chief disadvantage of these alloys when compared to the brasses?

22. What properties do the aluminum bronzes have that make them useful engineering alloys? What special property does aluminum give them?

23. In what way is the Cu-Al phase diagram at 11.8% Al similar to the Fe-Fe_3C diagram at 0.8% C?

24. How can a martensitic transformation be produced in a Cu–9.8% Al alloy?

25. What properties make copper-silicon bronzes useful for engineering alloys? Why are Cu-Si bronzes substituted for Cu-Sn bronzes for some applications?

26. Are copper-silicon bronzes precipitation-hardenable? Explain.

27. What properties make copper-beryllium alloys useful engineering alloys? What is their main disadvantage?

28. Why are copper-beryllium alloys with up to 2.5% Be precipitation-hardenable?

29. How are strengths of about 200 ksi attained in a Cu–2% Be–0.2% Co alloy?

30. What are the cupronickels?

31. What effect does 10 to 30% Ni have on the mechanical strength and electrical conductivity of copper?

32. What are some of the applications for the cupronickels?

33. What are the nickel-silver alloys? Why is the word "silver" a misnomer for these alloys?

34. What are the properties of the nickel-silver alloys that make them useful engineering alloys?

35. When a Cu–10 wt% Al alloy is quenched to room temperature from 920°C in the β-phase region (Fig. 6-30) an all β' martensitic structure is produced. Upon tempering this alloy at 356°C, an 80 percent increase in 0.2 percent yield strength is observed. That is, the alloy in the quenched and tempered condition is 80 percent stronger than in the quenched condition. How can this great increase during the first 5 min of tempering be explained? [See A. A. Hussein et al., *Met. Trans.* 9A(1978):1783.]

36. Precipitation hardening of a Cu–1.81 wt% Be–0.28 wt% Co alloy is affected by cold work (50 percent) before aging. When this alloy is solution-heat-treated, quenched, cold-worked 50 percent, and aged at 175°C, continuous precipitation is retarded as compared to the alloy without cold work before aging. When the alloy is solution-heat-treated, quenched, cold-worked 50 percent, and aged at 315°C, continuous precipitation is accelerated in the metal which is cold-worked as compared to the metal without cold work. How can this difference in precipitation behavior be explained? [See W. Bonfield and B. C. Edwards, *J. Mat. Sci.* 9(1974):415.]

37. Although commercial bronze (Cu–5 wt% Sn) is not considered an age-hardening alloy, increases in yield strength during aging have been noted. However, the age hardening occurs only after severe rolling (90 to 97 percent reduction) and not with < 70 percent reduction. What effect could explain this difference? [See T. C. Tiscone et al., *Met. Trans.* 1(1970):2010.]

REFERENCES

1. A. Butts (ed.): "Copper," Reinhold Publishing Co., City, 1954.
2. E. R. Petty: "Physical Metallurgy of Engineering Materials," American Elsevier, New York, 1968.
3. R. M. Brick, A. W. Pense, and R. B. Gordon: "Structure and Properties of Engineering Materials," McGraw-Hill, New York, 1977.
4. P. G. Shewmon: "Transformations in Metals," McGraw-Hill, New York, 1969.
5. *Metals Handbook*, vol. 1: "Copper and Copper Alloys," American Society for Metals, Metals Park, OH, 1961, pp. 960–1006.
6. "ASM Databook," American Society for Metals, Metals Park, OH, 1975, pp. 96–100.
7. "Standards Handbook," Parts 1 through 7, Copper Development Association, Inc., New York.
8. P. J. Macken and A. A. Smith: "The Aluminum Bronzes," Copper Development Association, London, 1966.
9. "The Nickel Silvers," Copper Development Association, London, 1965.
10. H. H. Uhlig: "Corrosion and Corrosion Control," 2d ed., John Wiley, New York, 1971.
11. D. J. Mack: "The Isothermal Transformation of a Eutectoid Aluminum Bronze," *Trans. AIME* 175(1948):240.
12. W. R. Hibbard: "The Kappa Eutectoid Transformation in the Copper-Silicon System," 180(1949):92.
13. E. N. Pugh, W. G. Montague, and A. R. C. Westwood: "On the Role of Complex Ions in the Season Cracking of Alpha Brass," *Trans. ASM* 58(1965):665.
14. H. S. Campbell: "Stress Corrosion Cracking of Copper Alloys," *J. Inst. Metals* 101(1973):232.
15. S. Murphy and C. J. Ball: "The Recrystallization of Tough-Pitch Copper," *J. Inst. Metals* 100(1972):225.
16. T. C. Tiscone, G. Y. Chin, and B. C. Wonsiewicz: "Precipitation in Rolled Phosphor Bronze," *Met. Trans.* 1(1970):2011.
17. J. Hedworth and G. Pollard: "Influence of Stacking Fault Energy on the Creep Behavior of α-Brasses," *Met. Sci. J.* 5(1971):41.
18. V. A. Phillips and L. E. Tanner: "G.P. Zones in Aged Cu-1.97% Be Crystal," *Acta Met.* 21(1973):441.
19. W. Bonfield and B. C. Edwards: "Precipitation Hardening in Cu-1.81% Be-0.28% Co Alloy," Parts 1, 2, and 3, pp. 398–422, *J. Mater. Sci.* 9(1974):398.
20. P. Wilkes and M. M. Jackson: "An Electron-Microscope Study of Precipitation in Cu-Be Alloys," *Met. Sc. J.* 3(1969):130.

21. G. Thomas: "Dislocation Substructures, Stacking Fault Energies and Yield Stresses of Alpha Brasses," *J. Aust. Inst. Metals* 8(1963):80.
22. P. R. Swann and J. Nutting: "The Influence of Stacking-Fault Energy on the Modes of Deformation of Polycrystalline Copper Alloys," *J. Inst. Metals* 50(1961–62):133.
23. P. E. Flewitt and J. M. Towner: "The Decomposition of Beta Prime in Copper-Zinc Alloys," *J. Inst. Metals* 95(1967):273.
24. G. R. Purdy: "Widmanstätten Precipitation from Non-Ideal Solid Solutions: α in β-Cu-Zn," *Met. Sci. J.* 5(1971):81.

SEVEN

STAINLESS STEELS

Stainless steels are selected as engineering materials mainly because of their excellent corrosion resistance, which is principally due to their high chromium contents. Small amounts of chromium, for example about 5 percent, add some corrosion resistance to iron, but in order to make a stainless steel "stainless," at least 12% Cr in iron is required. According to classical theories, chromium makes the iron surface "passive" by forming a surface oxide film which protects the underlying metal from corrosion. In order to produce this protective film, the stainless steel surface must be in contact with oxidizing agents.

The addition of nickel to stainless steels improves their corrosion resistance in neutral or weakly oxidizing media but adds to their cost. Nickel in sufficient amounts also improves their ductility and formability by making it possible for the austenitic (FCC) structure to be retained at room temperature. Molybdenum, when added to stainless steels, improves corrosion resistance in the presence of chloride ions, whereas aluminum improves high-temperature scaling resistance. The effects of these and other alloying elements are discussed in following sections.

In this chapter, the important alloy systems for stainless steels are treated first. Then, after a summary of the four main classes of stainless steels, the metallurgical structure and properties of each of these classes are discussed.

7-1 IRON-CHROMIUM ALLOYS

Since chromium is the basic alloying addition for all stainless steels, we consider first the iron-chromium binary phase diagram which is shown in Fig. 7-1. Two

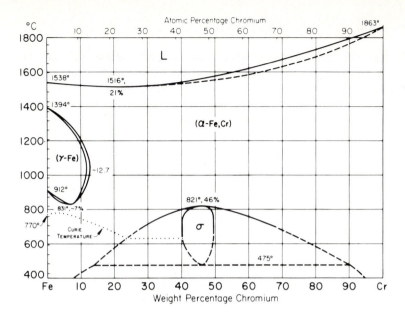

Figure 7-1 Iron–chromium phase diagram. *(After Metals Handbook, 8th ed., vol. 8, American Society for Metals, 1973, p. 291.)*

important features of this phase diagram are the γ *loop* and the presence of the σ *phase*.

Formation of the γ Loop

Chromium, since it has the same BCC structure as α ferrite, acts as a ferrite stabilizer and extends the α-phase field, while suppressing the γ-phase field. As a result, the "γ loop" is formed, which divides the iron-chromium diagram into FCC and BCC regions. Iron-chromium alloys with less than 12 to 13% Cr undergo an austenite-to-ferrite transformation on cooling from temperatures within the γ loop. Iron-chromium alloys with more than 12 to 13% Cr do not undergo the FCC to BCC transformation, and on cooling from high temperatures remain as solid solutions of chromium in α iron.

Formation of the σ Phase

The iron-chromium phase diagram at low temperatures is not a complete range of solid solutions; an intermediate phase called the "σ phase" forms below 821°C, centered at about 46% Cr (Fig. 7-1). The σ phase has a tetragonal crystal structure and is hard and brittle. It can be a source of difficulty in engineering alloys since its presence can lead to structures which are brittle or which possess variable mechanical properties. More will be said of the σ phase in Sec. 7-5, under the heading "Embrittlement Mechanisms."

7-2 IRON-CHROMIUM-CARBON ALLOYS

Carbon is an austenitic stabilizer and, when added to Fe-Cr alloys, enlarges the austenitic phase field. Figure 7-2 shows the effect of increasing the carbon content from 0.05 to 0.4 wt% on enlarging the austenitic phase field in Fe-Cr alloys. The austenitic phase boundary increases to a maximum of 18% Cr with 0.6% C. A further increase in carbon beyond 0.6 percent leads to the formation of free carbides. The general sequence of carbide formation in Fe-Cr alloys is probably

$$(Fe,Cr)_3C \rightarrow (Cr,Fe)_7C_3 \rightarrow (Cr,Fe)_{23}C_6$$

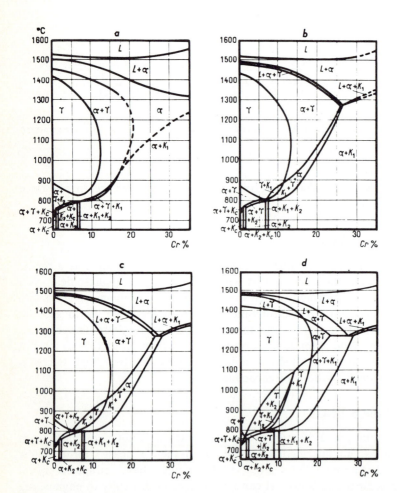

Figure 7-2 Iron–chromium phase diagrams for different carbon contents: (*a*) 0.05% C; (*b*) 0.1% C; (*c*) 0.2% C; (*d*) 0.4% C. [*After K. Bungardt, E. Kunze, and E. Horn, Archiv. Eisenhuttenw. 29(1958):193, as presented in L. Colombier and J. Hochmann, "Stainless and Heat Resisting Steels," Edward Arnold, Ltd. London, 1967, p. 11.*]

The cementite-type carbide $(Cr,Fe)_3C$ forms in alloys with up to 10% Cr and can contain up to 15% Cr. With higher percentages of Cr, the carbide $(Cr,Fe)_7C_3$ forms and has a minimum of 36% Cr. With even higher Cr/C ratios, the $(Cr,Fe)_7C_3$ carbide transforms into $(Cr,Fe)_{23}C_6$. The $(Cr,Fe)_{23}C_6$ carbide generally precipitates in the grain boundaries of some stainless steels heat-treated under certain conditions, while the $(Cr,Fe)_7C_3$ is dispersed with the grains. The phase fields where these carbides exist in the Fe-Cr diagrams for carbon contents of 0.05, 0.1, 0.2, and 0.4 percent are shown in Fig. 7-2. In these diagrams κ_c, κ_1, and κ_2 are the carbides $(Cr,Fe)_3C$, $(Cr,Fe)_{23}C_6$, and $(Cr,Fe)_7C_3$, respectively.

7-3 IRON-CHROMIUM-NICKEL-CARBON ALLOYS

When nickel is added to iron, it stabilizes the austenitic phase since nickel has the same FCC crystal structure as austenite. Nickel is therefore an austenitic stabilizer in iron and counteracts the opposing ferrite-forming effect of chromium in stainless steels. If sufficient nickel is added to low-carbon stainless steels, the austenitic structure can be produced at room temperature.

Figure 7-3a and b shows phase diagrams for Fe–18% Cr alloys with 4 and 8% Ni, respectively. At 4% Ni (Fig. 7-3a), the δ-ferrite zone is displaced to higher temperatures, and an austenitic structure can be produced upon cooling

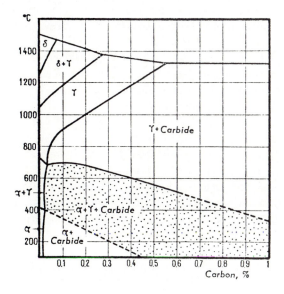

(a)

Figure 7-3 Phase diagrams for Fe–18% Cr-C alloys containing (a) 4% Ni and (b) 8% Ni. *(From L. Colombier and J. Hochmann, "Stainless and Heat Resisting Steels," Edward Arnold Ltd., London, 1967, p. 16.)*

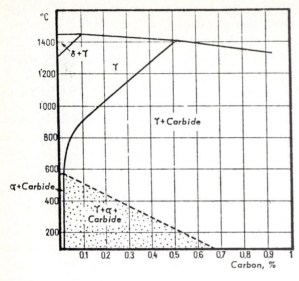

(b)

Figure 7-3 Continued

to room temperature. For example, a steel containing 0.2% C, 19.8% Cr, and 4.4% Ni, after quenching from 1100°C, can produce an austenitic structure. However, it is unstable and can be readily transformed by tempering at 650°C or by cold working. Upon increasing the nickel content of Fe–18% Cr alloys to 8 percent, the δ + γ phase field is restricted to very high temperatures and low carbon contents while, in contrast, the austenitic field is extended. Austenite is stabilized by nickel, and with 8% Ni a stable austenitic structure is produced in an Fe–18% Cr–8% Ni alloy at room temperature.

The solubility of carbon in austenite for the Fe–18% Cr–8% Ni alloy decreases rapidly with decreasing temperature (Fig. 7-3b). By rapidly quenching an Fe–18% Cr–8% Ni alloy containing about 0.08% C from temperatures above about 1000°C, the carbon is retained in solid solution. However, if this alloy is slow-cooled from the austenitic region, the rejected carbon will combine with chromium and iron to form carbides. This reaction takes place principally at the grain boundaries, where atomic diffusion is more rapid. This problem with austenitic stainless steels is discussed in Sec. 7-7.

7-4 CLASSIFICATION OF WROUGHT STAINLESS STEELS

On the basis of compositional and structural differences, wrought stainless steels are divided into the following four main groups:

1. *Ferritic stainless steels.* These alloys normally contain from 11 to 30% Cr, with their carbon contents kept below about 0.12 percent. Other alloying elements

are added in relatively small amounts to improve their corrosion resistance or other special properties such as machinability. Ferritic stainless steels, because of their low carbon contents, do not normally undergo the austenite-to-ferrite transformation and are therefore not considered heat-treatable. However, small amounts of carbon in many ferritic stainless steels produce some hardening if these steels are quenched from high temperatures. For weldability, improved ductility, and good corrosion resistance, the carbon and nitrogen levels in these alloys must be kept extremely low.

2. *Martensitic stainless steels.* These alloys contain from 12 to 17% Cr, with 0.1 to 1% C. They can be hardened by heat treatment to form martensite in the same way as can plain-carbon steels. Very high hardnesses are obtained if their carbon content is about 1 percent and the proper heat treatment is applied. Small amounts of other elements are added to improve corrosion resistance, strength, and toughness.

3. *Austenitic stainless steels.* These alloys are essentially ternary alloys containing from 6 to 22% Ni. Like the ferritic stainless steels, they cannot be hardened by heat treatment. However, they usually retain an austenitic structure at room temperature, are more ductile, and normally have better corrosion resistance than the ferritic stainless steels. In order to avoid intergranular corrosion, many of the austenitic stainless steels have to be specially heat-treated or have their chemical compositions modified.

4. *Precipitation-hardening stainless steels.* These alloys usually contain from 10 to 30% Cr, along with varying amounts of nickel and molybdenum. Precipitation-hardening phases are formed by additions of Cu, Al, Ti, and Cb. These alloys have high mechanical strengths, without significant loss of corrosion resistance for many applications. Even at high temperatures many of these alloys possess good strength properties.

7-5 FERRITIC STAINLESS STEELS

Chemical Compositions and Typical Applications

The ferritic stainless steels are essentially iron-chromium alloys containing 12 to 30% Cr. These alloys are called *ferritic* since their structure remains mostly ferritic (BCC α-iron type) at normal heat treatment conditions. Table 7-1*a* lists the chemical compositions and applications of some of the prominent standard ferritic stainless steels. These alloys are used mainly as general construction materials, in which their special corrosion- and heat-resistant properties are required. The ferritic stainless steels are of interest to the design engineer because they provide about the same corrosion resistance as the nickel-containing stainless steels but at a lower cost since no nickel is needed as an alloying element. However, the ferritic stainless steels have had more restrictive use than the austenitic steels because of their lack of ductility, their notch sensitivity, and their poor weldability. To overcome the ductility problem of the standard ferritic stainless steels, new ferritics with very low carbon and nitrogen contents have

Table 7-1a   Chemical compositions and typical applications of wrought ferritic stainless steels†

AISI type	Chemical composition wt%					Typical applications
	Cr	C (max)	Mo	Al	Other‡	
405	13.0	0.08		0.20		Nonhardenable grade for assemblies where air-hardening types such as 410 or 403 are objectionable. Annealing boxes; quenching racks; oxidation-resistant partitions.
409	11.0	0.08			Ti 6 × C	General-purpose construction stainless, Automotive exhaust systems; transformer and capacitor cases; dry fertilizer spreaders; tanks for agricultural sprays.
430	17.0	0.12				General-purpose nonhardenable chromium type. Decorative trim, nitric acid tanks; annealing baskets; combustion chambers; dishwashers; heaters; mufflers; range hoods; recuperators; restaurant equipment.
434	17.0	0.12	1.0			Modification of type 430 designed to resist atmospheric corrosion in the presence of winter road-conditioning and dust-laying compounds. Automotive trim and fasteners.
436	17.0	0.12	1.0		Cb 5 × C	Similar to types 430 and 434. Used where low "roping" or "ridging" required. General corrosion and heat-resistant applications such as automobile trim.
442	20.5	0.20				High chromium steel, principally for parts which must resist high service temperatures without scaling. Furnace parts; nozzles; combustion chambers.
446	25.0	0.20				High resistance to corrosion and scaling at high temperatures, especially for intermittent service; often used in sulfur-bearing atmosphere. Annealing boxes; combustion chambers; glass molds; heaters; pyrometer tubes; recuperators; stirring rods; valves.

† After "ASM Databook," *Metal Progress*, mid-June 1979, vol. 116, no. 1.
‡ S: 0.030% max; P: 0.045% max; Si: 1.00% max.

Table 7-1b Chemical compositions of some of the new ferritic stainless steels†

Name	% C	% Cr	% N	% Ti	% Mo
18-2	0.02	18		0.4	2
26-1S	0.03	26		0.5	1
E-Brite 26-1 (Airco)	0.002	26	0.01		1
29Cr-4Mo (Du Pont)	0.004	29	0.01		4

† From Ref. 6.

been developed and produced commercially (Table 7-1b). These alloys have improved corrosion resistance and are weldable.

Microstructures

The structure of ferritic stainless steels remains essentially ferritic (α-iron BCC type) at all normal heat treatment temperatures. Most ferritic stainless steels can be divided into two groups on the basis of chromium content:

Group 1. Ferritic stainless steels with 15 to 18% Cr and about 0.06% C. Example: type 430 alloy (17% Cr, 0.06% C).
Group 2. Ferritic stainless steels with 25 to 30% Cr and about 0.08% C. Example: type 446 alloy (25% Cr, 0.08% C).

The microstructure of the 15 to 18% Cr alloys (group 1) is almost entirely ferritic at temperatures below 900°C. Figure 7-4 shows a micrograph of type 430

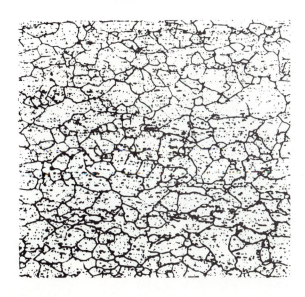

Figure 7-4 Type 430 (ferritic) stainless steel strip annealed at 788°C (1450°F). The structure consists of a ferrite matrix with equiaxed grains and dispersed carbide particles. (Etchant: picral + HCl; 100 × .) *(Courtesy of United States Steel Co., Research Laboratories.)*

Figure 7-5 Fe–17% Cr stainless steel after quenching in water from 1200°C. Structure shows islands of martensite in ferrite matrix. Note that the martensite is much harder than the ferrite. (Etch: Aqua Regia + Glycerol; 540 × .) *(From L. Colombier and J. Hochmann, "Stainless and Heat Resisting Steels," Edward Arnold Ltd., London, 1967, p. 49.)*

alloy (17% Cr, 0.06% C) alloy after annealing at 788°C. The structure consists of a chromium-rich solid solution of α ferrite, with most of the carbon in the form of intergranular and finely divided matrix (Fe,Cr) carbide precipitates. Very little of the carbon is in solid solution because of the low solubility of carbon in α ferrite. When alloys of this type are heat-treated at temperatures above 900°C, some austenite is formed and, upon subsequent water quenching, it is transformed to martensite. Figure 7-5 shows islands of martensite in a matrix of α

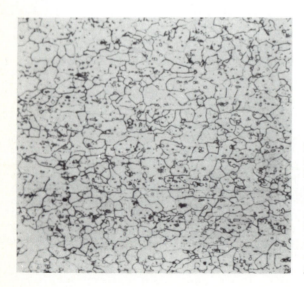

Figure 7-6 Type 446 (ferritic) stainless steel strip annealed at 802°C (1475°F). The structure consists of dispersed carbide particles in matrix of ferrite with equiaxed grains. (Electrolytic: HCl-methanol; 100 × .) *(After Metals Handbook, 8th ed., vol. 7, American Society for Metals, 1972, p. 146.)*

ferrite in a 17% Cr, 0.1% C ferritic stainless steel after water quenching from 1200°C.

The annealed microstructure of a cold-worked sheet of alloy 446 (25% Cr, 0.08% C), a group 2 ferritic stainless steel, is shown in Fig. 7-6. This Fe–25% Cr alloy has a coarser distribution of matrix (Fe,Cr) carbides than is found in a similarly annealed sheet of the Fe–18% Cr alloy (Fig. 7-4). However, in contrast to the Fe–18% Cr alloy, the Fe–25% Cr alloy can only form a relatively small amount of martensite after being heated above 950°C and then water quenched.

Embrittlement Mechanisms

Although the commonly used ferritic stainless steels cannot be strengthened significantly by the martensitic transformation, they are subject to some strengthening mechanisms that lead to embrittlement and associated low ductility. Distinction can be made among three types of embrittlement:

1. 475°C embrittlement
2. σ-phase embrittlement
3. High-temperature embrittlement

475°C embrittlement This type of embrittlement takes place when ferritic stainless steels are heated for long times in the 400 to 540°C range. This heat treatment results in an increase in tensile strength and hardness and a considerable decrease in ductility and impact values. The effect of 475°C embrittlement on increasing the strength and drastically reducing the ductility of a Fe–27% Cr ferritic stainless steel is shown in Fig. 7-7. The cause of the 475°C embrittlement is attributed to the precipitation of a chromium rich α' phase on dislocations (Ref. 11). Development of this precipitate with aging time at 482°C in a high-purity Fe–18% Cr steel is shown in Fig. 7-8. Since long-time isothermal treatments are required to produce 475°C embrittlement, this type of embrittlement does not normally interfere with welding and heat treatment of ferritic stainless steels.

σ-phase embrittlement The phase diagram for the Fe-Cr system shows that the σ phase should form at lower temperatures if conditions approaching equilibrium are attained. When Fe-Cr alloys containing about 15 to 70% Cr are heated in the 500 to 800°C range for prolonged periods, the σ phase will precipitate. Figure 7-9 shows intergranular σ phase which precipitated in an Fe–27% Cr alloy after heating for 131 days at 565°C. Since either very slow cooling rates through intermediate temperatures or prolonged aging at these temperatures is required to precipitate the σ phase, it does not normally interfere with welding and heat treatment of ferritic stainless steels.

High-temperature embrittlement When ferritic stainless steels with moderate to high carbon and nitrogen contents are heated above about 950°C and cooled to room temperature, they show severe embrittlement and loss of corrosion resis-

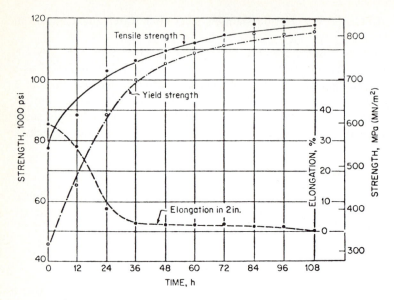

Figure 7-7 Effect of aging time at 475°C (885°F) on the room temperature aging properties of an Fe–27% Cr steel, air-melted. *(After H. D. Newell, Metal Progress, May 1946, p. 977.)*

tance. The cause of the high-temperature embrittlement is believed to be the precipitation of chromium-rich carbides and nitrides in the grain boundaries and/or dislocations (Ref. 1, pp. 5-13). Since BCC Fe-Cr alloys have such low solubility for carbon and nitrogen, chromium carbides form in these alloys unless their interstitials are kept to very low levels. This type of embrittlement is particularly damaging since it can occur in almost all operations necessary for a structural material. That is, processes such as welding, high-temperature heat treatment, and casting will lead to low ductility and corrosion resistance in these alloys. Thus, to circumvent the high-temperature embrittlement problem, new

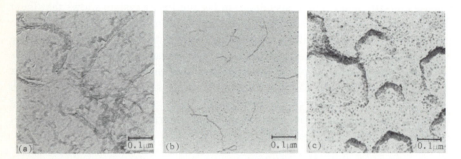

Figure 7-8 Structure of Fe–18% Cr (ferritic) stainless steel, vacuum melted, after aging at 482°C (900°F) for (*a*) 240 h, (*b*) 480 h, and (*c*) 2400 h. (Carbon replicas.) [*After P. J. Grobner, Met. Trans. 4(1973):251.*]

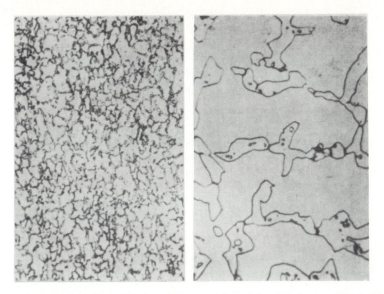

Figure 7-9 Fe–27% Cr ferritic stainless steel heated for 131 days at 565°C. Structure consists of ferrite matrix with dispersed carbides and intergranular sigma phase constituent. (Etchant: aqua regia; (*a*) 200 × , (*b*) 1000 × .) *(After H. D. Newell, Metal Progress, May 1946, p. 977.)*

ferritic stainless steels have been developed with high chromium and very low carbon and nitrogen contents. The new ferritics have been made possible through the use of vacuum and argon-oxygen decarburization and electron-beam and large-scale vacuum melting.

Mechanical properties

Representative tensile properties of some ferritic stainless steels in the annealed condition are shown in Table 7-2. Since these alloys are not completely hardened by solutionizing and quenching, they are used in the annealed condition, in which their structure consists of an equiaxed ferritic matrix with dispersed carbide particles. The standard ferritic stainless steels have slightly higher tensile and yield strengths and lower elongations than the low-carbon steels. The newer ferritics, because of their low carbon and nitrogen levels, have higher elongation values than the standard ferritics. The Charpy V-notch impact transition temperatures of the new ferritics are also much lower, as indicated in Table 7-3.

Figure 7-10 shows how the impact energy of an Fe–17% Cr alloy in the annealed condition is increased as its carbon and nitrogen contents are reduced. The ductile-brittle transition of this alloy is relatively low for all carbon levels from 0.002 to 0.061 percent when heated for 1 h at 815°C and water-quenched (Fig. 7-10*a*). However, when this alloy is heated 1 h at 815°C plus 1 h at 1150°C and water-quenched, the impact resistance decreases drastically such that very

Table 7-2 Typical room-temperature properties of annealed standard and nonstandard ferritic stainless steels†

Steel	Yield strength (0.2% offset)		Tensile strength		Elongation in 2 in (50.8 mm), %
	ksi	MN/m²	ksi	MN/m²	
Standard					
405	40	275.8	65	448.2	25
409	40	275.8	68	468.9	20
429	40	275.8	70	482.7	30
430	50	344.8	75	517.1	25
430F	55	379.2	80	551.6	25
430Se	55	379.2	80	551.6	25
434	53	365.4	77	530.9	23
436	53	365.4	77	530.9	23
442	45	310.3	80	551.6	20
446	50	344.8	80	551.6	20
Nonstandard					
18-2	43	296.5	68	468.9	37
26-1	50	344.8	70	482.7	30

† From Ref. 1, pp. 20–23.

Table 7-3 Charpy V-notch impact transition temperatures of some ferritic stainless steels†

Steel	Transition temperature‡	
	°F	°C
Standard		
Type 405	40	4.4
Type 409	70	21
Type 430	70–212	21–100
Type 446	250	121
Nonstandard		
Type 409 modified§	−70	−57
18Cr-2Mo	0	−17.7
26Cr-1Mo	0	−17.7

† From Ref. 1, pp. 20–23.
‡ Based on 25 ft–lb (33.9J) of energy absorbed.
§ Modified with up to 1% Ni.

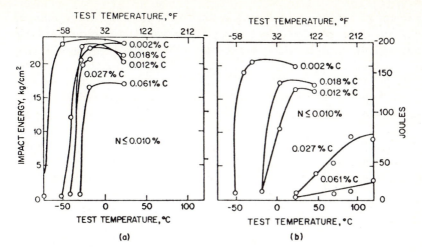

Figure 7-10 Transition curves for quarter-size charpy V-notch impact specimens of Fe–17% Cr–0.002 to 0.061% C ferritic stainless steels heat-treated at (*a*) 815°C for 1 h and water-quenched and (*b*) 815°C + 1150°C for 1 h and water-quenched. *(After M. Semchyshen, A. P. Bond, and H. J. Dundas, Toward Improved Ductility and Toughness, Climax Molybdenum Co., Greenwich, Conn., 1971), as presented in "Handbook of Stainless Steels," McGraw-Hill, New York, 1977, p. 5-24.)*

low levels of carbon are necessary to reduce the ductile-brittle transition temperature to low temperatures (Fig. 7-10*b*).

Corrosion Properties

The ferritic stainless steels are in general nonhardenable and show their best corrosion resistance in the annealed condition. The *general corrosion resistance* of these alloys increases as their chromium contents increase, with a chromium level of about 23 to 28 percent providing the best corrosion resistance if the alloy is solution-annealed. Although the *pitting resistance* of ferritic stainless steels increases to some degree with increasing chromium, molybdenum additions have been found to be especially beneficial.[1] To provide pitting resistance, the ferritics should contain at least 23 to 24% Cr and over 2% Mo (Ref. 1, p. 15–2). The precipitation of chromium carbides during welding or heat treatment decreases the pitting resistance of these alloys.

Ferritic stainless steels containing even small amounts of carbon and nitrogen are susceptible to *intergranular corrosion*. The mechanism of intergranular corrosion in ferritic stainless steels involves the precipitation of chromium carbides and nitrides at the grain boundaries. These precipitates cause chromium concentrations in regions adjacent to the grain boundaries to be lowered below the critical 12 percent level needed for corrosion resistance.

[1] M. A. Streicher, *Corrosion* 30(1974):77.

The solubility of carbon and nitrogen in ferrite is much lower than in austenite at a given temperature. Since precipitation reactions in ferritic stainless steels occur rapidly at higher temperatures (i.e., 600 to 800°C) because of the high diffusion rates of carbon and nitrogen in ferrite, precipitates cannot be prevented from forming even by rapid quenching. Hence, even small amounts of carbon and nitrogen are detrimental for corrosion resistance. Figure 7-11a and b shows that levels of 0.012% C and 0.022% Ni are high enough to cause an Fe–17% Cr alloy to be susceptible to intergranular corrosion after heating at 925°C and quenching. By lowering the carbon content to 0.002 percent and the nitrogen to 0.0095 percent, the Fe–17% Cr alloy is found to be very corrosion-resistant to intergranular corrosion after high-temperature heat treatment (Fig. 7-11c). Recently, as has been pointed out previously, newer ferritics have been produced commercially with extremely low levels of carbon and nitrogen (Table 7-1b).

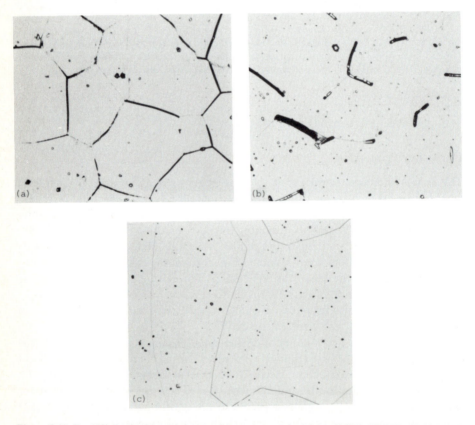

Figure 7-11 Fe–17% Cr (ferritic stainless steels) water-quenched from 925°C (1700°F). Held for 90 min at 1.2V vs. standard calomel electrode in 1N sulfuric acid at 24°C. (a) 0.0126% C, 0.0089% N; (b) 0.0025% C; 0.022% N; (c) 0.0021% C; 0.0095% N. 340 × . Note that the concentration of both carbon *and* nitrogen must be kept very low to get improved corrosion resistance. [*After A. P. Bond, Trans. AIME 245(1969):2127.*]

Another method for reducing the intergranular susceptibility of standard ferritic stainless steels is to *stabilize* these alloys with titanium or columbium. By forming titanium or columbium carbides at higher temperatures, the intergranular corrosion of these alloys is improved at lower temperatures. For example, alloy 409 is stabilized by a titanium addition of six times its carbon content and alloy 436 by a columbium addition of five times its carbon content.

7-6 MARTENSITIC STAINLESS STEELS

Chemical Compositions and Typical Applications

The martensitic stainless steels are essentially iron-chromium alloys containing 12 to 17% Cr with sufficient carbon so that a martensitic structure can be produced by quenching from the austenitic phase region. These alloys are called martensitic since they are capable of developing a martensitic structure with an austenitizing and quenching heat treatment. Table 7-4 lists the chemical compositions and typical applications of some of the predominantly used martensitic stainless steels. Since the composition of the martensitic stainless steels is adjusted to optimize strength and hardness, the corrosion resistance of these alloys is relatively poor as compared to the ferritic and austenitic stainless steels.

The chemical compositions of the martensitic stainless steels are relatively limited since a minimum of 12% Cr is required for corrosion resistance. At this chromium level, the maximum amount of carbon that can be added is about 0.15 percent or the excess carbon will precipitate carbides near the grain boundaries and lower the chromium content there below the critical 12 percent. For higher hardnesses, e.g., for cutlery, the carbon level is raised to the 0.60 to 1.1 percent range (types 440A, B, and C), along with an increase in chromium content up to the 16 to 18 percent range. Fortunately, the γ loop of the iron-chromium phase diagram is expanded by the increased carbon content to about 18% Cr so that these high-chromium high-carbon alloys can be austenitized and quenched to form a martensitic structure. The amount of alloying elements that can be added to martensitic stainless steels is limited since these elements, like carbon, depress the M_s temperature and, if the M_s is depressed too low, austenite will be retained at room temperature. Thus, the other alloying elements added to these alloys are restricted to several percent nickel, as in the 414 and 431 alloys, and to about 1% Mo, along with 1% W and 0.25% V, in the 422 alloy.

Microstructures

The microstructures of martensitic stainless steels are principally determined by their chromium and carbon contents and by heat treatment. Chromium restricts the range over which the austenitic phase in Fe-Cr alloys is stable, and thus with a carbon content of about 0.1 percent the chromium level cannot exceed about

Table 7-4 Chemical compositions and typical applications of wrought martensitic steels†

AISI type	% Cr	% C	% Ni	% Mo	% V	% W	% Other‡ Typical applications
403	12.2	0.15 max					"Turbine quality" grade. Steam turbine blading and other highly stressed parts including jet engine rings.
410	12.5	0.15 max					General-purpose heat-treatable type. Machine parts; pump shaft; bolts; bushings; coal chutes; cutlery; finishing tackle; hardware; jet engine parts; mining machinery; rifle barrels; screws; valves.
414	12.5	0.15 max	1.8				Higher carbon modification of type 410. Cutlery; surgical instruments; valves; wear-resisting parts; glass molds; hand tools; vegetable choppers.
420	13	Over 0.15					High-hardenability steel. Springs; tempered rules; machine parts; bolts; mining machinery; scissors; ships belts; spindles; valve seals.
422	12	0.22		1.0	0.25	1.0	High strength and toughness at service temperatures up to 1200°F. Steam turbine blades; fasteners.
431	16	0.20 max	1.8				Special-purpose hardenable steel used where particularly high mechanical properties are required. Aircraft fittings; beater bars; paper machinery; bolts.
440A	17	0.72					Hardenable to higher hardness than type 420; with good corrosion resistance. Cutlery; bearings; surgical tools.
440B	17	0.85					Cutlery grade. Cutlery; valve parts; instrument bearings.
440C	17	1.07					Yields highest hardnesses of hardenable stainless steels. Balls, bearings; races; nozzles; balls and seats for oil well pumps; valve parts.

† After "ASM Databook," *Metal Progress*, mid-June 1979, vol. 116, no. 1.
‡ S:0.030% max; P: 0.040% max; Si: 1.00% max.

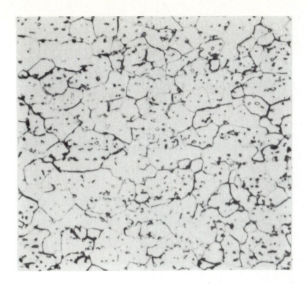

Figure 7-12 Type 410 (martensitic) stainless steel strip annealed at 815°C (1500°F), furnace-cooled to 595°C (1100°F), and air-cooled to room temperature. Matrix consists of equiaxed ferrite grains with randomly dispersed carbides. (Vilella's reagent; 500 × .) *(After Metals Handbook, 8th ed., vol. 7, American Society for Metals, 1972, p. 142.)*

13 percent if full hardening is to be achieved. When the carbon content of the Fe-Cr alloys is increased to about 0.4 to 0.6 percent, the γ loop in the Fe-Cr phase diagram is expanded so that as much as 18% Cr can be alloyed in martensitic steels with full hardening still being attained (Fig. 7-2d).

In the annealed condition, the optical microstructure of 410 alloy (12% Cr, 0.1% C) consists of a matrix of equiaxed ferrite grains with randomly dispersed carbides (Fig. 7-12). In the air-quenched and tempered condition, the structure of this alloy consists of martensite with precipitated carbide particles (Fig. 7-13). The microstructure of 440C alloys in the air-quench hardened condition consists of a high density of primary carbides in a martensitic matrix (Fig. 7-14). The very high hardness developed in this alloy is due to the large number of (Fe,Cr) carbides in the martensitic matrix.

Heat Treatment

The heat treatment of martensitic stainless steels for increased strength and hardness is basically the same as that for plain-carbon or low-alloy steels. That is, the alloy is austenitized, cooled at a rate fast enough to produce a martensitic structure, and then tempered to increase toughness and relieve stresses.

Austenitizing Figure 7-15 shows the effects of increased austenitizing temperatures on the quenched hardness of four Fe–12% Cr alloys with 0.016 to 0.14% C contents. In general, the maximum hardness is achieved when the austenitizing temperature is between 980 and 1090°C, and increases as the carbon content increases. The hardness decrease caused by heating above about 1100°C is probably due to the formation of δ ferrite, whereas heating below 900°C will not produce sufficient austenite (see the Fe-Cr-C phase diagrams of Fig. 7-2).

Figure 7-13 Type 410 (martensitic) stainless steel strip hardened by rapidly air cooling from 980°C (1800°F) to room temperature, and tempered 4 h at 205°C (400°F). Structure consists of martensite with precipitated carbide particles. Oblique illumination. (Viella's reagent; 500 × .) *(After Metals Handbook, 8th ed., vol. 7, American Society for Metals, 1972, p. 142.)*

Figure 7-14 Type 440C (martensitic) stainless steel hardened by austenitizing at 1010°C (1850°F) and air-cooled. Structure consists of primary carbides in martensite matrix. (Etchant: HCl + Picral; 500 × .) *(Courtesy of the Allegheny Ludlum Steel Co.)*

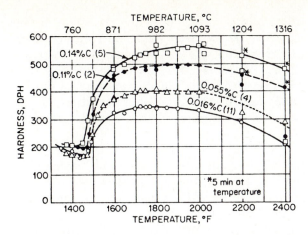

Figure 7-15 Hardness of four Fe–12% Cr stainless steels containing 0.016 to 0.14% carbon. After quenching from a series of temperatures. Held 1 h at 1093°C (2000°F) or lower and 30 min at 1205°C (2200°F). Before quenching. [*After R. L. Rickett, W. F. White, C. S. Walton, and J. C. Butler, Trans. ASM, 44(1952):138; as presented in "Handbook of Stainless Steels," McGraw-Hill, New York, 1977, p. 20-19.*]

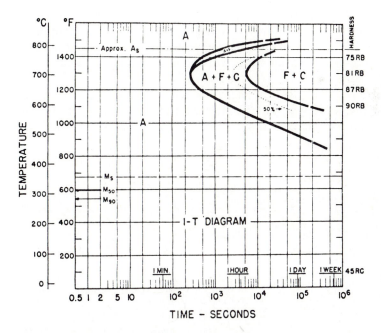

Figure 7-16 Isothermal transformation diagram for type 410 stainless steel (Fe–12% Cr–0.1% C). Austenitized at 980°C (1800°F). Grain size 6–7. (A, austenite; F, ferrite; C, carbide; M, martensite.) *(Isothermal Transformation Diagrams, United States Steel Co., 1963, p. 48.)*

Rate of cooling The high hardenability contribution of the chromium in the Fe–12% Cr type of alloys eliminates the necessity for a water quench after austenitizing, and allows for a slower cooling rate (e.g., air cooling) to produce a martensitic structure. The influence of chromium on the hardenability of the type 410 alloy is shown in the IT diagram of Fig. 7-16. A comparison of this IT diagram with that for 1035 plain-carbon steel emphasizes the effect of the chromium in shifting the beginning of the austenite → ferrite + carbide transformation to longer times.

Tempering As with other alloyed martensitic steels, a tempering treatment is necessary to increase toughness and ductility. Figure 7-17 shows the effect of tempering temperature on the hardness of Fe–12% Cr stainless steels containing 0.055 to 0.14% C. It should be noted from this plot that these steels do not soften to any great extent until the tempering temperature is above 480°C. Also, when the tempering temperature reaches about 430°C, a slight secondary hardening effect is observed. This is most probably due to the early stages of the formation of the $(Fe,Cr)_{23}C_6$ phase, which becomes stable at about this temperature and gradually replaces the $(Fe,Cr)_3C$ phase.

Mechanical Properties

The tensile properties of martensitic stainless steels can be controlled to some extent through heat treatment. These alloys, because of their high chromium contents, can be air-quenched to form martensite, and then subsequently tempered. Table 7-5 lists representative tensile properties of some martensitic stainless steels in the quenched (air-hardened) and tempered condition and also in the annealed condition.

Figure 7-18 shows the effect of tempering temperature (for 1 h) on the mechanical properties of a type 410 stainless steel (12% Cr–0.1% C). Tensile strengths of this alloy follow the same trend as the hardnesses of the 12% Cr

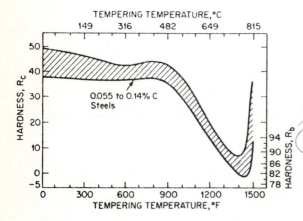

Figure 7-17 Effect of temperature on the hardness of Fe–12% Cr, 0.055 to 0.14% carbon (martensitic) stainless steels tempered for 2 h. [*After R. L. Rickett, W. F. White, C. S. Walton and J. C. Butler, Trans. ASM 44(1952):138; as presented in the "Handbook of Stainless Steels," McGraw-Hill, New York, 1977, p. 20-19.*]

Table 7-5 Typical tensile properties of AISI standard martensitic stainless steels in the annealed and tempered conditions†

AISI type	Tempering temperature		Yield strength (0.2% offset)		Tensile strength		Elongation in 2 in (50.8 mm), %	Reduction of area, %
	°F	°C	ksi	MN/m²	ksi	MN/m²		
403, 410,	None (annealed)		40	275.8	75	517.1	30	65
416, 416Se	400	204	145	999.8	190	1310.1	15	55
	600	315	140	965.3	185	1275.6	15	55
	800	426	150	1034.3	195	1344.5	17	55
	1000	538	115	792.9	145	999.8	20	65
	1200	648	85	586.1	110	758.5	23	65
	1400	760	60	413.7	90	620.6	30	70
414	None (annealed)		95	655	120	827.4	17	55
	400	204	150	1034.3	200	1379	15	55
	600	315	145	999.8	190	1310.1	15	55
	800	426	150	1034.3	200	1379	16	58
	1000	538	120	827.4	145	999.8	20	60
	1200	648	105	724	120	827.4	20	70
420	None (annealed)		50	344.8	95	655	25	55
	400	204	200	1379	255	1758.2	10	35
	600	315	195	1344.5	250	1723.8	10	35
	800	426	200	1379	255	1758.2	10	35
	1000	538	145	999.8	170	1172.2	15	40
	1200	648	85	586.1	115	792.9	20	55
431	None (annealed)		95	655	125	861.9	20	60
	400	204	155	1068.7	205	1413.5	15	55
	600	315	150	1034.3	195	1344.5	15	55
	800	426	155	1068.7	205	1413.5	15	60
	1000	538	130	896.4	150	1034.3	18	60
	1200	648	95	655	125	861.9	20	60
440A	None (annealed)		60	413.7	105	724	20	45
	600	315	245	1689.3	265	1827.2	5	20
440B	None (annealed)		62	427.5	107	737.8	18	35
	600	315	270	1861.7	280	1930.6	3	15
440C	None (annealed)		70	482.7	110	758.5	13	25
	600	315	275	1896.1	285	1965.1	2	10

† After Ref. 1, p. 20-18.

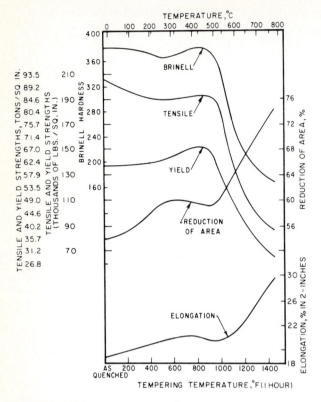

Figure 7-18 Effect of tempering temperature on the mechanical properties of Fe–12% Cr type 410 stainless steel. *(After the "Making, Shaping, and Treating of Steel," 9th ed., United States Steel Co., 1971, p. 1178.)*

steels previously discussed. The slight increase in tensile strength before the rapid decline at about 450°C is again believed due to secondary hardening by $(Fe,Cr)_{23}C_6$ precipitation.

Figure 7-19 shows the effect of tempering temperature (for 1 h) on the Charpy V-notch impact properties of some Fe–12% Cr stainless steels. The highest impact values (toughness[1]) are obtained around 260°C. Above 260°C, the impact values decrease, reaching a minimum between about 450 to 550°C. In practice, the tempering range between 475 to 550°C is avoided because of the poor toughness of all martensitic stainless steels tempered in this region (Ref. 1, p. 6–20). This decrease in impact strength corresponds to the secondary hardness peak in the tempering curve, and is attributed to carbide (or nitride) precipitation, a large part of which takes place at grain boundaries. Temper

---

[1] Toughness is the ability of a steel to deform plastically under the influence of a notch.

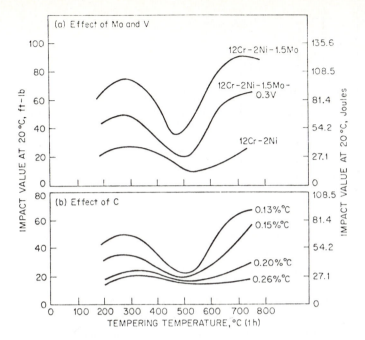

Figure 7-19 Effect of tempering temperature on impact properties (Charpy V-notch) of Fe–12% Cr stainless steels. *(After Irvine, Crowe, and Pickering, as presented in "The Making, Shaping, and Treating of Steel," 9th ed., United States Steel Co., 1971, p. 1178.)*

embrittlement mechanisms operative in low-alloy steels could also be partly responsible (see Sec. 4-10).

Corrosion Properties

The corrosion resistance of the martensitic stainless steels is relatively poor compared to the austenitic and ferritic stainless steels. Most martensitic stainless steels contain just the minimum 12% Cr required for passivity in moist air since, if more chromium were added, the formation of ferrite would be promoted at the expense of austenite, which is necessary for the formation of martensite. The chemical composition of martensitic stainless steels is designed for strength and hardness as well as for corrosion resistance, and therefore the chemical balance for corrosion resistance in these alloys is poor. Only limited amounts of other alloying elements such as nickel can be added because the transformation of austenite to martensite will be inhibited.

The martensitic stainless steels are usually tempered after quenching, but care must be taken to avoid the 370 to 600°C range where impact strengths are low (Fig. 7-19). In this critical region, the corrosion resistance of these alloys is also reduced, as Fig. 7-20 shows for alloys 410 and 414.

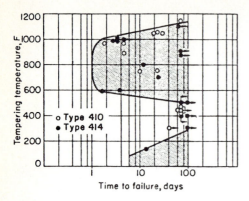

Figure 7-20 Effect of tempering temperature on the stress-corrosion characteristics of types 410 and 414 stainless steels at high stress. Data apply to a stress level of 80,000 psi for tests in a salt fog cabinet. *(After Metals Handbook, 8th ed., vol. 2, American Society for Metals, 1964, p. 249.)*

7-7 AUSTENITIC STAINLESS STEELS

Chemical Compositions and Typical Applications

Austenitic stainless steels are essentially ternary iron-chromium-nickel alloys containing 16 to 25% Cr and 7 to 20% Ni. These alloys are called *austenitic* since their structure remains austenitic (FCC, γ-iron type) at all normal heat treatment temperatures. Some of the nickel in these alloys can be replaced by manganese and their structure will still remain austenitic. Table 7-6 lists the chemical compositions and typical applications of some of the more commonly used austenitic stainless steels.

Austenitic stainless steels make up about 65 to 70 percent of the total U.S. stainless steel production (Ref. 1, p. 4-2). These alloys have this dominant position mainly because of their high corrosion resistance and formability, and thus they possess highly desirable properties for many engineering applications. Types 302 and 304 are the most widely used stainless steels, finding application at elevated temperatures as well as ambient temperatures. Type 316, which contains 2.5% Mo and essentially the same base as type 304, has higher corrosion resistance and enhanced elevated-temperature strength. Alloys with increased chromium levels (23 to 25 percent) such as types 309 and 310 find use primarily for elevated-temperature applications.

Microstructures

The austenitic stainless steels retain the austenitic (FCC) structure at room temperature after high-temperature annealing principally because of the austenitic stabilizing effect of their high nickel contents. However, Mn, C, and N also contribute to the retention and stabilization of the austenitic structure. The addition of nickel to iron-chromium alloys widens the region over which austenite is stable (Fig. 7-3) and decreases the M_s temperature. In an 18% Cr–8% Ni stainless steel, an austenitic structure can be retained at room temperature after cooling from the annealing temperature (e.g., 1050°C). The austenite,

Table 7-6 Chemical compositions and typical applications of wrought austenitic stainless steels†

AISI type	Chemical compositions, wt%						Typical applications
	Cr	Ni	C (max)	Mn	Mo	Other ‡	
301	17	7	0.15				High work-hardening rate; used for structural applications where high strength plus high ductility is required. Railroad cars; trailer bodies; aircraft structurals; fasteners; automobile wheel covers, trim; pole-line hardware.
302	18	9	0.15				General-purpose austenitic stainless steel. Trim; food-handling equipment; aircraft cowlings; antennas; springs; cookware; building exteriors; tanks; hospital, household appliances; jewelry; oil refining equipment; signs.
304	19	9	0.08				Low-carbon modification of type 302 for restriction of carbide precipitation during welding. Chemical and food processing equipment; brewing equipment; cryogenic vessels; gutters; downspouts; flashings.
304L	19	10	0.03				Extra-low-carbon modification of type 304 for further restriction of carbide precipitation during welding. Coal hopper linings; tanks for liquid fertilizer and tomato paste.
309	23	13.5	0.20				High-temperature strength and scale resistance. Aircraft heaters; heat-treating equipment; annealing covers; furnace parts; heat exchangers; heat-treating trays; oven linings; pump parts.
310	25	20.5	0.25				Higher elevated-temperature strength and scale resistance than type 309. Heat exchangers; furnace parts; combustion chambers; welding filler metals; gas turbine parts; incinerators; recuperators.
316	17	12	0.08		2.5		Higher corrosion resistance than types 302 and 304; high creep strength. Chemical and pulp handling equipment; photographic equipment; brandy vats; fertilizer parts; ketchup cooking kettles; yeast tubs.
316L	17	12	0.03		2.5		Extra-low carbon modification of type 316. Welded construction where intergranular carbide precipitation must be avoided. Type 316 applications requiring extensive welding.

Table 7-6, Continued

AISI type	Chemical compositions, wt%						Typical applications
	Cr	Ni	C (max)	Mn	Mo	Other ‡	
321	18	10.5	0.08			Ti 5 × C	Stabilized for weldments subject to severe corrosive conditions, and for service from 800 to 1600°F. Aircraft exhaust manifolds; boiler shells; process equipment; expansion joints; cabin heaters; fire walls; flexible couplings; pressure vessels.
347	18	11	0.08			Cb 10 × C	Similar to type 321 with higher creep strength. Airplane exhaust stacks; welded tank cars for chemicals; jet engine parts.
201	17	4.5	0.15	6			High work-hardening rate; low-nickel equivalent of type 301. Flatware; automobile wheel covers, trim.
202	18	5	0.15	8.75			General purpose low-nickel equivalent of type 302. Kitchen equipment; hub caps; milk handling.

† After "ASM Databook," *Metal Progress*, vol. 116, no. 1, mid-June 1979.
‡ Mn: 2.00% max; S: 0.030% max; P: 0.045% max; Si: 1.00% max.

however, in some Fe-Cr-Ni stainless steels (e.g., type 301) is not thermodynamically stable at room temperature due to lower Cr and Ni contents than 18% Cr and 8% Ni. If alloys of this type are plastically deformed at room temperature or slightly below it, some of the austenite can be transformed to martensite. The "Mechanical Properties" part of this section discusses the effect of cold work on producing this transformation.

Most of the commonly used stainless steels contain significant amounts of carbon. For example, there is usually about 0.1% C in type 302 alloy and about 0.06% C in type 304. Since the solubility of carbon in austenitic stainless steels, like that in the 18% Cr–8% Ni alloy, decreases so rapidly with decreasing temperature, chromium carbides can precipitate if these alloys are slowly cooled (Fig. 7-21). For example, if alloy type 304 (19% Cr–9% Ni) is slowly cooled from about 1050°C to room temperature, chromium carbides will precipitate in the grain boundaries in the 850 to 400°C range.

During slow cooling through the critical temperature range of 850 to 400°C, an insufficient number of chromium atoms diffuse in from the grain matrix to replace those chromium atoms removed at the grain boundary region by the precipitated chromium carbides. Consequently, the regions adjacent to the grain boundaries have their chromium contents lowered to less than the critical 12 percent needed for corrosion resistance, and the alloy thus becomes susceptible

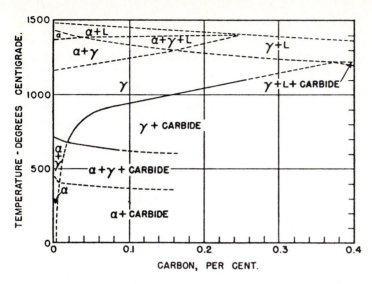

Figure 7-21 Effect of carbon on the constitution of Fe–18% Cr–8% Ni stainless steels. *(After "Making, Shaping, and Treating of Steel," 9th ed., United States Steel Co., 1971, p. 1178.)*

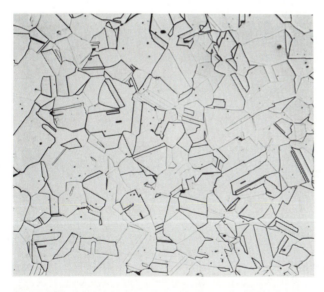

Figure 7-22 Type 304 (austenitic) stainless steel strip annealed 5 min at 1065°C (1950°F) and air-cooled. Structure consists of equiaxed austenite grains. Note annealing twins. (Etchant: HNO_3–acetic–HCl–glycerol; 250 × .) *(Courtesy of Allegheny Ludlum Steel Co.)*

to intergranular corrosion. Austenitic stainless steels in this condition are said to be *sensitized* since they are susceptible to intergranular corrosion. This phenomenon is discussed in greater detail later, under the heading "Corrosion Properties" in this section. Such austenitic stainless steels, therefore, must be annealed at a temperature high enough to put the chromium carbides into solid solution but low enough to prevent excessive grain growth. Most stainless steels are annealed in the 1050 to 1120°C range. After the high-temperature anneal, they must be cooled rapidly to prevent chromium carbides from precipitating.

Figure 7-22 shows the microstructure of type 304 austenitic stainless steel after annealing at 1050°C and air cooling. Figure 7-23 shows an electron transmission micrograph of a thin foil of this alloy after 2 h at 1060°C and water quenching. In the electron micrograph there is no precipitate visible in the grain boundaries. If, however, this alloy is reheated 2 h at 600°C, which is in the critical range for precipitation, carbide precipitates that are almost continuous are visible in the grain boundaries (Fig. 7-24).

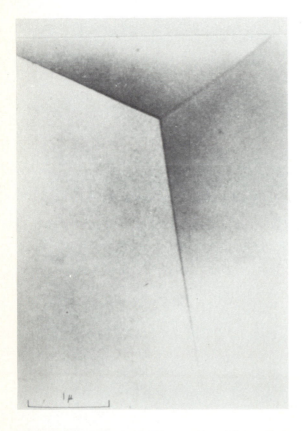

Figure 7-23 Structure of type 304 austenitic stainless steel after 2 h at 1060°C and water quenching. Transmission electron micrograph from thin foil. Note absence of grain boundary carbide precipitates. (25,000 × .) [*After K. T. Aust, J. S. Armijo, and J. H. Westbrook, Trans. ASM 59(1966):544.*]

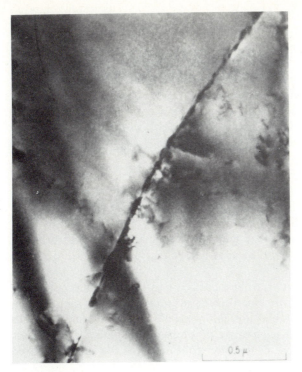

Figure 7-24 Structure of type 304 austenitic stainless steel after 2 h at 1060°C, water quenching, followed by 2 h at 600°C and water quenching. Transmission electron micrograph from thin foil. Note the almost continuous precipitate of carbides along grain boundary. (50,000 × .) [*After K. T. Aust, J. S. Armijo, and J. H. Westbrook, ASM Trans. 59(1966):544.*]

It is not always possible to cool rapidly alloys such as type 304 stainless steel after high-temperature annealing, and this can cause corrosion problems. For example, welding stainless steel in the field where it must slow-cool may mean that a subsequent annealing treatment to redissolve the chromium carbide precipitates will no longer be possible. In order to prevent intergranular precipitation caused by slow cooling, variations in chemical composition which have the carbon combined with other elements have been developed.

In type 321 alloy, titanium in the amount of five times the carbon content is added to the alloy. By heating this alloy at 870°C for sufficient time, the titanium will combine with the carbon to form titanium carbide (TiC). This is called a *stabilizing* treatment since chromium carbides are prevented from precipitating if a subsequent heat treatment in which slow cooling through the critical temperature range occurs.

Another variation of this method of stabilizing the 18-8 stainless steels is to add columbium (niobium) so that columbium carbides (CbC) will form in preference to chromium carbides. In type 347 alloy, an amount of columbium equal to 10 times the carbon content is added to the 18-8 stainless steel. This alloy must also be stabilized at 870°C to combine the carbon with the columbium.

Still another method of preventing chromium carbide precipitation in the grain boundaries of austenitic stainless steels is to reduce the carbon content to a sufficiently low level. In types 304L and 316L, a maximum of 0.03% C is allowed. For service temperatures below about 425°C, these alloys are usually used in preference to those containing titanium and columbium carbides. The economics and application dictate which of the alloys will be used to prevent intergranular corrosion in service.

Mechanical Properties

Since the austenitic stainless steels have an austenitic (FCC) microstructure at room temperature, they cannot be hardened to any great extent by heat treatment. They can, however, be considerably strengthened by cold work. For example, the yield strength of type 301 alloy can be increased from 40 to 200 ksi by cold working.

The austenitic stainless steels can be classified into two groups according to the stability of the austenite in the microstructure: *stable austenitic* and *metastable austenitic* steels. The stable austenitic steel's microstructure remains austenitic after cold working. The structure of the metastable austenitic stainless steel is transformed to some degree by cold working so that a mixed martensitic-austenitic structure is developed.

The difference between the strain-hardening behavior at room temperature of a metastable austenitic stainless steel (type 301) and a stable one (type 304) is shown by the engineering stress-strain curves in Fig. 7-25. The type 304 steel exhibits normal strain-hardening behavior and shows a parabolic curve which is indicative of normal strain hardening throughout the application of the stress. However, the metastable type 301 shows an accelerated strain-hardening effect

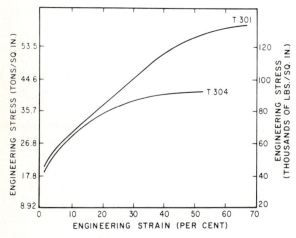

Figure 7-25 Engineering stress and strain curves for types 301 and 304 stainless steels. *(From "Making, Shaping, and Treating of Steel," 9th ed., United States Steel Co., 1971, p. 1181.)*

after about 10 to 15 percent plastic deformation. This accelerated work hardening is due to the formation of martensite from the unstable austenite.

Table 7-7 lists the room-temperature tensile properties of selected austenitic stainless steels in the annealed condition. The effect of small differences in carbon content on the yield strength can be seen by comparing the yield strength of alloy 304 with 304L. Type 304, with a carbon content of about 0.08 percent, has a yield strength of 42 ksi, whereas type 304L, with a lower carbon content of 0.03 percent, has a yield strength of only 39 ksi. The difference between the stable and metastable austenitic stainless steels is sharply indicated by the differences in their annealed tensile strengths. For example, metastable steel type 301 has a tensile strength of 110 ksi whereas stable steel type 304 has only a tensile strength of 84 ksi. For a given amount of cold work, the metastable steels (e.g., type 301) show higher tensile and yield strengths and

Table 7-7 Typical room-temperature tensile properties of annealed standard austenitic stainless steels†

AISI type	Yield strength (0.2% offset)		Tensile strength		Elongation in 2 in (50.8 mm), %
	ksi	MN/m²	ksi	MN/m²	
201	55	379.2	115	792.9	55
202	55	379.2	105	724	55
301	40	275.8	110	758.5	60
302	40	275.8	90	620.6	50
302B	40	275.8	95	655	55
303	35	241.3	90	620.6	50
303Se	35	241.3	90	620.6	50
304	42	289.6	84	579.2	55
304L	39	268.9	81	558.5	55
305	38	262	85	586.1	50
308	35	241.3	85	586.1	50
309	45	310.3	90	620.6	45
309S	45	310.3	90	620.6	45
310	45	310.3	95	655	45
310S	45	310.3	95	655	45
314	50	344.8	100	689.5	40
316	42	289.6	84	579.2	50
316L	42	289.6	81	558.5	50
317	40	275.8	90	620.6	45
321	35	241.3	90	620.6	45
347	40	275.8	95	655	45
348	40	275.8	95	655	45
384	35	241.3	75	517.1	55
385	30	206.9	72	496.4	55

† After Ref. 1, p. 20-16.

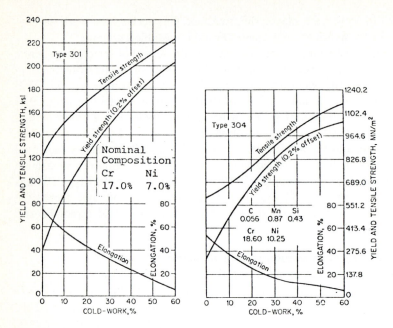

Figure 7-26 Effect of cold-working on the mechanical properties of type 301 (metastable) and type 304 (stable) stainless steels. *(Allegheny Ludlum Steel Co. data.)*

elongations than the stable steels (e.g., type 304), as is shown in Fig. 7-26. The higher strengths of the metastable steels are again attributed to the transformation of some unstable austenite to martensite.

Corrosion Properties

General corrosion In general, the austenitic stainless steels are considered to have the best overall corrosion resistance of all the stainless steels and to be the most resistant to industrial atmospheres and acid media. Bright-polished surfaces maintained free of dust and dirt will remain bright under most natural conditions (Ref. 1, p. 15-2). As the corrosion conditions become more severe (e.g., higher temperatures and stronger acids), more alloy content above that in type 304 is required.

Pitting corrosion The addition of over 2% Mo to austenitic stainless steels increases resistance to *pitting*. Type 316 is a popular alloy in this class and contains 2.5% Mo. For aggressive pitting media (i.e., high chloride contents), higher nickel and molybdenum contents other than those in type 316 alloy are necessary. Cleaner steels, with fewer inclusions and impurities, in general have better pitting resistance but are more costly to produce.

Intergranular corrosion A major disadvantage of some austenitic stainless steels such as the popular type 304 alloy is that they are susceptible to *intergranular corrosion* if they are heated in the *sensitizing range*, 400 to 850°C (see "Microstructures" in this section). The degree of susceptibility to intergranular corrosion is dependent on the composition of the alloy and the time at temperature in the *sensitizing range*. Severe environments such as highly oxidizing acids are used to accelerate intergranular corrosion for laboratory testing purposes. For example, severe integranular corrosion of annealed type 304 alloy occurs in boiling nitric-dichromate solutions, as shown in Fig. 7-27. The attack after 4 h of exposure is shown in Fig. 7-27a, and that after 8 h in Fig. 7-27b.

Another accelerated laboratory corrosion test is carried out in an oxidizing solution called Strauss' solution. The effects of 45 h of exposure to this solution on a type 304 alloy which has been sensitized for 150 h at various temperatures is shown in Fig. 7-28. These micrographs show that the most severe attack occurred in the 730 to 815°C range, which is indicated in the time-temperature plot of Fig. 7-29. It is important, therefore, that austenitic stainless steels like type 304 be quenched through the 870 to 600°C range to avoid sensitization.

As previously discussed (see "Microstructures" in this section), by lowering the carbon content below solid-solution levels, intergranular corrosion is largely avoided in austenitic steels. For example, type 304L alloy has a maximum of

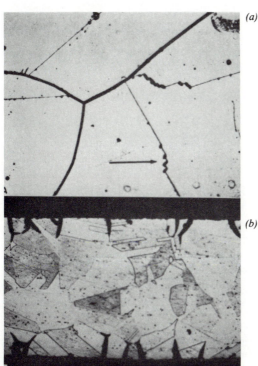

(a)

(b)

Figure 7-27 Typical intergranular attack of solution-treated type 304 stainless steel in boiling nitric-dichromate solution. (*a*) 4-h exposure; 500 × . (*b*) 8-h exposure; 150 × . [*After K. T. Aust, J. S. Armijo, and J. H. Westbrook, Trans. ASM 59(1966):544.*]

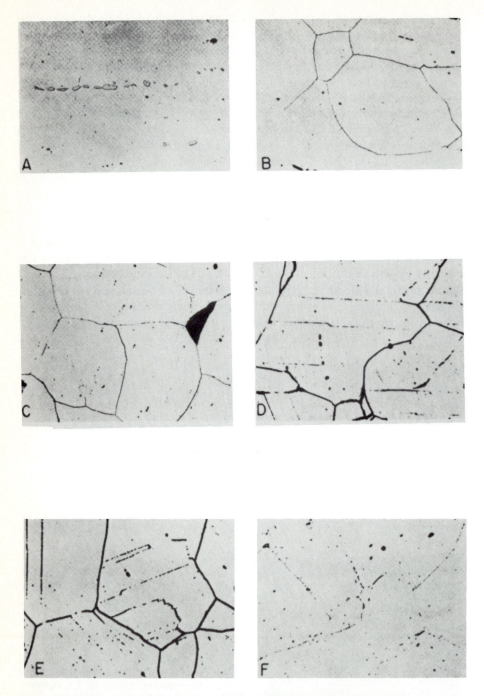

Figure 7-28 Etching of polished surface of type 304 (0.038% C) stainless steel after 45-h exposure to strauss solution. Specimens were sensitized for 150 h at (*a*) 480°C, (*b*) 565°C, (*c*) 650°C, (*d*) 730°C, (*e*) 815°C, (*f*) 900°C. [*After R. Stickler and A. Vinckier, Trans. ASM 54(1961):362.*]

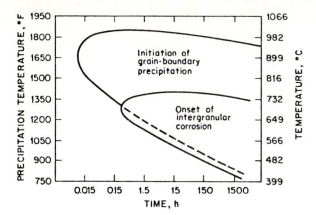

Figure 7-29 Relationship between precipitation of $M_{23}C_6$ and intergranular corrosion in type 304 austenitic stainless steel. *(After R. Stickler and A. Vinckier, Trans. ASM 54(1961):362; as modified in the "Handbook of Stainless Steels," McGraw-Hill, New York, 1977, p. 4-39.)*

0.03% C. Another way of avoiding it is to tie up the carbon with titanium or columbium to form titanium or columbium carbides as is done in alloy types 321 and 347 (again, see "Microstructures" in this section).

7-8 PRECIPITATION-HARDENING STAINLESS STEELS

Precipitation-hardening stainless steels were first developed during the 1940s and since then have become of increasing importance for a variety of applications where their special properties can be utilized. The most important of these properties are ease of fabrication, high strength, relatively good ductility, and excellent corrosion resistance. In this presentation, the two groups of precipitation-hardenable stainless steels which are the most commonly used are discussed. These groups are the *semiaustenitic* and *martensitic* types.

Semiaustenitic Type

These alloys are called *semiaustenitic* since they are essentially austenitic in the annealed condition (solution-heat-treated) but can be transformed to martensite by relatively simple thermal or thermomechanical heat treatments. In order to make this type of alloy, very close control must be maintained between the austenitic and ferritic balance. If the austenite and/or ferrite is too high, the austenite will be too stable to transform to martensite. If the austenite balance is too low, a stable austenite in the annealed condition that resists partial or complete transformation to martensite cannot be produced.

The chemical composition of some semiaustenitic precipitation-hardening stainless steels are listed in Table 7-8. The 17-7PH steel has approximately the same Cr and Ni contents as the type 301 austenitic stainless steel, but has the

Table 7-8 Nominal chemical compositions of selected semiaustenitic precipitation-hardenable stainless steels†

Grade	% C	% Mn	% Si	% Cr	% Ni	% Mo	% Al	% N
17-7PH‡	0.07	0.50	0.30	17.0	7.1		1.2	0.04
PH 15-7MO‡	0.07	0.50	0.30	15.2	7.1	2.2	1.2	0.04
PH 14-8Mo§	0.04	0.02	0.02	15.1	8.2	2.2	1.2	0.005
AM-350¶	0.10	0.75	0.35	16.5	4.25	2.75		0.10
AM-355¶	¶0.13	0.85	0.35	15.5	4.25	2.75		0.12

† After Ref. 1, p. 7–9.
‡ 17-7PH and PH 15-7Mo are registered trademarks of the Armco Steel Corporation.
§ PH 14-8Mo is a trademark of the Armco Steel Corporation.
¶ AM-350 and AM-355 are trademarks of the Allegheny Ludlum Steel Corporation.

addition of 1.2% Al for precipitation hardening. The other semiaustenitics listed in Table 7-8 have 2 to 3% Mo as a substitute for some of the chromium and nickel. Alloys AM-350 and AM-355 also contain about 0.1% N.

These alloys are most often supplied from the mill in the annealed (solution-heat-treated) condition designated "condition A." In it, the structure consists of a matrix of austenite with stringers of δ ferrite (Fig. 7-30). In

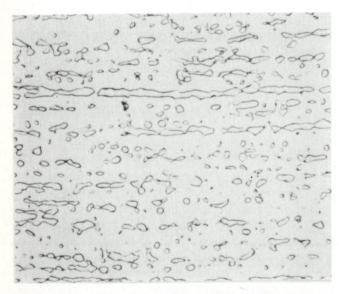

Figure 7-30 Type 17-7PH precipitation hardening stainless steel that was mill annealed [solution treated at 1065°C (1950°F)] to put it in condition A. Structure consists of austenite matrix with stringers of delta ferrite. (Etchant: HNO₃–acetic, then 10% oxalic; 1000 × .) *(Courtesy of Armco Steel Co.)*

condition A, these alloys may be fabricated almost as easily as if they were true austenitic stainless steels.

After fabrication in the soft condition, which is an advantage of these alloys, the austenite is conditioned to allow transformation to martensite. The conditioning treatment consists of heating the austenite to a high enough temperature to remove carbon from solid solution and precipitate it in the form of chromium carbide ($Cr_{23}C_6$). Precipitation occurs first at the ferrite-austenite interfaces, as shown in Fig. 7-31. Removing the carbon and some of the chromium from the austenite matrix makes the austenite unstable, and upon cooling to the M_s temperature the austenite transforms to martensite.

The 17-7PH steel is conditioned at 760°C (1400°F) and is cooled to about 16°C (60°F) to produce the T condition (Fig. 7-31). Conditioning at a higher temperature of 950°C (1750°F) results in fewer carbides being precipitated (Fig. 7-32), and thus the steel must be cooled to a lower temperature [about −73°C (−110°F)] to transform the austenite to martensite. Note that, since carbon and chromium lower the M_s temperature, removing *less* carbon and chromium will leave the alloy with a *lower* M_s temperature.

The final step in the heat treatment of the semiaustenitic precipitation-hardening stainless steels is *precipitation hardening*, which is carried out in the 480 to 650°C range (900 to 1200°F). The effects of precipitation hardening cannot be observed in the optical microscope (Fig. 7-33), but in the electron microscope a veining condition is observed when these alloys are in the precipitation-hardened condition (Fig. 7-34). During precipitation hardening, the

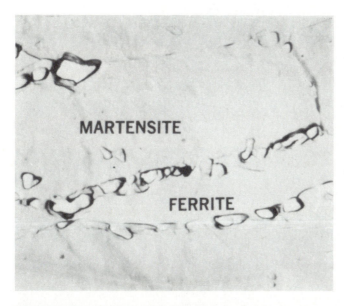

Figure 7-31 Type 17-7PH (precipitation hardening) stainless steel in condition T (transformed) showing carbides at ferrite–martensite interface. (Etchant: Vilella's; 18,000 × . Electron micrograph, plastic replica.) *(Courtesy of Armco Steel Co.)*

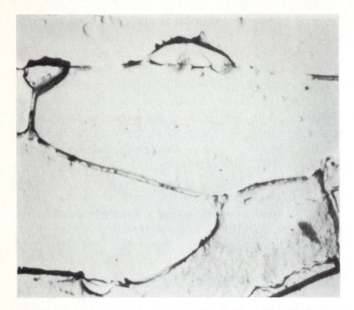

Figure 7-32 Type 17-7PH (precipitation hardening) stainless steel in condition R-100 (transformed) showing carbides and grain boundaries. Note: even though this alloy was more heavily etched than the one in Fig. 7-31, fewer carbides are observed. (Etchant: (1) nitric–acetic, electrolytic; (2) 10% oxalic, electrolytic; 18,000 × ; electron micrograph, plastic replica.) *(Courtesy of Armco Steel Co.)*

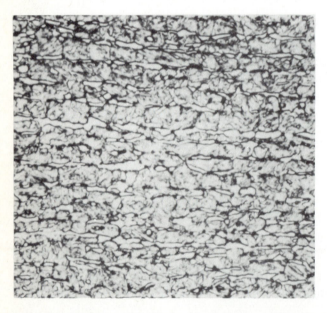

Figure 7-33 17-7PH (precipitation hardened) stainless steel in the transformed and precipitation hardened condition. At this magnification the effects of precipitation cannot be detected. (Etchant: (1) nitric–acetic, electrolytic; (2) 10% oxalic, electrolytic; 1000 × .) *(Courtesy of Armco Steel Co.)*

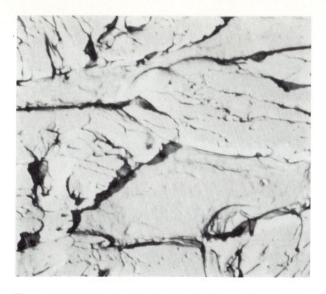

Figure 7-34 17-7PH (precipitation hardening) stainless steel in the transformed and precipitation hardened conditions TH-1050 and RH-950. Note veining in the martensite. (Etchant: (1) nitric–acetic, electrolytic; (2) 10% oxalic, electrolytic; 18,000 × ; electron micrograph, plastic replica.) *(Courtesy of Armco Steel Co.)*

aluminum in the martensite combines with some of the nickel to produce precipitates of NiAl and Ni₃Al, which strengthen the alloy considerably (Table 7-9).

Figure 7-35 summarizes the standard heat treatments used for the 17-7PH and PH 15-7Mo alloys. Typical mechanical properties of these alloys in the most common aged conditions are given in Table 7-9.

Table 7-9 Typical mechanical properties of selected semiaustenitic precipitation-hardenable stainless steels†

Grade	Condition	Form	0.2% yield strength		Ultimate tensile strength		Elongation in 2 in (50.8 mm), %	Hardness, R_c
			ksi	MPa	ksi	MPa		
17-7PH	TH-1050	Sheet	185	1276	200	1379	9	43
PH 15-7Mo	RH-950	Sheet	225	1551	240	1655	6	48
PH 15-7Mo	CH-900	Sheet	260	1793	265	1827	2	49
PH 14-8Mo	SRH-950	Sheet	215	1482	230	1586	6	48
AM-350	SCT-850	Sheet	175	1207	206	1420	12	46
AM-355	SCT-850	Sheet	181	1248	219	1510	13	48

† After Ref. 1, p. 7–19.

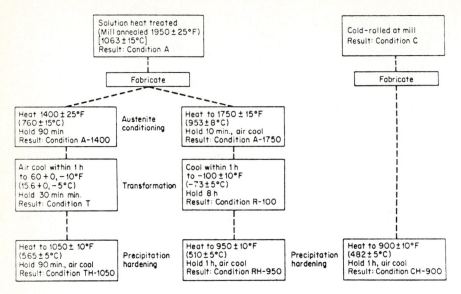

Figure 7-35 Standard heat treatments for Armco 17-7PH and PH 15-7Mo. *(After Handbook of Stainless Steels, McGraw-Hill, New York, 1977, p. 7-8.)*

Martensitic Type

From a weight-use standpoint, the martensitic precipitation-hardenable stainless steels are used more than any other type. Because of their relatively high hardness in the solution-annealed condition, these steels are used principally in the form of bar, rod, wire, and heavy forgings, and only to a minimum extent in the form of sheet. The austenitic and ferritic balance in these steels is such that, after solution heat treatment and cooling to room temperature, they are in the martensitic condition.

Since the development of Stainless W in the 1940s, a whole series of martensitic precipitation-hardening stainless steels have been developed. Table 7-10 lists the chemical compositions of four of these alloys. The principal hardening element for the 17-4PH steels is copper, which forms a highly dispersed copper phase. In Custom 450, a lower copper content is used with an addition of molybdenum.

Stainless W and 17-4PH steels have a two-phase structure consisting of some stringers of δ ferrite (usually less than 10 percent) in a martensitic matrix. The ferrite causes poor through-thickness properties at all strength levels, especially in heavy sections. Newer martensitic precipitation hardening-stainless steels such as 15-5PH and Custom 450 are produced essentially ferrite-free, and thus have improved through-thickness properties.

Table 7-10 Nominal chemical compositions of selected martensitic precipitation-hardenable stainless steels

Grade	% C	% Mn	% Si	% Cr	% Ni	% Mo	% Al	% Cu	% Ti	% Cb
Moderate strength										
17-4PH‡	0.04	0.30	0.60	16.0	4.2			3.4		0.25
15-5PH‡	0.04	0.30	0.40	15.0	4.5			3.4		0.25
Custom 450§	0.03	0.25	0.25	15.0	6.0	0.8		1.5		0.3
Stainless W¶	0.06	0.50	0.50	16.75	6.25		0.2		0.8	
High strength										
PH 13-8Mo‡	0.04	0.03	0.03	12.7	8.2	2.2	1.1			
Custom 455§	0.03	0.25	0.25	11.75	8.4			2.5	1.2	0.3

† After Ref. 1, p. 7–14.
‡ 17-4PH, PH 13-8Mo, and 15-5PH are registered trademarks of the Armco Steel Corporation.
§ Custom 450 and Custom 455 are trademarks of the Carpenter Technology Corporation.
¶ Stainless W is a trademark of the United States Steel Corporation.

These steels are heat-treated by solution annealing, cooling to room temperature, and aging. The 17-4PH steel is customarily solution-annealed at about 1040°C (1400°F), and then aged for maximum hardness at 450 to 510°C (850 to 950°F). The effect of treatments on the tensile and yield strengths of 17-4PH steel is shown in Fig. 7-36. Table 7-11 lists typical mechanical properties for selected martensitic precipitation-hardening stainless steels.

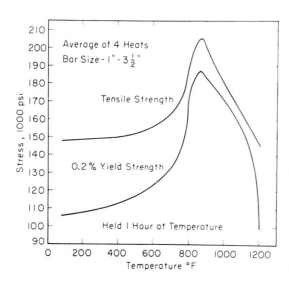

Figure 7-36 The effect of aging temperature on the tensile and yield strengths of 17-4PH precipitation hardenable stainless steels. *(After Armco Steel Co., as presented in "Precipitation from Solid Solution," American Society for Metals, 1959, p. 268.)*

Table 7-11 Typical mechanical properties of selected martensitic precipitation-hardenable stainless steels†

Name	Condition	Form	0.2% yield strength ksi	0.2% yield strength MPa	Ultimate tensile strength ksi	Ultimate tensile strength MPa	Elongation in 2 in (50.8 mm) or 4D, %	Reduction of area, %	Hardness, R_c
					Moderate strength				
17-4PH	H-925	Bar	175	1207	190	1310	14	54	42
15-5PH	H-925	Bar	175	1207	190	1310	14	54	42
Custom 450	H-900	Bar	184	1269	196	1351	14	60	42
Stainless W	H-950	Bar	180	1241	195	1345	10		42
					High strength				
PH 13-8Mo	H-950	Bar	210	1448	225	1551	12	50	47
Custom 455	H-900	Bar	235	1620	245	1689	10	45	49

† After Ref. 1, p. 7–114.

PROBLEMS

1. In order to make a "stainless steel" stainless, about what weight percent chromium in iron is necessary?

2. By what mechanism does chromium protect the stainless steel surface?

3. What is the "γ loop" in the iron-chromium phase diagram?

4. Is chromium a ferritic or austenitic stabilizer? Explain.

5. What is the σ phase in Fe-Cr alloys? Why is it considered detrimental in engineering alloys?

6. Is carbon a ferritic or austenitic stabilizer in Fe-Cr-C alloys? Explain.

7. List the general sequence of carbide formation in Fe-Cr alloys and indicate the chemical composition for which each carbide is stable.

8. Describe how increasing the carbon content of Fe-Cr alloys from 0.05 to 0.4 percent changes the Fe-Cr phase diagram.

9. Is nickel a ferritic or austenitic stabilizer in Fe-Ni alloys? Explain.

10. How does the addition of 4 and 8% Ni affect the Fe–18% Cr phase diagram?

11. What makes the Fe–18% Cr–8% Ni alloy so special?

12. Can small amounts of carbon (i.e., about 0.08 percent) be retained in solid solution in an Fe–18% Cr–8% Ni alloy by (a) rapid cooling from high temperature, or (b) slow cooling from high temperature? Explain.

13. List the four main types of wrought stainless steels and describe their principal characteristics.

14. Why are the ferritic stainless steels of interest to the design engineer?

15. Why does the structure of ferritic stainless steels remain essentially ferritic at all normal heat treatment temperatures?

16. Describe the two main types of ferritic stainless steels.

17. Why are the ferritic stainless steels used in the annealed condition? Why are they not heat-treated to produce martensite?

18. List the three types of embrittlement found in ferritic stainless steels and describe each of these.

19. What is the mechanism believed to be the cause of the 475°C embrittlement in ferritic stainless steels?

20. What is the cause of the high-temperature embrittlement problem in stainless steels?

21. How can the high-temperature embrittlement problem in ferritic stainless steels be circumvented?

22. What techniques are used to lower the carbon and nitrogen contents in the "new" ferritic stainless steels?

23. How does the lowering of the carbon and nitrogen contents of Fe–17% Cr alloys affect their ductile-brittle transition temperatures?

24. How does the lowering of the carbon and nitrogen contents affect the corrosion resistance of ferritic stainless steels?

25. What is the mechanism which makes carbon-containing ferritic stainless steels susceptible to intergranular corrosion?

26. What alloying addition is commonly added to ferritic stainless steels to improve their pitting resistance?

27 Why is carbon an essential alloying addition to martensitic stainless steels?

28. How is it that in martensitic stainless steels with higher carbon contents, i.e., 0.6 to 1.1 percent, a martensitic structure can be obtained with chromium contents of 16 to 18 percent?

29. What is believed to be the cause of the lowered impact values between 370 and 600°C in tempered Fe–12% Cr stainless steels?

30. What is believed to be the cause of the increase in hardness at about 450°C when tempering an Fe–12% Cr martensitic stainless steel?

31. Why is the corrosion resistance of the martensitic stainless steels relatively poor compared to that of austenitic or ferritic ones?

32. What properties make the austenitic stainless steels very attractive for engineering materials?

33. What other alloying additions besides nickel contribute to the retention of an austenitic structure in stainless steels?

34. What is the mechanism which causes intergranular corrosion in austenitic stainless steels?

35. What term is used to describe austenitic stainless steels that are susceptible to intergranular corrosion?

36. Why must austenitic stainless steels containing about 0.08% C be rapidly cooled after annealing at 1050 to 1120°C?

37. How can additions of titanium and columbium prevent intergranular corrosion in austenitic stainless steels?

38. What is a stabilizing treatment with respect to austenitic stainless steels?

39. What other method can be used to prevent the precipitation of carbides in austenitic stainless steels?

40. What causes the accelerated strain-hardening effect in type 301 austenitic stainless steel as compared to that observed for type 304 alloy?

41. In what temperature range are austenitic stainless steels the most subject to intergranular corrosion? Why this temperature range?

42. What are the advantages of precipitation-hardening stainless steels as engineering materials?

43. Why is the chemical composition of semiaustenitic precipitation-hardening stainless steels so critical?

44. Describe the heat treatment sequence necessary to harden the 17-7PH type of stainless steels.

45. What is the precipitation-hardening mechanism in the 17-7PH stainless steels?

46. Why is the presence of δ ferrite detrimental to the strength properties of martensitic precipitation-hardening stainless steels?

47. What is believed to be the hardening precipitate in the 17-4PH alloys?

REFERENCES

1. D. Peckner and I. M. Bernstein (eds.): "Handbook of Stainless Steels," McGraw-Hill, New York, 1977.
2. L. Colombier and J. Hochmann: "Stainless and Heat Resisting Steels," Edward Arnold Publishers Ltd., London, 1967.
3. "Advances in the Technology of Stainless Steels and Related Alloys," ASTM Spec. Tech. Publ. 369, 1965.
4. F. B. Pickering: "Physical Metallurgy of Stainless Steel Developments," *Int. Metals Rev.*, Dec. 1976, p. 227.
5. "Toward Improved Ductility and Toughness," Climax Molybdenum Co., Greenwich, CN, 1971.
6. R. A. Lula: "Ferritic Stainless Steels: Corrosion + Economy," *Metal Progress* 110 no. 2, (1976):24.
7. Symposium on "New Developments in Stainless Steels," *Trans. AIME* 245(1969):2117.
8. *Metals Handbook*, vol. 7: "Atlas of Microstructures," American Society for Metals, Metals Park, OH, 1972.
9. H. E. McGannon (ed.): The Making, Shaping, and Treating of Steel," 9th ed. United States Steel Co., Pittsburgh, 1971, p. 1163.
10. K. T. Aust: "Intergranular Corrosion of Austenitic Stainless Steels," *Trans. AIME* 245(1969):2117.
11. P. J. Grobner: "The 885°F(475°C) Embrittlement of Ferritic Stainless Steels," *Met. Trans.* 4(1973):251.
12. K. T. Aust, J. S. Armijo, E. F. Koch, and J. H. Westbrook: "Intergranular Corrosion and Electron Microscopic Studies of Austenitic Stainless Steels," *Trans. ASM* 60(1967):363.
13. R. Stickler and A. Vinckier: "Morphology of Grain-Boundary Carbides and its Influence on Intergranular Corrosion of 304 Stainless Steel," *Trans. ASM* 54(1961):362.
14. A. P. Bond: "Mechanisms of Intergranular Corrosion in Ferritic Stainless Steels," *Trans. AIME* 245(1969):2127.
15. J. J. Demo: "Weldable and Corrosion-Resistant Ferritic Stainless Steels," *Met. Trans.* 5(1974):2253.
16. I. A. Franson: "Mechanical Properties of High Purity Fe-26Cr-1Mo Ferritic Stainless Steel," *Met. Trans.* 5(1974):2257.
17. C. S. Carter et al.: "Stress Corrosion Properties of High Strength Precipitation Hardening Stainless Steels," *Corrosion* 27(1971):190.
18. B. R. Banerjee, J. J. Hauser, and J. M. Capenos: "Structure and Properties of PH15-7Mo Stainless," *Trans. ASM* 57(1964):856.
19. B. Weiss and R. Stickler: "Phase Instabilities During High Temperature Exposure of 316 Austenitic Stainless Steels," *Met. Trans.* 3(1972):851.
20. "Stainless Steel 77," Climax Molybdenum Co., Greenwich, CN, 1978.
21. F. B. Pickering: "The Metallurgical Evolution of Stainless Steels," American Society for Metals, Metals Park, OH, 1979.

CAST IRONS

Cast irons are a family of ferrous alloys with a wide diversity of properties and, as their name implies, they are intended to be cast into the desired shape rather than being worked in the solid state. Unlike steels which contain less than 2% C and usually less than 1% C, cast irons normally contain from about 2 to 4% C and 1 to 3% Si. Other alloying metallic and nonmetallic elements are added in order to control and vary specific properties. Besides chemical composition, other important factors which affect their properties are the solidification process, solidification rate, and subsequent heat treatments. Cast irons make excellent casting alloys, have a wide range of strengths and hardness, and in most cases are easy to machine. They are alloyed to produce superior wear, abrasion, and corrosion resistance. Their widespread use is primarily the result of their comparatively *low cost* and *versatile engineering properties*. In spite of vigorous competition from new materials, cast irons have proven to be the most economical and suitable materials for thousands of engineering applications.

8-1 CLASSIFICATION OF CAST IRONS

Four basic types of cast irons can be differentiated from each other by the distribution of the carbon in their microstructures. Since their chemical compositions overlap, they cannot be distinguished by chemical analyses. These four basic metallurgical types are *white iron, gray iron, malleable iron,* and *ductile iron. High-alloy cast irons* constitute a fifth type of cast iron. Table 8-1 lists the range of chemical compositions of typical unalloyed cast irons, and Fig. 8-1 shows the approximate ranges of their carbon and silicon contents in comparison with the steels.

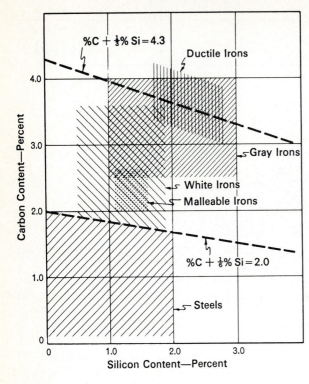

Figure 8-1 The approximate range in carbon and silicon contents of ferrous alloys. [*After C. F. Walton (ed.), "Gray and Ductile Iron Castings Handbook," Gray and Ductile Founders' Society, Inc., Cleveland, 1971.*]

White Cast Iron

If the chemical composition of the cast iron is in the white-cast-iron range (Table 8-1) and the solidification rate is fast enough, white cast iron will be produced. In white cast iron, the carbon in the molten iron remains combined with iron in the form of iron carbide or cementite, which is a hard brittle compound (Fig. 8-2). White cast iron is therefore relatively hard and brittle, and

Table 8-1 Chemical composition ranges for typical unalloyed cast irons†

Element	Gray iron, %	White iron, %	Malleable iron (cast white), %	Ductile iron, %
Carbon	2.5–4.0	1.8–3.6	2.00–2.60	3.0–4.0
Silicon	1.0–3.0	0.5–1.9	1.10–1.60	1.8–2.8
Manganese	0.25–1.0	0.25–0.80	0.20–1.00	0.10–1.00
Sulfur	0.02–0.25	0.06–0.20	0.04–0.18	0.03 max
Phosphorus	0.05–1.0	0.06–0.18	0.18 max	0.10 max

† After Ref. 1, p. 94.

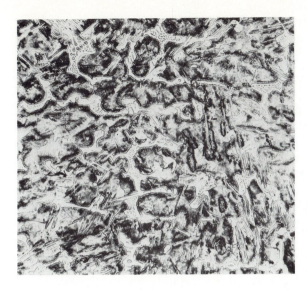

Figure 8-2 Microstructure of white cast iron. The white constituent is iron carbide. The gray areas are unresolved pearlite. (Etch: 2% nital; X100.) (*Courtesy of Central Foundry.*)

shows a "white" crystalline fractured surface. White iron has high compressive strength and excellent wear resistance.

Gray Cast Iron

If the chemical composition of the cast iron is in the gray-cast-iron range and the solidification rate is correct, the carbon in the iron separates or *graphitizes* during solidification to form separate graphite flakes (Fig. 8-3). Gray cast irons are the most fluid of the ferrous alloys and, as a result, intricate and thin sections can be produced. These irons have excellent machinability at hardness

Figure 8-3 Pearlitic gray cast iron in the annealed condition; structure shows graphite flakes as dark etched constituent. (Etch: 2% nital; X100.) (*Courtesy of Central Foundry.*)

levels to provide good wear resistance. The fractured surface appearance of gray cast iron has a gray sootish color, and hence the term "gray cast iron."

Malleable Cast Iron

This type of cast iron has most of its carbon in the form of irregularly shaped nodules of graphite (Fig. 8-4). Malleable cast iron is first cast as white iron of a suitable composition. Then, during an annealing treatment often called *malleab- lizing*, the graphite nucleates and grows from the cementite of the white iron to form nodules. A wide range of mechanical properties can be obtained in malleable iron by varying the annealing heat treatment. However, since rapid solidification is needed to first form the white iron, the metal thickness of malleable iron castings is limited.

Ductile Cast Iron

Ductile iron has its free carbon in the form of spheres instead of flakes. For this reason it is sometimes referred to as *nodular cast iron* in the United States and as *spherulitic graphite iron* (SG iron) in England. Figure 8-5 shows the microstruc- ture of ductile cast iron in the annealed condition. The spheroidal graphite in these irons is obtained by adding a very small amount of magnesium to the molten iron before casting. The chemical compositions of ductile irons are similar to gray cast irons, but with low levels of minor elements such as sulfur and phosphorus. Ductile iron has a good range of yield strengths along with reasonable ductility and, in contrast to malleable cast iron, can be cast into a wide range of sizes with thin or very thick sections.

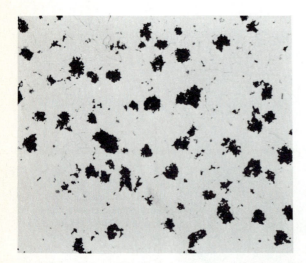

Figure 8-4 Ferritic malleable cast iron in the annealed condition; structure shows irregularly shaped graphite nodules in ferritic matrix. (Etch: 2% nital; X100.) (*Courtesy of Central Foundry*.)

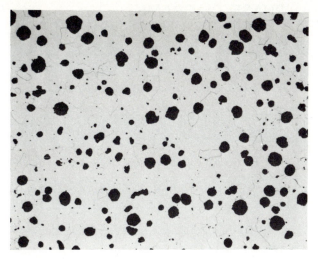

Figure 8-5 Ferritic ductile cast iron in the annealed condition. Structure shows regular spheres of graphite as dark etched constituent. (Etch: 2% nital; X100.) (*Courtesy of Central Foundry*.)

High-Alloy Cast Irons

This group of cast irons includes the high-alloy white irons, highly alloyed gray irons, and highly alloyed ductile irons. The alloy cast irons are grouped separately because they have special properties considerably different from those of unalloyed or low-alloyed cast irons, such as high abrasive wear resistance, heat resistance and corrosion resistance. They are usually specified by their chemical compositions, but mechanical property requirements may also be included.

8-2 THE IRON-CARBON-SILICON SYSTEM

Cast irons contain an appreciable amount of silicon (about 1 to 3 percent) as well as high carbon contents, and must therefore be considered ternary Fe-C-Si alloys. Since the presence of silicon in Fe-C alloys promotes graphitization, cast irons may solidify in either the iron-iron carbide system or the iron-graphite system or even both. Long holding times at high temperatures, slow cooling, and the presence of certain alloying elements favor the formation of graphite nuclei, and thus promote the change from the metastable iron-carbide phase to the stable graphite phase. On the other hand, rapid cooling and the presence of certain alloying elements can assist in preventing the nucleation of graphite and in retaining the iron-carbide phase.

Figure 8-6 shows the effects of additions of 2 and 4% Si in modifying the iron-carbon phase diagram. Silicon additions of 2 and 4 percent decrease the eutectoid carbon content to 0.6 and 0.4 percent, respectively. Also, silicon

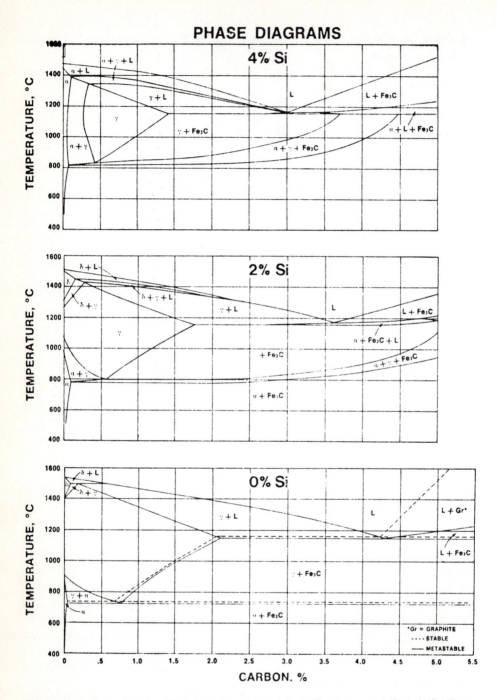

PHASE DIAGRAMS

4% Si

2% Si

0% Si

*Gr = GRAPHITE
···· STABLE
— METASTABLE

CARBON, %

Figure 8-6 Vertical sections of the iron-carbon-silicon ternary alloy system at 0, 2, and 4% silicon. (*Courtesy of the American Foundrymen's Society.*)

320

additions of 2 and 4 percent lower the maximum solid solubility of carbon in austenite to 1.7 and 1.4 percent, respectively. Accordingly, silicon additions lower the carbon content of the pearlite in these alloys. Further, the addition of silicon to the Fe-C system causes the eutectic and eutectoid reactions to take place over a range of temperatures and at higher temperatures than in the Fe-C system. The temperature range increases as the amount of silicon increases.

8-3 GRAY CAST IRON

Since gray cast iron has so many useful characteristics, it is still the favorite of design engineers for casting simple and intricate shapes in both small and huge sizes. Today approximately 75 percent by weight of all castings are made with gray iron because of its performance advantages and low cost. The flake graphite in gray iron provides it with some of its special properties such as excellent machinability at hardness levels that produce superior wear resistance, the ability to resist galling with restricted lubrication, and excellent vibration damping. Gray iron is comparable to higher-strength steels for applications where compressive strength, dimensional stability, and accurate alignment under stress are required.

Classes of Gray Cast Iron

Gray cast irons are usually classified by the minimum tensile strength attained with a given section size. Most gray irons are classified by the ASTM specification A48, which has classes ranging from 20,000 to 60,000 psi tensile strength (Table 8-2). Other specifications are used for special products. The strength of gray iron depends mainly on the structure of the matrix and the size, distribution, and type of graphite flakes.

On the basis of carbon content of the ternary section of the Fe-C-Si phase diagram (Fig. 8-6), gray cast irons can be classified as hypoeutectic or hypereutectic. For example, a gray cast iron with 2% Si has its eutectic composition at about 3.6% C. Any gray cast iron with less than 3.6% C and 2% Si would be

Table 8-2 Classes of gray cast irons according to ASTM Specification A48†

Class	Minimum tensile strength, psi
20A	20,000
30A	30,000
40A	40,000
50A	50,000
60A	60,000

† Tensile specimen has 0.88-in nominal diameter.

classified as hypoeutectic, while one with more than 3.6% C and 2% Si would be hypereutectic.

Slow Solidification of an Hypoeutectic Gray Cast Iron

Consider the *slow* solidification of a hypoeutectic (3% C–2% Si) gray cast iron with reference to Fig. 8-7. Under slow (equilibrium) solidification conditions, solid primary austenitic dendrites begin to fo m at the liquidus [point (1)] (about 1250°C) and continue to grow into the liqu d until the beginning of eutectic solidification, which occurs at about 1150°C at point (2). The eutectic will then

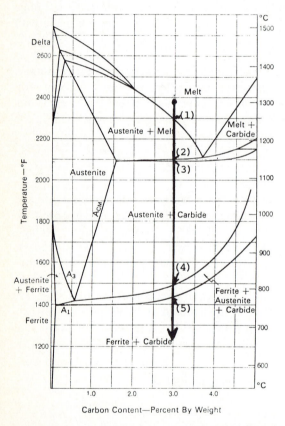

Figure 8-7 The section of the iron-graphite-silicon. Ternary equilibrium diagram at 2% silicon. (1) to (2): Austenite dendrites begin to form and grow until the temperature at (2) is reached. (2) to (3): Eutectic freezing occurs between temperatures (2) to (3). If iron is freezing as gray iron, the eutectic will be a mixture of austenite and graphite. (3) to (4): As the temperature is decreased in the austenite + carbide region, carbon will be rejected from the austenite as graphite and precipitate on the graphite flakes in the eutectic. (4) to (5): Equilibrium cooling through the eutectoid range will result in the transformation of austenite to ferrite and the precipitation of the remaining carbon on the graphite flakes. [*After C. F. Walton (ed.), "Gray and Ductile Iron Castings Handbook," Gray and Ductile Founders' Society, Inc., Cleveland, 1971, p. 364.*]

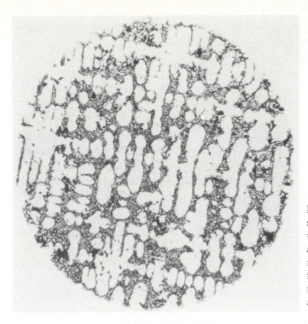

Figure 8-8 Microstructure of a slowly cooled Fe-C-Si hypoeutectic alloy. Note that the ferrite areas were formed from the original austenite dendrites and the ferrite-graphite eutectic was formed from the original austenite-graphite eutectic.

solidify over a small temperature range [points (2) to (3)], after which the alloy will consist of primary austenitic dendrites and an austenitic-graphite eutectic. At the solidus, the austenite is saturated with about 1.7% C.

Slow cooling below the solidus from points (3) to (4) is accompanied by the rejection of carbon from the austenite and its precipitation at the existing graphite in the eutectic. The excess carbon continues to precipitate until the eutectoid temperature of about 800°C is reached. Cooling through the eutectoid range [points (4) to (5)] causes the austenite to transform to ferrite and the excess carbon to be precipitated on the existing graphite flakes. The final microstructure then consists of ferritic regions originating from the primary austenitic dendrites, along with other regions of mixed ferrite and graphite flakes which originate from the austenitic-graphite eutectic. This type of structure is shown in Fig. 8-8. In commercial cast irons, the solidification process is much more complex because of the effects of the presence of many other elements and the introduction of other variables such as solidification rate and section size.

Effects of Chemical Composition on the Microstructure of Gray Cast Iron

Carbon and silicon are the major alloying elements in gray cast iron and have the greatest effect on its microstructure. However, all elements influence that microstructure to some degree. Elements which promote *graphitization* (the formation of graphite) are called *graphite stabilizers*. Silicon is a strong graphite stabilizer and is the most important single compositional factor promoting graphitization in gray cast irons. Graphitization is the process whereby either free carbon precipitates in the iron or the iron carbide (Fe_3C) decomposes into free carbon (graphite) and iron according to the reaction

$$Fe_3C \rightarrow 3Fe + C \text{ (graphite)}$$

Other elements can stabilize the iron carbide, and are called *carbide stabilizers*. Chromium, manganese, and sulfur are examples of carbide-stabilizing elements.

Carbon and silicon Both carbon and silicon promote the formation of graphite in gray cast iron, and thus, as the percentages of these elements are increased, the formation of gray iron will be favored over white, as indicated in Fig. 8-9. If the amount of carbon and silicon is decreased below critical levels, white cast iron will be formed. A mottled cast iron consisting of mixed white and gray can be produced as an intermediate structure.

Carbon in gray cast iron can exist in the form of graphite or as iron carbide. If graphitization is complete, the gray cast iron will have graphite flakes with a ferritic matrix. However, if 0.5 to 0.8 percent of the carbon is combined in the form of Fe_3C, the matrix of the cast iron will be pearlitic, as shown in Fig. 8-3.

The silicon content of gray cast iron ranges from 1.0 to 3.5 percent by weight. Increasing the silicon content of Fe-C-Si alloys shifts the eutectic composition to the left (Fig. 8-6). This eutectic shift can be expressed by the following equation:

$$\% \text{ Eutectic carbon(Fe-C-Si alloy)} = 4.3 - 0.33 \times \% \text{ Si (in alloy)}$$

Many properties of gray cast irons can be related to a term called the *carbon equivalent* (CE). The carbon equivalent takes into account both the carbon and silicon contents of the cast iron by the following relationship:

$$\text{Carbon equivalent} = \% \text{ C (in iron)} + \tfrac{1}{3}\% \text{ Si (in iron)}$$

Since the eutectic composition of the binary Fe-C system is 4.3% C, a carbon equivalent of about 4.3 indicates that the alloy is approximately of eutectic composition. An alloy with *less* than a CE of 4.3 would be hypoeutectic, and one with *more* than 4.3 would be hypereutectic.

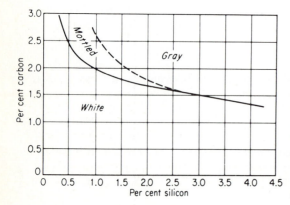

Figure 8-9 The effect of carbon and silicon percentages on the type of cast iron formed. [*After C. R. Loper, Jr., and R. W. Heine, Trans. AFS 68 (1960): 313, as presented in R. W. Heine, C. R. Loper, and P. C. Rosenthal, "Principles of Metal Casting" McGraw-Hill, New York, 1967, p. 579.*]

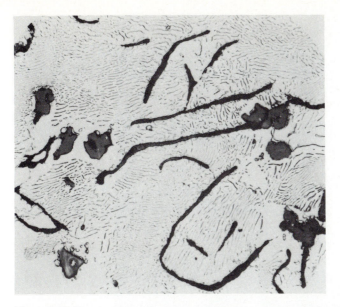

Figure 8-10 Gray cast-iron microstructure showing compact angular gray particles of manganese sulfide. Since MnS solidifes above the solidification temperature of iron, this compound forms as separate particles. (Etch: 2% nital; X250.) [*After C. F. Walton (ed.), "Gray and Ductile Iron Castings Handbook," Gray and Ductile Iron Founders' Society, Cleveland, 1971, p. 104.*]

Sulfur and manganese Sulfur is present to some degree in all cast irons. For ductile cast irons, the sulfur content must be kept very low in order to allow the formation of spheroidal graphite upon the addition of magnesium. However, for the other cast irons, the influence of sulfur must be considered relative to its reaction with manganese.

Without manganese in the cast iron, sulfur will combine with iron to form iron sulfide (FeS), which segregates into the grain boundaries during freezing. When manganese is present in the cast iron, MnS, or complex manganese-iron-sulfides, precipitate during the entire solidification process. As a result, a random dispersion of angular manganese sulfide particles is created. Figure 8-10 shows some of these geometrically shaped MnS particles. These particles have little influence on the castability and use properties of commercial cast irons.

The effect of both sulfur and manganese alone in cast irons is to restrict graphitization and promote pearlite formation. Thus, either sulfur or manganese alone in cast irons is a carbide-stabilizing element. When both are present, however, their carbide-stabilizing effects are nullified. If a pearlitic structure is desired, for example in a gray cast iron, excess manganese above that necessary to combine with the sulfur as MnS is added to the cast-iron melt.

Phosphorus Cast irons containing sufficient phosphorus, especially gray irons, can form an eutectic of iron and iron phosphide called *steadite*. Steadite, which

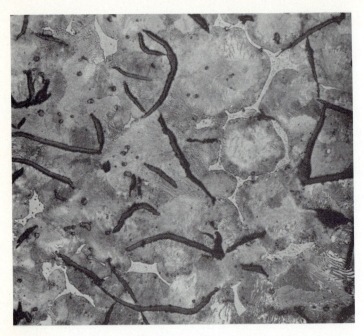

Figure 8-11 Gray cast-iron microstructure showing the phosphorus constituent steadite which forms as separate particles at lower phosphorus contents. [*After C. F. Walton (ed.), "Gray and Ductile Iron Castings Handbook," Gray and Ductile Iron Founders' Society, Cleveland, 1971, p. 105.*]

has a low melting point (between 954 and 980°C), solidifies at a relatively low temperature and segregates at the boundaries of the solidification cells. At the 0.2% P level, which is typical of many cast irons, steadite solidifies at the junction of three cells to form concave triangular-shaped constituents such as those shown in Fig. 8-11.

At higher phosphorus levels, steadite forms much larger constituents. Since iron phosphide is hard and brittle, an increase in the amount of steadite (i.e., above 0.3% P) in a cast iron can increase its hardness and brittleness and decrease machinability. However, this hard compound also increases wear resistance, and so its presence is desirable for some applications.

Graphitization during Solidification

During the solidification of gray cast iron, various sizes, shapes, and distributions of graphite flakes can develop. Five basic types of graphite flakes (types A to E) have been established by ASTM and AFS as standards and are shown in Fig. 8-12. Type A flakes with a random distribution and small size are considered desirable. However, in practice interdendritic segregated and cellular types are often present. The type E flakes with interdendritic segregation and preferred orientation are often found in hypoeutectic cast irons. Many researchers

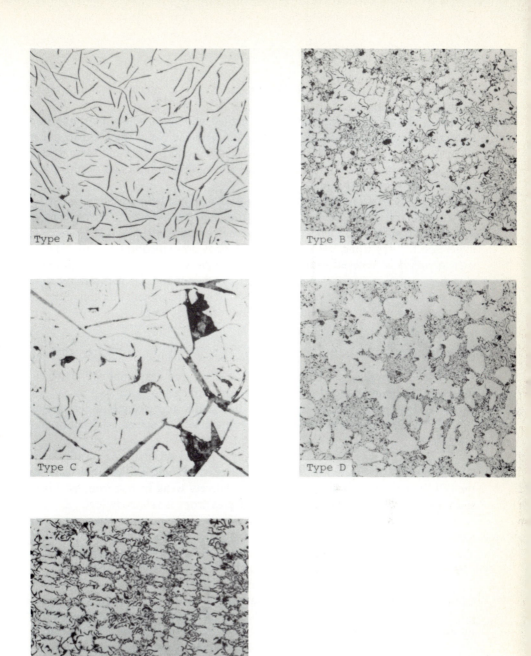

Figure 8-12 The five types of flake graphite as established by ASTM specification A 247. Type A—uniform distribution and random orientation. Type B—rosette groupings and random orientation. Type C—superimposed flake size and random orientation. Type D—interdendritic segregation and random orientation. Type E—interdendritic segregation and preferred orientation. (*After Metals Handbook, 8th ed., vol. 7, American Society for Metals, 1972, Metals Park, OH, p. 82.*)

attribute the formation of type D graphite flakes to an undercooling effect during solidification.

Large flakes which are randomly oriented will form when the nucleation and solidification rates are low and graphitization is easy. Small flakes occur when the nucleation rate is high due to moderate undercooling and there is time for graphitization. Severe undercooling prevents graphitization and results in the formation of white cast iron.

The addition of *inoculants*[1] to molten gray iron just before casting can affect the eutectic cell size, graphite pattern, and metallic matrix. If ferrosilicon or some other graphitizing agent is added in small amounts (0.05 to 0.25 percent) to hypereutectic gray irons, the formation of type A graphite and fine eutectic cells is favored.[2] It is believed that the inoculation provides nuclei for the graphite eutectic formation, and thereby prevents undercooling of the solidification temperature. Hypoeutectic gray irons respond well to inoculation, but little or no effect is obtained for eutectic or hypereutectic gray irons.

Microstructures

Typical microstructures of classes 20, 30, and 40 gray cast irons are shown in Fig. 8-13. The matrix metal provides the basic strength of gray cast irons. The harder and stronger the matrix metal is, the harder and stronger the gray iron will be. The graphite flakes have a weakening effect on strength by acting like notches.

The class 20 gray irons have essentially a matrix of ferrite, as shown in the microstructures of Fig. 8-13a and b. However, dark bands of pearlite occur at the cell boundaries. The ferritic matrix is relatively weak in cast iron, as it is in steel, and thus the strength of this class of gray irons is relatively low.

Class 30 gray irons are stronger (30,000 psi min) and their microstructures have a matrix of mixed pearlite and ferrite, as shown in Fig. 8-13c and d. However, the graphite flakes are rather coarse, although of the desirable type A. The mixed pearlitic-ferritic matrix is stronger than the straight ferritic type.

The microstructure of a class 40 gray cast iron is shown in Fig. 8-13e and f. In this case, rapid solidification has created a matrix of fine pearlite and type D graphite flakes. Numerous carbide particles can also be seen, and were created by the rapid solidification.

The rate of solidification has a major effect on the microstructure, and hence upon the mechanical properties, of gray cast iron. Slow cooling rates lead to a coarsening of the graphite flakes and of the pearlitic lamellae. Very slow cooling favors a ferritic matrix. Thus, slow cooling results in lower strength structures, as shown in Fig. 8-13a and b. Rapid solidification favors a pearlitic

---

[1] An inoculant may be defined as an addition to molten iron which produces effects far out of proportion to any change in chemical composition.

[2] H. D. Merchant, L. I. Toriello, and J. F. Wallace, *AFS Trans.* 69(1961):117.

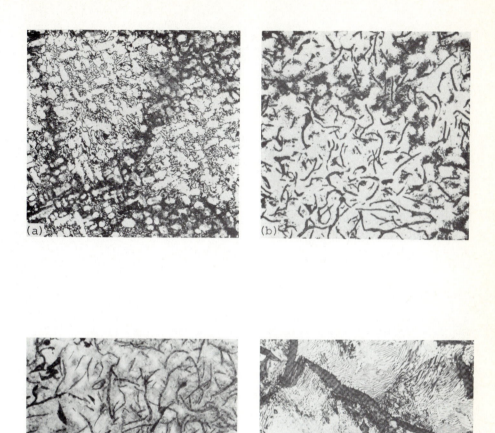

Figure 8-13 Microstructures of Class 20, 30, and 40 gray cast irons. (*a*) Class 20 gray iron stress relieved for 1 h between 607 and 621°C. Structure: as cast consisting of type D graphite flakes in a matrix of ferrite, dark bands of pearlite occur at cell boundaries. (3% picral; X100.) (*b*) Class 20 gray iron annealed at 788°C 1 h per inch of thickness; furnace cooled to 427°C, air cooling. Type A graphite in matrix free ferrite and pearlite; dark bands of pearlite at cell boundaries. (5% nital; X100.) (*c*) Class 30 gray iron as-cast in sand mold structure: Type A graphite flakes in a matrix of 20% free ferrite and 80% pearlite (dark constituent). (3% nital; X100.) (*d*) Class 30 gray iron as cast. Structure type A graphite flakes in a matrix of pearlite (alternating lamellae). (3% nital; X500.) (*e*) Class 40 gray iron. Structure: Type D graphite flakes in a matrix of fine pearlite, with numerous carbide particles (light) due to rapid solidification. (2% nital; X100.) (*f*) Class 40 gray iron same as (*e*) but at higher magnification; structure shows details of the fine pearlite in the matrix. (2% nital; X750.) [*After Metals Handbook, 8th ed., vol. 7, American Society for Metals, 1972, p. 82–83.*]

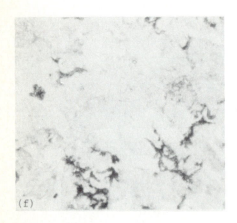

Fig. 8-13 *continued*

matrix and even some iron carbides, resulting in higher strengths. This type of microstructure is shown in Fig. 8-13*e*.

Engineering Properties

Mechanical properties The mechanical properties of gray cast irons result mainly from a combination of the effects of chemical composition and solidification cooling rate. The rate at which gray iron solidifies has a direct influence on the size, shape, and distribution of the graphite in gray cast irons. The cooling rate after solidification is completed acts in a similar manner to a heat treatment. Thus, whether a ferritic or pearlitic matrix is obtained is basically determined by the cooling rate after solidification.

Carbon and silicon are the most important elements which determine the mechanical properties of gray cast iron. In general, as the carbon equivalent is decreased, the strength of the gray iron is increased. The effect of carbon equivalence and section size on the tensile strength of gray cast iron is shown in Fig. 8-14. When tensile strengths above about 50,000 psi are required, alloying additions of chromium, nickel or molybdenum are needed except for very thin sections. Heat treatment by rapid cooling and tempering can also be used to increase the strength of gray cast iron.

The relatively low strength of gray cast iron is due to the interlacing network of graphite flakes, as shown in the scanning-electron micrograph of Fig. 8-15. The higher the graphite content and the coarser the flakes, the more the strength of the cast iron is reduced. The brittle graphite flakes weaken the matrix and act as internal notches to initiate cracks.

The influence of different graphite flake types and austenitic dendrites on the mechanical properties of gray irons at various carbon equivalences has recently been investigated by Ruff and Wallace (Ref. 10). They found that increasing amounts of primary austenitic dendrites improved the tensile strength of the gray irons. Under conditions of comparable amounts of dendrites and similar carbon equivalents, larger percentages of type A graphite and refined eutectic cells appeared to increase tensile strength by reducing the effective span of the graphite flakes. Their best tensile properties were obtained from a

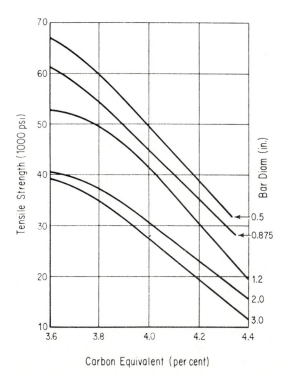

Figure 8-14 Effect of carbon equivalent and section size on the tensile strength of gray cast-iron bars. [Carbon equivalent = %C + 0.3 (%Si + %P).] (*After J. F. Wallace, Foundry, Dec. 1963, p. 40.*)

Figure 8-15 Scanning electron micrograph of hypereutectic gray iron with matrix etched to show position of type B graphite in space. (*After Metals Handbook, 8th ed., vol. 7, American Society for Metals, Metals Park, OH, 1972, p. 82.*)

combination of large amounts of long, primary austenitic dendrites, nearly 100% type A graphite, refined eutectic cells, and a pearlitic matrix. This structure was obtained by adding nitrogen to a base melt containing small amounts of titanium.

Wear resistance Gray cast iron has outstanding resistance to the sliding friction type of wear. For this reason it is used in applications involving sliding surfaces such as cylinder bores and piston rings in internal combustion engines and in sliding ways in machine tools. Gray cast iron has excellent resistance to galling

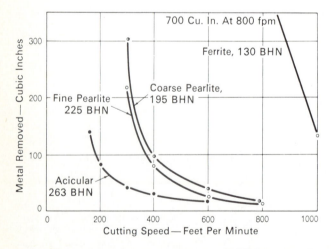

Figure 8-16 The influence of matrix microstructure and cutting speed on the tool life in machining of gray iron. [*After C. F. Walton, (ed.), "Gray and Ductile Iron Castings Handbook," Gray and Ductile Iron Founders' Society, Cleveland, 1971, p. 519.*]

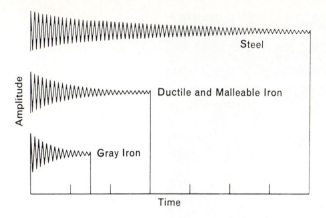

Figure 8-17 The relative ability of ferrous metals to dampen vibrations. The energy absorbed per cycle, or specific damping capacity of these can differ by more than ten times. [*After C. F. Walton (ed.), "Gray and Ductile Iron Castings Handbook," Gray and Ductile Iron Founders' Society, Cleveland, 1971, p. 155.*]

and seizing, which is explained by the lubricating effect of the graphite flakes and retention of oil in the graphite areas.

Machinability Gray cast iron is one of the best machinable ferrous alloys, as indicated in Fig. 8-16. The finer pearlitic matrices, which are stronger and harder, machine at lower speeds. However, a pearlitic matrix has the best combination of machinability and wear resistance for gray cast irons.

Damping capacity Damping capacity is defined as the ability of a material to absorb energy caused by vibrations and thus to dampen them. Gray iron, especially the type with high percentages of flake graphite, rapidly dampens vibrations, as shown in Fig. 8-17. The damping capacity of gray iron is sometimes its greatest advantage for some applications. Gear covers, cylinder blocks, and heads are some of the ways in which the damping capacity of gray iron is utilized.

8-4 DUCTILE CAST IRON

Ductile cast iron consists of graphite spheroids dispersed in a matrix of ferrite, pearlite, or both. During the solidification of ductile iron, most of the carbon forms as graphite spheroids, in contrast to the graphite flakes formed in gray cast iron. The usual as-cast microstructure of ductile iron consists of graphite nodules which are surrounded by free ferrite ("Bull's eye" structure) in a matrix of pearlite (Fig. 8-18).

Ductile iron has an unusual combination of properties because its graphite occurs as nodules rather than flakes. It has advantages of gray cast irons such as

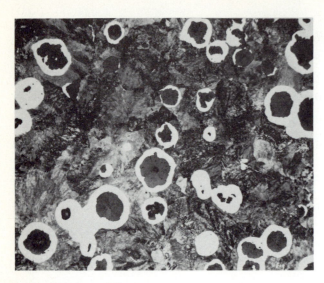

Figure 8-18 Grade 80-55-06 pearlitic ductile iron, as-cast. Structure consists of graphite nodules surrounded by envelopes of free ferrite (bull's eye structure) in a matrix of pearlite. (Etch: 2% nital; X100.) (*Courtesy of Central Foundry*.)

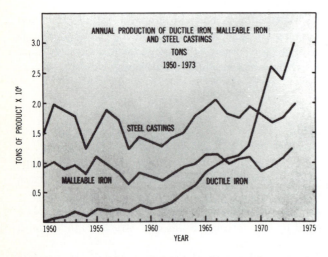

Figure 8-19 Annual production of ductile iron, malleable iron, and steel castings from 1950 to 1973. (*After J. Schuyten, Casting Engineering, Summer, 1975*.)

low melting point, good fluidity and castability, excellent machinability, and good wear resistance. But it also has high strength, ductility, toughness, and hot workability.

As a result of its favorable properties, ductile iron has shown a phenomenal growth in popularity since it was first discovered in 1948.[1] Figure 8-19 shows how its production has greatly increased over the past years. In comparison, malleable-cast-iron and steel-castings production has remained relatively constant.

The principal types and applications for ductile cast irons are listed in Table 8-3. As in the case of gray cast irons, the various grades of ductile cast irons are designated by their tensile strengths. Different grades are produced by changing the matrix microstructure. Some ductile irons are produced and utilized in the as-cast condition, whereas the higher-strength regular grades require heat treatment after casting. There is generally no difference in chemical analyses of the regular grades. Sometimes, however, small chemical additions and changes in foundry procedure are necessary to obtain the desired microstructure.

Solidification of Ductile Cast Iron

To produce ductile cast iron, a small addition of magnesium (about 0.1 percent) is added to molten iron which has 3.0 to 4.0% C and 1.8 to 2.8% Si. The function of the magnesium is to deoxidize and desulfurize the molten iron. If sulfur and oxygen are absorbed on the graphite/melt interface during solidification, graphite flakes such as those found in gray cast iron will be formed. In order to produce the graphite nodules of ductile cast iron, the sulfur and oxygen impurities of the molten iron must be removed. In the absence of these impurities, the normal growth of the graphite leads to a spherulitic morphology.

Figure 8-20 shows the graphite crystal structure. In order for nodules to form, there must be a combination of an unstable interface and basal plane growth, as indicated in Fig. 8-20a. Adsorbed impurities such as sulfur and oxygen poison the growth sites on the basal planes and thereby stabilize the basal plane/metal interface. As a result, graphite flakes (Fig. 8-20c) are formed instead of spheroidal graphite (Fig. 8-20b).

Solidification of ductile iron is a process similar to the solidification of gray iron except that the graphite grows in radial directions and assumes a nodular morphology. Nuclei for spheroidal graphite nodules in ductile iron are probably the same for flake graphite in gray iron except that the products of the nodularizing additions may also serve as nuclei. These products serving as nuclei may be magnesium sulfide or magnesium silicate, identified as $3MgO \cdot 2SiO_2 \cdot 2H_2O$. The action of these products as nuclei in ductile iron could help explain the much higher eutectic cell counts observed in ductile iron as compared to gray iron.

[1] H. Morrogh, *AFS Trans.* 56(1948):72.

Table 8-3 Common grades and typical applications of ductile cast irons†

Type TS-YS-% elongation	Tensile strength, psi	Yield strength, psi	Typical Elonga- tion, %	Hardness, Bhn	Heat treatment	Typical microstructure	Typical applications
60-40-18	60,000	40,000	18	137–170	Annealed	All ferritic	Pressure castings such as valve and pump bodies.
65-45-12	65,000	45,000	12	149–229	—	Ferritic	Machinery castings subject to shock and fatigue loading
80-55-06	80,000	55,000	6	179–255	—	Ferritic and pearlitic	Crankshaft gears and rollers
100-70-03	100,000	70,000	3	229–302	Normalized	All Pearlitic	High strength gears, automotive and machine components
120-90-02	120,000	90,000	2	250–350⁺	Quench and temper	Tempered martensitic	Pinions, gears, rollers and slides

† After Ref. 1, p. 100.

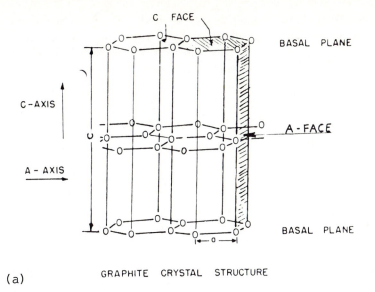

(a)   GRAPHITE   CRYSTAL   STRUCTURE

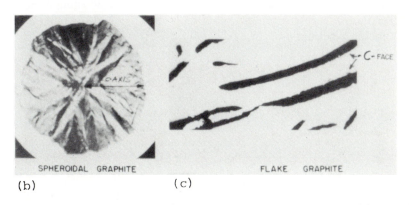

(b)   (c)

Figure 8-20 Graphite structure relationships to spheroidal and flake graphite growth directions. (*After P. F. Weiser, C. E. Bates, and J. F. Wallace, "Mechanisms of Graphite Formation in Iron-Silicon-Carbon Alloys," Malleable Founders Society, Cleveland, 1967, p. 100.*)

Experimental evidence indicates that graphite spheroids grow directly from the melt in molten ductile iron, as they do in gray iron. These spheroids grow in the direction of the graphite basal pole with the basal plane in contact with the melt, but may soon be surrounded with an austenitic shell. Further growth occurs by the diffusion of carbon through the shell. As the carbon *must* diffuse through the shell, the growth of the spheroids is slower than that of gray iron eutectic solidification. Thus, the liquid melt is present in a wider temperature range and to a lower temperature in ductile iron than it is in gray iron.

The number of graphite spheroids is determined at an early stage of the solidification process. As the ductile iron is cooled to lower temperatures, the carbon precipitates as graphite on the existing spheroids at temperatures down to the eutectoid range. As with gray iron, the cooling rate through the eutectoid range determines the matrix structure. The bull's eye microstructure shown in Fig. 8-18 is typical of the structure of ferritic-pearlitic as-cast ductile irons.

Effect of Chemical Composition on Structure and Properties of Ductile Irons

Carbon and silicon The carbon content of ductile irons ranges from 3.0 to 4.0 percent, although much narrower limits of 3.6 to 3.9 percent are common. The higher carbon content of ductile cast iron above that of gray iron is necessary to develop the high density of graphite nodules. If the carbon equivalent becomes too high (i.e., above 4.6), carbon flotation may occur, as is indicated in Fig. 8-21. The silicon content of ductile iron is in the 1.8 to 2.8 percent range, but narrower limits of 2.2 to 2.7 percent are common. Silicon affects the carbon equivalent, and thus, as the silicon is increased, so are the number of nodules. A low silicon content in ductile iron increases the chilling tendency and, if low enough, may cause the formation of excess carbides in thin sections. Silicon also strengthens the ferrite in ductile cast iron.

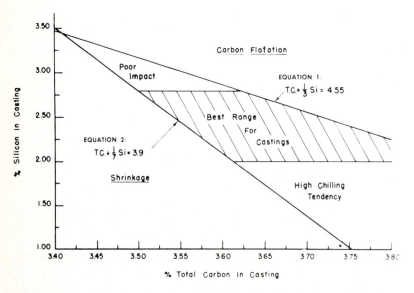

Figure 8-21 Carbon and silicon range of ductile iron. (*After H. E. Henderson, Gray, Ductile and Malleable Iron Castings-Current Capabilities, ASTM STP 455, 1969, p. 37; Reprinted, with permission, from the American Society for Testing and Materials, 1916 Race St., Philadelphia, PA 19013; Copyright.*)

Sulfur The sulfur content of ductile iron is usually kept below 0.03 percent. An increase in sulfur content means that additional magnesium must be added to spheroidal graphite. The sulfur content after the magnesium treatment is usually about 0.015 percent.

Phosphorus Since phosphorus forms the brittle eutectic structure steadite, it adversely affects impact properties and ductility. A maximum of about 0.10 percent is specified, but it is usually kept below 0.05 percent.

Other elements Close control must be maintained over elements such as lead, titanium, aluminum, antimony, and zirconium, which are known to promote graphite flakes. Other elements which favor the formation of pearlite and/or iron carbide, such as arsenic, boron, chromium, tin, and vanadium, must also be avoided. Alloy ductile irons are made with additions of manganese, nickel, and molybdenum, and will be discussed later.

Heat Treatment and Microstructures

The tensile strength of unalloyed ductile cast iron can be varied from about 60 to 120 ksi by the proper choice of heat treatment. This range of strength is achieved by variation in matrix structure from all ferrite, to ferrite and pearlite, to martensite, or to tempered martensite.

Although the processing conditions can be adjusted to produce some ductile iron castings to specifications without heat treatment, it is common practice to heat-treat most ductile iron castings to obtain the desired properties. The principal types of heat treatments are listed here:

Stress relieving This treatment removes internal stresses in castings by heating at 538 to 675°C for 1 h, plus 1 h per inch of thickness.

Annealing This treatment increases ductility and produces the best machinability. Ductile iron in the as-cast condition is usually of the 80-55-06 grade with envelopes of ferrite around the graphite nodules and with a pearlitic matrix, as shown in Figs. 8-18 and 8-22a. Many different annealing treatments are given to ductile iron castings, but two of the commonly used ones are the following:

1. *Single-step anneal.* The iron is heated to 788°C for 6 h and then furnace-cooled. This treatment decomposes most of the pearlite of the as-cast structure, as shown in Fig. 8-22b.
2. *Two-stage anneal.* The iron is heated to 900°C at 130°C per hour held 4 h, cooled at 22°C per hour to 691°C, held 6 h, and then furnace-cooled. The resulting microstructure consists of graphite nodules in a ferritic matrix (Fig. 8-22c). Figure 8-22d shows the secondary graphite produced by annealing and which surrounds a primary graphite nodule.

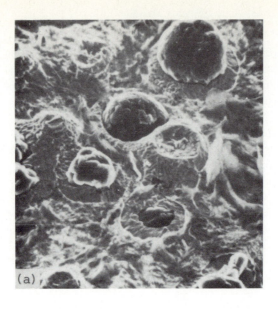

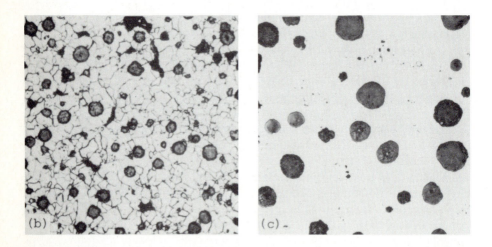

Figure 8-22 Microstructures of ductile cast irons. (*a*) Scanning electron micrograph of as-cast pearlitic ductile iron with matrix etched away to show secondary graphite, and bull's-eye ferrite, around primary graphite nodules. (3:1 methyl acetate-liquid bromine; X475.) (*b*) Grade 60-40-18 ferrite ductile iron. As-cast pearlitic ductile iron was annealed 6 h at 788°C and furnace-cooled. Most of the original pearlite was decomposed, resulting in a matrix of free ferrite (light) and 5% pearlite (black irregular). (3% nital; X100.) (*c*) Grade 60-45-12 ferritic ductile iron. Heated to 900°C at 139°C/h, held 4 h, cooled at 22°C/h to 691°C, held 6 h, furnace-cooled. Graphite nodules in ferrite matrix. (2% nital lightly etched; X140.) (*d*) Grade 60-45-12 nodular ductile iron given same treatment as *c*, which decomposed pearlite and free carbide in original matrix, producing secondary graphite around primary graphite nodule, in an all-ferrite matrix. (2% nital; X750.) (*After Metals Handbook, 8th ed., vol. 7, American Society for Metals, 1972, pp. 88–89.*)

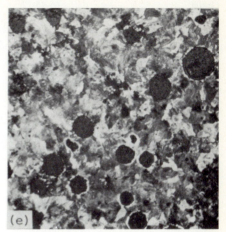

Fig. 8-22 *Continued*

Normalizing and tempering This treatment can be used to develop higher strengths such as in the 80-55-06 and 100-70-03 grades. Normalizing is usually carried out by holding for about 1 h at 900°C followed by air cooling. After normalizing, the iron can be tempered for 1 h at 566°C. Figure 8-22*e* shows a typical normalized and tempered ductile iron microstructure for a grade 80-55-06 iron.

Quenching and tempering Ductile iron can be oil-quenched and tempered to produce a tempered martensitic structure which has higher strength. Sometimes austempering and martempering treatments are also used.

Engineering Properties

In general, ductile iron combines the processing advantages of gray cast iron with the engineering advantages of steel. No other ferrous material can equal the versatility of ductile cast iron. It has good fluidity and castability, excellent machinability, and good wear resistance. In addition, ductile cast iron has a number of properties similar to those of some steels in terms of high strength, toughness, ductility, hot workability, and hardenability.

Mechanical properties Table 8-4 lists the mechanical properties of four popular grades of ductile iron. Figure 8-23 shows how heat treatment can decrease or increase the strength of as-cast ductile iron in the range of 60 to 120 ksi. This great variation in strength is possible since the graphite nodules do not have as much effect on the strength properties in this type of iron as do the graphite flakes in gray iron.

Table 8-4 Mechanical properties of ductile iron†

Grade	65-45-12	80-55-06	100-70-03	120-90-02
Hardness, Bhn	167	192	235	331
Tension:				
Tensile strength 10^3 psi	67.3	81.1	118.6	141.3
0.2 % yield strength 10^3 psi	48.2	52.5	98.2	125.3
% elongation in 2 in	15.0	11.2	4.5	1.5
Modulus 10^4 psi	24.4	24.5	23.5	23.8
Poisson's ratio	0.29	0.31	0.28	0.28
Compression:				
0.2 % yield strength 10^3 psi	52.5	56.0	87.5	133.5
Modulus 10^3 psi	23.6	23.9	22.7	23.8
Poisson's ratio	0.31	0.31	0.27	0.27
Torsion:				
Shear strength 10^3 psi	68.9	73.1	87.3	126.9
0.0375 % yield strength 10^3 psi	30.0	28.0	47.3	71.3
Shear modulus 10^3 psi	9.3	9.0	8.7	9.2

† After Ref. 6, p. 45.

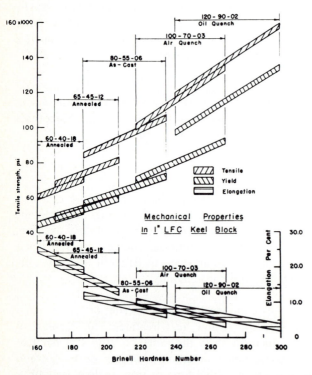

Figure 8-23 Mechanical properties of ductile iron. (*After Gray, Ductile and Malleable Iron Castings —Current Capabilities, ASTM STP 455, 1969, p. 37; Reprinted, with permission, from the American Society for Testing and Materials, 1916 Race St., Philadelphia, PA 19013; Copyright.*)

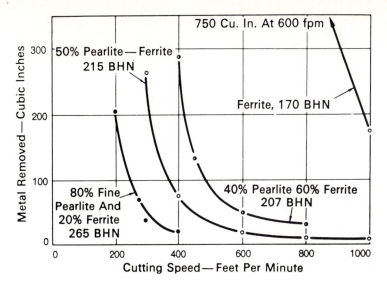

Figure 8-24 The effect of matric microstructure and cutting speed on tool life in the machining of ductile iron. [*After C. F. Walton (ed.), "Gray and Ductile Iron Castings Handbook," Gray and Ductile Iron Founders' Society, Cleveland, 1971, p. 518.*]

Wear resistance The spheroidal graphite in ductile iron is able to retain oil and thus prevent galling and seizing of moving parts particularly during start up. Ductile iron has a wear resistance equivalent to gray iron.

Machinability Ductile iron has superior machinability to gray iron for equivalent hardness, as can be seen by comparing Fig. 8-24 to Fig. 8-16.

8-5 MALLEABLE CAST IRON

Types of Malleable Cast Iron

Malleable cast iron is an important engineering material since it has the desirable properties of castability, machinability, toughness and ductility, corrosion resistance for certain applications, adequate strength, and uniformity due to the heat treatment of all castings.

The structure of malleable cast iron is obtained by heat treatment of castings which have the white-cast-iron structure. That is, the iron carbide white-iron structure of Fig. 8-2 is converted to a malleable structure (Fig. 8-4) by an appropriate annealing treatment. Typical chemical composition ranges of white cast irons which are subsequently heat-treated to produce malleable cast irons are given in Table 8-5.

Table 8-5 Typical chemical compositions for malleable cast irons†

	ASTM No. 32510	ASTM No. 35018
% C	2.30–2.65	2.00–2.45
% Si	0.9–1.40	0.90–1.30
% Mn	0.25–0.55	0.21–0.55
% P	0.18	Less than 0.18
% S	0.05–0.18	0.05–0.18

† After Ref. 2.

Table 8-6 Mechanical properties and typical applications for malleable cast irons†

Designa-tion	Mechanical properties				Typical applications
	Tensile Strength,* psi	Yield Strength,* psi	Elongation in 2 in,* %	Brinell hardness§	
Ferritic					
35018	53,000	35,000	18	110–156	Iron grillework, railroad car hardware, hand tools, high-pressure parts, hardware for oil industry
32510	50,000	32,500	10	110–156	Gear cases and housings, chain links, auto hinges, brackets, mounting pads, brake shoes, wheel hubs
Pearlitic					
40010	60,000	40,000	10	149–197	C clamps, diesel engine brackets, levers, transmission cases, artillery shells, gears, farm implement parts
45008	65,000	45,000	8	156–197	
45006	65,000	45,000	6	156–207	
50005	70,000	50,000	5	179–229	
60004	80,000	60,000	4	197–241	Pistons for diesel engines, differential axle cases, rocker arms, clutch hubs, transmission gears, universal joint yokes, crankshafts, idler gears and shafts
70003	85,000	70,000	3	217–269	
80002	95,000	80,000	2	241–285	
90001	105,000	90,000	1	269–321	

† After "ASM Databook," published in *Metal Progress*, mid-June 1974 vol. 106, no. 1.

* ASTM minimum (A220-68).

§ Hardnesses are listed for informational purposes only; they are typical but not part of the ASTM specification.

The term malleable iron includes *ferritic* (or "standard") *malleable iron* and *pearlitic malleable iron*. In commercial usage, the term "malleable iron" normally refers to the ferritic type.

As in the case of gray and ductile irons, malleable cast irons are usually specified by their minimum tensile strengths. The tensile properties and applications for ferritic and pearlitic unalloyed malleable cast irons are listed in Table 8-6. The lower tensile properties of the ferritic malleable irons are due to the weaker ferritic matrix. The higher-strength "pearlitic" malleable cast irons have a tempered martensitic structure and not a pearlitic one.

Heat Treatment and Microstructures

Ferritic malleable cast irons By applying an annealing treatment, the brittle as-cast white-iron structure is converted to a more malleable and tough structure consisting of graphite nodules in a ferritic matrix. This treatment is called *malleablization*. The initial structure of white cast iron consists of massive iron carbides, pearlite, and some eutectic areas (Fig. 8-26a). The malleablization heat treatment converts the structure to one of a ferritic matrix and tempered graphite nodules (Fig. 8-4). The time-temperature cycle of a typical malleablization treatment is given in Fig. 8-25.

The annealing heat treatment for ferritic malleable cast iron consists of the following three steps:

1. First, the graphite is *nucleated*. This occurs principally in the heating to temperature stage and in the early part of the holding at temperature (Fig. 8-25).
2. The second step, in which the iron is held at 870 to 954°C, involves the elimination of the massive carbides and the conversion of their carbon to graphite. This is the initial stage of graphitization.
3. The third and last step of the heat treatment involves the slow cooling of the iron through the allotropic transformation range of the iron so that a ferritic matrix completely free of pearlite and carbide is created. This resultant structure is shown in Fig. 8-26b.

A number of structural changes occur during malleablization:

Nucleation of graphite In order to produce satisfactory ferritic malleable cast iron, a sufficient number of nuclei must develop. The nuclei first form within the

Figure 8-25 Cycle of temperature and time for malleablizing white iron. (Actual duration of the cycle may be much less or longer than indicated.) (*After R. W. Heine, C. R. Loper, and P. C. Rosenthal, "Principles of Metal Casting," McGraw-Hill, New York, 1967, p. 662.*)

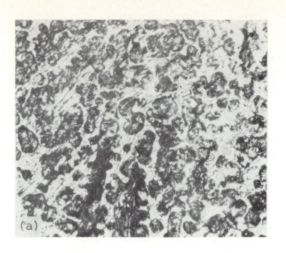

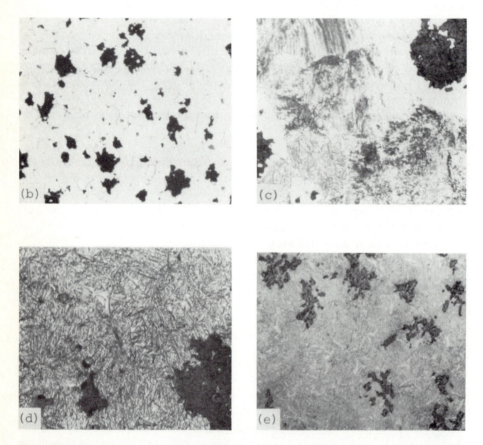

Figure 8-26 Microstructures of white cast iron and malleable cast irons. (*a*) Hypoeutectic white iron (nominal 2.5% C, 1.5% Si) as-cast showing a dendritic pattern of fine pearlite (dark area), and an interdendritic mixture of massive and acicular free cementite (light) with some pearlite. (1% nital; X100.) (*b*) ASTM A602, grade M3210 ferritic malleable iron two-stage annealed by holding 4 h at

pearlite of the white cast iron and at the interfaces of the iron carbide and austenite, or at nonmetallic inclusions. Some of the factors which influence the nucleation of the graphite are

1. *Chemical composition.* A high silicon content favors nucleation.
2. *Heating rate.* Rapid heating to holding temperature decreases the number of nuclei developed.
3. *Section size.* Thin sections develop more nuclei.
4. *Pretreatment.* A preliminary holding of the iron at 315 to 650°C increases nucleation upon subsequent malleablization.

First-stage graphitization In this stage the iron carbide of the white cast iron dissolves in the austenite and the carbon diffuses to the graphite nuclei previously formed and precipitates on them, causing nodules to grow. The process is completed when the carbide disappears. The time required to complete it is dependent on the number of nuclei present, the rate of dissolution of the iron carbide, and the diffusion rate of the carbon. Carbide-forming elements such as chromium and manganese delay the completion of the process. The chromium content should be kept low and the manganese should be in proper balance with the sulfur to prevent excess manganese or sulfur from stabilizing the iron carbide.

Second-stage graphitization In this stage, which involves slow cooling through the transformation range of iron, the cooling rate must permit the austenite to transform to ferrite and precipitate the rejected carbon as graphite. If the cooling rate is too fast, pearlite will be formed. The slow cooling rate must be maintained to at least 650°C to prevent the formation of pearlite.

Pearlitic-malleable cast irons Increased strength and wear resistance with reduced ductility are obtained by producing a structure of a matrix of pearlite or tempered martensite with tempered graphite nodules.

To produce a pearlitic matrix in the malleable iron, it is first annealed for about 13 h at 970°C and then *air-cooled*. Slow cooling will produce envelopes of ferrite, as shown in Fig. 8-26c. Air cooling at faster rates will produce less ferrite and a finer pearlitic structure, as shown in Fig. 8-26d.

954°C, cooling to 704°C in 6 h, air cooling. Type III graphite (temper carbon) nodules in a matrix of granular ferrite; small gray particles are MnS. (2% nital; X100.) (c) Grade 45008 malleable iron that was first stage annealed by austenitizing for 13-1/2 h at 971°C and then air cooled slowly. Iron was tempered for 2 h at 677°C. Nodules of type III temper carbon graphite (black) in bull's eye ferrite (white); the pearlitic matrix has been slightly spheroidized by the tempering. (2% nital; X500.) (d) Pearlitic malleable iron that was first annealed by holding 13.5 h at 943°C, and then quenched in oil at 82°C. Nodules of type III temper-carbon graphite (black), and particles of manganese sulfide (gray) in a matrix of tempered martensite. (2% nital; X500.) (e) Centrifugally cast pearlitic malleable iron (3.1% C, 1.1% Si, 0.75% Mn) first stage annealed by holding 1 h at 1093°C and air cooling. Iron was austenitized 1 h at 871°C, oil-quenched, and tempered 1 h at 482°C. Nodules of temper-carbon graphite (black) in a matrix of tempered martensite (gray). (5% nital; X100.) (*After Metals Handbook, 8th ed., vol. 7, American Society for Metals, 1972, pp. 95–97.*)

A martensitic matrix can be produced by first annealing at 943°C for about 13 h and then quenching in oil. The tempering temperatures used range from 260° to 727°C, depending on the desired properties. Figure 8-26e shows the microstructure of a tempered martensitic malleable cast iron.

Engineering Properties

The mechanical properties of ferritic and pearlite malleable cast irons are listed in Table 8-6. By varying the matrix structure, tensile strengths ranging from 50 to 100 ksi can be obtained with corresponding elongations from 8 to 1 percent. Thus, an advantage of malleable cast irons, like ductile irons, is that they can be produced in a wide range of strengths by carefully controlled heat treatments.

Malleable iron is one of the most machinable of the ferrous alloys. Since all castings are heat-treated, a high degree of structural uniformity, and hence machinability, is obtained. Pearlitic malleable iron can be machined to a high-quality finish, and also has good wear resistance and excellent capability for surface or through-hardening by flame or induction heating. Ferritic malleable iron does not have special wear resistance greater than that of ordinary soft ferrous alloys.

8-6 ABRASION-RESISTANT ALLOY CAST IRON

Chilled Cast Iron

For some purposes, it is desirable to produce a casting with a hard, abrasive-resistant surface layer and with a tougher inner core. This type of casting is normally produced with a hard white-cast-iron surface and a softer gray-cast-iron core. The white-cast-iron structure is usually produced by casting the molten metal against a metal or graphite *chill* to provide rapid cooling of the solidifying cast iron. Rapid cooling tends to produce a white-(iron carbide) cast-iron structure, whereas slower cooling allows graphitization to occur and thus the formation of a gray-iron structure. Since these cast-iron castings are made with this duplex structure by using "chills," they are called *chilled iron castings*. The depth of the chilled layer can be controlled to some extent by varying the composition of the cast iron.

Figure 8-27 shows the microstructure of a nickel-chromium abrasive-resistant cast iron as-cast against a chill. The cast iron near the chill has a fine dendritic pattern of pearlite and interdendritic iron carbide (Fig. 8-27a). The white cast iron 2 in from the chill has a coarse dendritic pattern of pearlite and interdendritic iron carbide (Fig. 8-27b). At 4 in from the chill, the cast iron is of the gray structure, with type B graphite flakes in a matrix of fine pearlite (Fig. 8-27c).

White Cast Iron

White cast irons may be defined as cast irons in which the excess carbon occurs as iron carbides instead of graphite flakes or nodules. Unalloyed white cast irons

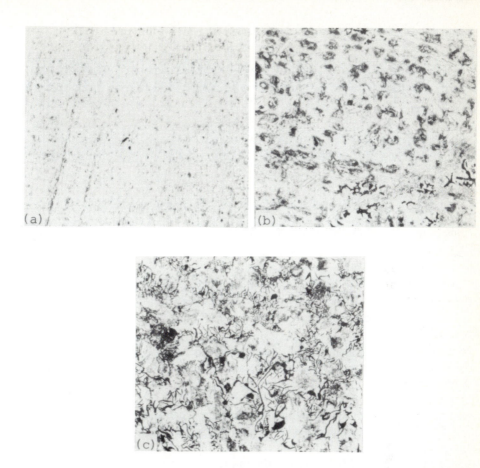

Figure 8-27 Microstructures of nickel-chromium abrasion-resistant cast irons as-cast against a chill. (*a*) White iron near chill, showing fine dendritic pattern of pearlite (gray) and interdendritic carbide (white). (*b*) White iron 2 in from chill, showing coarse dendritic pattern of pearlite (gray) and interdendritic carbide (white). (*c*) Gray iron 4 in from chill, showing type B graphite flakes (black) in a matrix of fine pearlite (gray) with some free ferrite (light). (2% nital plus 5% picral; X100.) (*After Metals Handbook, 8th ed., vol. 7, American Society for Metals, 1972, p. 99.*)

usually have a fine pearlitic matrix. By the addition of alloying elements such as nickel, chromium, or molybdenum, the matrix can be changed to martensitic (or bainitic) or austenitic. Table 8-7 lists the compositions and minimum Brinell hardnesses of some typical unalloyed and alloyed white cast irons.

Chromium white cast irons Chromium is added to white cast irons in amounts of 1 to 4 percent to increase hardness and improve abrasion resistance. Chromium is a strong carbide stabilizer which increases the tendency to form white cast iron and to suppress the formation of graphite, especially graphite resulting from the slow cooling of heavy sections. Chromium is used in amounts of 12 to 35 percent to increase corrosion and oxidation resistance as well as abrasion

Table 8-7 Chemical composition and hardness of typical white cast irons†

	% C	% Si	% Mn	% Cr	% Ni	% P, max	% S, max	Bhn, min
Cupola white iron	3.30–3.60	0.40–1.00	0.50–0.70	. . .	. . .	0.30	0.15	400
Cupola white iron (1 % Cr)	3.30–3.60	0.40–1.00	0.50–0.70	0.80–1.00	. . .	0.30	0.15	444
Malleable white iron	2.20–2.50	1.00–1.60	0.30–0.50	. . .	. . .	0.15	0.15	321
Martensitic nickel-chromium iron	3.00–3.60	0.40–0.70	0.40–0.70	1.40–3.50	4.00–4.75	0.40	0.15	550
Martensitic nickel-chromium high strength iron	2.90 max	0.40–0.70	0.40–0.70	1.40–3.50	4.00–4.75	0.40	0.15	525
High-chromium white iron	2.25–2.85	0.25–1.00	0.50–1.25	24.0–30.0	. . .	0.40	0.15	500

† After Ref. 7, p. 396.

resistance. The microstructure of a high-chromium (28.0% Cr) abrasion-resistant cast iron in the as-cast condition is shown in Fig. 8-28.

Nickel-chromium white cast irons Nickel and chromium both increase strength, oxidation, and corrosion resistance of cast irons, but have opposite stabilizing tendencies with respect to graphitization. Nickel stabilizes graphite, whereas chromium stabilizes iron carbide. The two elements are added together in white cast iron so that their effects on graphitization are counterbalanced.

Small amounts of nickel and chromium up to about 2% Ni and 1% Cr refine the pearlitic matrix of white cast irons. For sections up to 1 in with 3.3% C, 0.60% Si, and 0.50% Mn, a martensitic matrix is produced with about 3.25% Ni and 1.25% Cr. Figure 8-29 shows the microstructure of a Ni-Cr white cast iron

Figure 8-28 ASTM A532 Type III, high chromium (28% Cr) abrasion-resistant cast iron as-cast. White iron showing interdendritic network of iron-chromium carbide (white) and dendritic pattern of martensite (variegated gray). (3% nital; X100.) (*After Metals Handbook, 8th ed., vol. 7, American Society for Metals, 1972, p. 99.*)

Figure 8-29 ASTM A532, Type I, grade 1 nickel-chromium abrasion-resistant cast iron (3.3% C, 0.55% Si, 2.0% Cr, 4.2% Ni, 0.75% max Mo) as-cast in a 1-in-diam. bar. White iron showing a dendritic pattern of austenite (black), and an interdendritic eutectic of austenite (black dots) and carbide (white). The austenite transforms to martensite during abrasive service. (3% nital; X100.) (*After Metals Handbook, 8th ed., vol. 7, American Society for Metals, 1972, Metals Park, OH, p.99.*)

Table 8-8 Chemical composition and mechanical properties of several corrosion-resistant alloy cast irons†

	Types of cast iron		
Analysis	High silicon (Duriron) (Durichlor)	High chromium	High nickel (Ni-resist)
% Carbon	0.4–1.0	1.2–2.5	1.8–3.0
% Silicon	14–17	0.5–2.5	1.0–2.75
% Manganese	0.4–1.0	0.3–1.0	0.4–1.5
% Nickel	. . .	0–5	14–30
% Chromium	. . .	20–35	0.5–5.5
% Copper	. . .	. . .	0–7
% Molybdenum	0–3.5	. . .	0–1
Brinell hardness number	450–500	290–400	100–230
Tensile strength, 1000 psi	13–18	30–90	25–45
Compressive strength, 1000 psi	. . .	100–	100–160
Charpy-type impact,‡ ft·lb .	2–4	20–35	60–150

† After Ref. 1, p. 270.

‡ 1.2 in-diameter unnotched bar broken on 6-in supports (plain gray iron has 25 to 35 ft·lb).

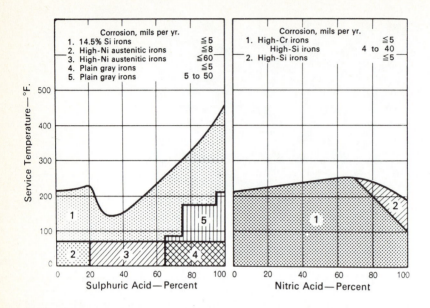

Figure 8-30 Useful life of plain and high-alloy iron castings in sulfuric and nitric acids. (Type 5 includes turbulence). [*After C. F. Walton (ed.), Gray and Ductile Iron Castings Handbook, Gray and Ductile Iron Founders' Society, Cleveland, 1971, p. 328.*]

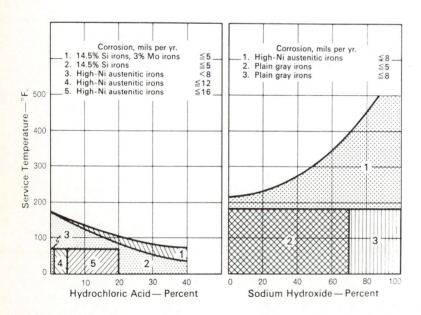

Figure 8-31 Useful life of high-alloy iron castings in hydrochloric acid and plain and austenitic irons in caustic. [*After C. F. Walton (ed.), Gray and Ductile Iron Castings Handbook, Gray and Ductile Iron Founders' Society, Cleveland, 1971, p. 329.*]

with 4.2% Ni and 2.0% Cr which has an austenitic matrix. When this cast iron is used in abrasive service, the austenite transforms to martensite.

8-7 CORROSION-RESISTANT CAST IRON

The corrosion resistance of alloy cast irons depends principally on their chemical composition and microstructure. The dominating factors are the chemical composition and structure of the matrix. There are three distinct groups of highly alloyed cast irons which have enhanced corrosion resistance for specific environments. These are (1) high-silicon irons, (2) high-chromium irons, and (3) high-nickel irons. The chemical composition ranges and mechanical properties of some of the important alloy cast irons are listed in Table 8-8.

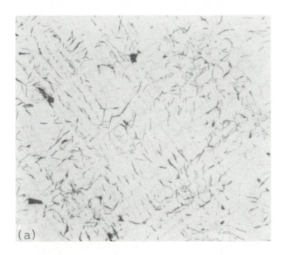

(a)

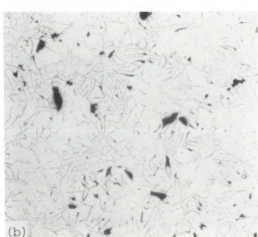

(b)

Figure 8-32 Microstructures of high-silicon corrosion-resistant cast irons. (*a*) ASTM A518 high-silicon (14.5% Si) corrosion-resistant cast iron as-cast. Gray iron showing type A graphite flakes (dark) in a matrix of iron-silicon ferrite solid solution (light). (HNO_3 plus HF, in glycerol; X100.) (*b*) High-silicon corrosion-resistant cast iron (0.9% C, 14.5% Si, 1.0% Mn, 4.5% Cr) as-cast. Gray iron with types A and E graphite flakes (dark) and interdendritic $(Fe, Cr)_3C$ (light, outlined) in matrix of dendritic Fe-Si-Cr ferrite. (HNO_3 plus HF, in glycerol; X100.) (*After Metals Handbook, 8th ed., vol. 7, American Society for Metals, Metals Park, OH, 1972, p. 99.*)

High-Silicon Irons

With a high-silicon content of from 12 to 18 percent, cast irons become very resistant to corrosive acids. With a silicon content of 14.5 percent or higher, these cast irons have a very high resistance to boiling 30% sulfuric acid (Fig. 8-30). High-silicon irons with 16.5% Si are resistant to boiling sulfuric and nitric acids at almost all concentrations. However, because of their high silicon content, they have poor mechanical properties such as low thermal and mechanical shock resistance, are difficult to cast, and are virtually unmachinable (Ref. 7, p. 402). The distributions of graphite flakes in the microstructures of two high-silicon cast irons are shown in Fig 8-32*a* and *b*.

High-Chromium Irons

High-chromium cast irons with 15 to 30% Cr are white cast irons. Chromium imparts abrasion resistance and resistance to oxidation. High-chromium cast irons are resistant to oxidizing acids, particularly nitric acid (Fig. 8-30*b*), and are useful for work with weak acids under oxidizing conditions, with many organic acid solutions, and with salt solutions. The mechanical properties of chromium

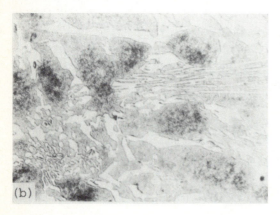

Figure 8-33 Microstructures of high-chromium corrosion-resistant cast irons. (*a*) High-chromium corrosion-resistant cast iron (3.09% C, 0.52% Si, 17.8% Cr, 3.3% Mo, 0.46% V) as-cast. White iron showing eutectic (Cr, Fe)$_7$C$_3$ (light, outlined), both interdendritic and as clusters, and some pearlite (gray) in a matrix of Fe-Cr ferrite solid solution. (Vilella's reagent; X250.) (*b*) High-chromium corrosion-resistant cast iron as in *a* but normalized by holding at 1010°C and air cooling, and tempered at 260°C. White iron showing eutectic (Cr, Fe)$_7$Cr$_3$ (light, outlined), both interdendritic and as radiating clusters, in a matrix of tempered martensite. (Vilella's reagent; X250.) (*After Metals Handbook, 8th ed., vol. 7, American Society for Metals, Metals Park, OH, 1972, p. 100.*)

cast irons are better than those of the high-silicon cast irons (Table 8-8). The high-chromium irons respond to heat treatment when the carbon and chromium contents are suitably adjusted. However, the machining of these alloys is very difficult. Figure 8-33 shows the microstructures of two high-chromium corrosion-resistant cast irons, one of which is in the as-cast condition and the other in the as-cast and heat-treated condition.

High-Nickel Irons

High-nickel austenitic cast irons are widely used and are generally known as *Ni-Resist* cast irons. Austenitic gray cast irons containing from 14 to 30 % Ni are fairly resistant to mildly oxidizing acids, including sulfuric acid at room temperature (Fig. 8-30*a*). High-nickel cast irons are more resistant to alkalis than unalloyed cast irons. Ni-Resist is particularly useful for alkalis at high temperatures (Fig. 8-31*b*).

High-nickel irons, because of their austenitic matrix, are the toughest of all cast irons with flake graphite. They have excellent machinability and good foundry properties, although their tensile strengths are relatively low (20 to 45

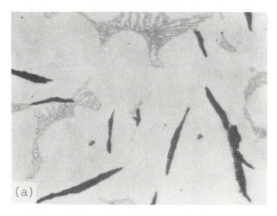

(a)

(b)

Figure 8-34 Microstructures of high-nickel corrosion-resistant cast irons. (*a*) High-nickel (30% Ni, 3% Cr) corrosion-resistant cast iron (ASTM A436, type 3) as-cast. Microstructure is gray iron showing type A graphite flakes (dark constituent) and some interdendritic (Fe, Cr)$_3$C (gray, outlined) in a matrix of high-nickel austenite. (2% nital plus 5% picral; X250.) (*b*) High-nickel corrosion-resistant cast iron (2.7% C, 2.8% Si, 1.4% Mn, 20.0% Ni, 2.4% Cr) as-cast. Structure: gray iron showing interdendritic graphite (dark) of types D and E, and interdendritic carbide (light, outlined) in a matrix of austenite (light). (5% nital; X500.) (*After Metals Handbook, 8th ed., vol. 7, American Society for Metals, Metals Park, OH, 1972, p. 100.*)

ksi) due to the flake graphite. High-nickel ductile irons have higher strength and ductility because they have nodular graphite. The microstructures of two high-nickel (flake-graphite-type) corrosion-resistant cast irons are shown in Fig. 8-34.

8-8 HEAT-RESISTANT ALLOY CAST IRONS

Heat-resistant alloy gray and ductile cast irons are Fe-C-Si alloys with additions of silicon (above 3 percent), chromium, nickel, molybdenum, or aluminum to improve their high-temperature properties. The chemical compositions and mechanical properties of some of the industrially important heat-resistant alloy cast irons are listed in Table 8-9.

At temperatures above 425°C, the mechanical properties of cast irons gradually decrease as the temperature rises and the iron undergoes the chemical changes of *growth* and *oxidation*.

Growth is a permanent increase in volume that occurs at elevated temperatures in some cast irons, particularly in gray cast iron. It is caused principally by (1) the expansion that accompanies the reaction of Fe_3C changing to graphite and iron and (2) the oxidation of the iron after graphite is oxidized away as carbon monoxide.

Table 8-9 Chemical compositions and mechanical properties of heat-resistant alloy cast irons†

	Types of cast iron				
Analysis	High silicon (silal)	High Chromium	High nickel (Ni-Resist)	Nickel-chromium silicon	High aluminum
% Carbon	1.6–2.5	1.8–3.0	1.8–3.0	1.8–2.6	1.3–2.0
% Silicon	4.0–6.0	0.5–2.5	1.0–2.75	5.0–6.0	1.3–6.0
% Manganese	0.4–0.8	0.3–1.5	0.4–1.5	0.4–1.0	0.4–1.0
% Nickel	. . .	0–5	14–30	13–32	
% Chromium	. . .	15–35	1.75–5.5	1.8–5.5	
% Copper	. . .	. . .	0–7	0–10	
% Molybdenum	. . .	. . .	0–1	0–1	
% Aluminum	. . .	. . .	. . .	. . .	20–25
Brinell hardness number	170–250	250–500	130–250	110–210	180–350
Tensile strength, 1000 psi	25–45	30–90	25–45	20–45	13–16
Compressive strength, 1000 psi	90–150	100–	100–160	70–100	
Charpy-type impact‡ ft·lb.	15–23	20–35	60–150	80–150	

† After Ref. 1.

‡ 1.2 -in-diameter unnotched bar broken on 6-in supports (plain gray iron has 25 to 35 ft·lb).

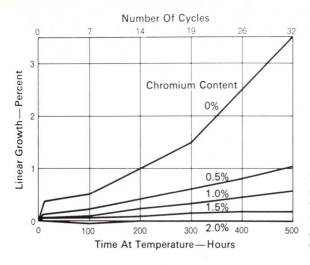

Figure 8-35 The effect of chromium on the growth of gray iron when subjected to cyclic heating to 800°C. [*After C. O. Burgess and A. E. Shrubsall, Trans. AFS, 50 (1942): 405.*]

Figure 8-36 Microstructures of heat-resistant cast irons. (*a*) Heat-resistant cast iron ASTM A319, (3.5% min C, 0.66–0.95% Cr) as-cast. Gray iron with Type A graphite flakes in matrix of pearlite. (2% nital plus 4% picral; X500.) (*b*) Same iron as above but after extended high-temperature service. Nearly all the cementite in the original matrix of pearlite has been spheroidized so that the matrix is now ferrite. (2% nital; X500.) (*After Metals Handbook, 8th ed., vol. 7, American Society for Metals, Metals Park, OH, 1972, p. 100.*)

Oxidation can also occur at the surface of cast-iron castings after sufficient exposure at high temperatures. If the surface oxide scale either is porous or flakes off at high temperatures, continued oxidation of the metal will occur. Eventually, the strength of the material will decrease due to loss of material.

Chromium Irons

Chromium is added to heat-resistance cast irons because it assists in stabilizing carbides and forms a protective oxide on the metal surface. Even small additions of chromium (0.5 to 2.0 percent) reduce growth in gray irons subjected to cyclic heating at 800°C, as is shown in Fig. 8-35. After extended high-temperature service, the pearlitic matrix of an as-cast 0.8% Cr heat-resistant cast iron is transformed to ferrite and its cementite spheroidized, as shown in Fig. 8-36a and b. Higher chromium additions of 15 to 35 provide excellent oxidation and growth resistance for temperatures up to about 980°C (Fig. 8-37). However, these high-chromium irons have a white-iron structure. Even though they have good strength properties, therefore, they also have limited machinability.

High-Silicon Irons

Silicon contents of less than 3.5 percent increase the rate of growth of gray cast iron by promoting graphitization. However, silicon contents of 4 to 8 percent

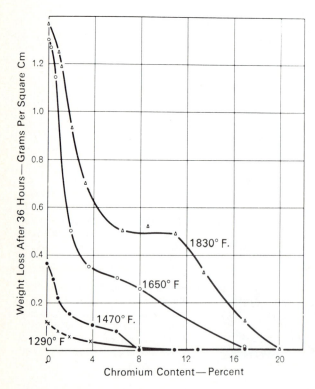

Figure 8-37 The effect of chromium on the oxidation weight loss by scaling of alloy cast irons at several temperatures in air. (*After C. F. Walton (ed.), Gray and Ductile Iron Castings Handbook, Gray and Ductile Iron Founders' Society, Cleveland, 1971, p. 272.*)

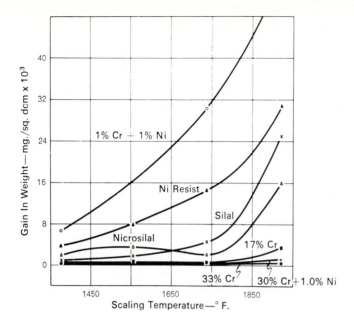

Figure 8-38 Oxidation of several alloy irons at temperature for 200 h in air. [*After C. F. Walton (ed.), Gray and Ductile Iron Castings Handbook, Gray and Ductile Iron Founders' Society, Cleveland, 1971, p. 272.*]

greatly reduce both oxidation (scaling) and growth. Silicon increases the scaling resistance of cast iron by forming a light surface oxide that is impervious to oxidizing atmospheres. Silicon also raises the ferrite-to-austenite transformation temperature to about 900°C so that the expansion and contraction due to the transformation can be avoided up to 900°C. The high-temperature oxidation resistance of the high-silicon cast iron *Silal* as compared to other alloy cast irons is shown in Fig. 8-38. Figure 8-39 shows the microstructure of a high-silicon heat-resistant cast iron in the as-cast condition.

High-Nickel Irons

Austenitic cast irons containing 18% or more Ni up to 7% Cu, and 1.75 to 4% C are used for applications where heat *and* corrosion resistance are required. The *Ni-Resist* cast irons have good resistance to high-temperature scaling and growth up to about 815°C for most oxidizing atmospheres. In sulfur-containing atmospheres, however, the nickel content of these alloys restricts their use to temperatures below about 500°C.

The austenitic nickel cast irons have considerably greater toughness and shock-resistance than the other heat-resistant silicon and chromium alloy irons.

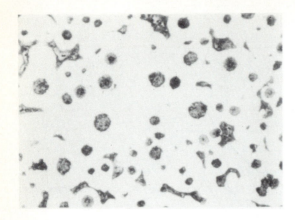

Figure 8-39 High-silicon heat-resistant cast iron (3.5% C, 3.5% Si, 0.7% Mn) as-cast. Structure shows nodules of temper-carbon graphite in a matrix of 15% pearlite (irregular gray constituent) and 85% free ferrite (light). (3% nital; X100.) (*After Metals Handbook, 8th ed., vol. 7, American Society for Metals, Metals Park, OH, 1972, p. 100.*)

The high-nickel cast irons with nodular graphite are considerably stronger and have higher ductility than the flake-graphite nickel alloy irons.

PROBLEMS

1. In what three principal ways do cast irons differ from steels?

2. What are some of the advantages of cast irons as engineering materials?

3. Describe the microstructures of the four major types of cast irons.

4. How does an addition of 2% Si to the iron-carbon system affect (*a*) the eutectoid composition and (*b*) the maximum solid solubility of carbon?

5. What properties make gray cast iron an extremely useful engineering material?

6. Describe the microstructural changes which occur when an Fe–3% C–2% Si gray cast iron is slowly cooled from the liquid state to room temperature. Use the Fe–C–2% Si phase diagram to indicate the changes.

7. Why are gray cast irons classified according to strength?

8. What element is most important for promoting graphitization in gray cast irons?

9. Which elements are iron-carbide stabilizers?

10. What is the carbon equivalent for gray cast irons?

11. How do sulfur and manganese interact in cast irons?

12. Why is manganese sulfide more desirable than iron sulfide in cast irons?

13. What is the morphology of MnS? How can this shape be explained in terms of solidification processes?

14. What is steadite?

15. What are the five main types of graphite flakes found in gray cast iron?

16. What type of microstructure is associated with a 20-ksi (min) gray cast iron? A 40-ksi one?

17. What is the chief cause of the relatively low tensile strengths of gray cast iron?

18. Why does gray cast iron have such an excellent damping capacity?

19. What are the engineering advantages of ductile cast iron?

20. Using a graphite crystal-structure model, illustrate how spheroidal graphite forms during the solidification of cast iron.

21. What mechanism is believed responsible for the creation of graphite flakes instead of graphite nodules?

22. How are graphite nodules produced in ductile cast irons?

23. Why is carbon content of ductile cast irons usually in the 3.5 to 3.8 percent range? What occurs during melting if the carbon is too high?

24. Why are the sulfur and phosphorus levels kept very low in ductile cast irons?

25. How can the strength and microstructure of ductile cast iron be changed by heat treatment?

26. What are the engineering advantages of malleable cast irons. What is their chief disadvantage?

27. Describe the structural changes which occur during malleablization to produce ferritic malleable cast iron?

28. How is the malleablization process modified to produce pearlitic malleable cast iron?

29. What are chilled cast irons? What is their main engineering application?

30. Describe the structure of high-chromium abrasion-resistant white cast iron?

31. What are the three main types of corrosion resistance cast irons? What are the dominant factors which determine the corrosion resistance of these alloys?

32. What special corrosion-resistance properties are obtained with (*a*) high-chromium cast irons, (*b*) high-nickel cast irons, (*c*) high-silicon cast irons?

33. What alloy additions are used to make heat-resistant cast irons?

34. What chemical changes occur in gray cast irons at high temperatures?

35. What property advantages are obtained with the following types of heat-resistant cast irons: (*a*) high-silicon, (*b*) high-chromium, and (*c*) high-nickel?

REFERENCES

1. C. F. Walton (ed.): "Gray and Ductile Iron Castings Handbook," Gray and Ductile Iron Founders' Society, Inc., Cleveland, 1971.
2. R. W. Heine, C. R. Loper, and P. C. Rosenthal: "Principles of Metal Casting," 2d ed., McGraw-Hill, New York, 1967.
3. H. T. Angus: "Cast Iron: Physical and Engineering Properties," 2d ed., Butterworths, London, 1976.
4. P. F. Wieser, C. E. Bates, and J. F. Wallace: "Mechanism of Graphite Formation in Iron-Silicon-Carbon Alloys," Malleable Founders Society, Cleveland, 1967.
5. B. Lux, I. Minkoff, and F. Mollard (eds.): "The Metallurgy of Cast Iron," Georgi Publishing Co., St. Saphorin, Switzerland, 1975.
6. H. E. Henderson: "Gray, Ductile, and Malleable Iron Castings—Current Capabilities," American Society for Testing and Materials, STP 455, 1969.
7. *Metals Handbook*, vol. 1: "Cast Iron," American Society for Metals, Metals Park, OH, 1961, pp. 349–406.
8. *Metals Handbook*, vol. 7: "Atlas of Microstructures of Industrial Alloys," American Society for Metals, Metals Park, OH, 1972.
9. J. F. Wallace and L. Leonard: "Cast Irons and Steels," *Machine Design*, Mar. 12, 1964.
10. G. F. Ruff and J. F. Wallace: "Control of Structure and its Effect on Mechanical Properties of Gray Iron," *AFS Trans.* 84(1976):705.
11. J. Schuyten: "Ductile Iron: an Overview of Where and How It Is Used," *Metal Progress*, Nov. 1975, p. 73.
12. W. Scholz and M. Semchyshen: "Effect of Molybdenum on the Hardenability of Ductile Cast Iron," *Modern Castings*, Jan. 1968, p. 1.
13. J. F. Wallace: "Effects of Minor Elements on the Structure of Cast Irons," *AFS Trans.* 83(1975):363.

14. R. D. Maier and J. F. Wallace: "Literature Search on Controlling the Shape of Temper Carbon Nodules in Malleable Iron," *AFS Trans.* 84(1976):687.
15. J. Dodd: "High Strength, High Ductility Ductile Irons," *Modern Castings*, May 1978.
16. W. C. Johnson and H. B. Smartt: "The Role of Interphase Boundary Adsorption in the Formation of Spherical Graphite in Cast Iron," *Met. Trans.* 8A(1977):553.
17. J. O. T. Adewara and C. R. Loper: "Effect of Pearlite on Crack Initiation and Propagation in Ductile Iron," *AFS Trans.* 84(1976):513.
18. J. O. T. Adewara and C. R. Loper: "Crack Initiation and Propagation in Fully Ferritic Ductile Iron," *AFS Trans.* 84(1976):527.
19. S. Yamamoto et al.: "A Proposed Theory of Nodularization of Graphite in Cast Irons," *Met Sci.* 9(1975):360.
20. R. W. Heine and R. J. Warshal: "Primary Solidification and Spiking in White Cast Iron," *Trans. ASM* 59(1966):163.
21. W. H. Moore: "Structure and Properties of White Cast Iron," *Casting Engineering*, May/June 1970, p. 13.
22. F. Maratray: "Choice of Appropriate Compositions for Chromium-Molybdenum White Irons," *AFS Trans.* 79(1971):121.

TOOL STEELS

From a use standpoint, *tool steels* are utilized in working and shaping basic materials such as metals, plastics, and wood into desired forms. From a composition standpoint, tool steels are carbon or alloy steels which are capable of being hardened and tempered. Some desirable properties of tool steels are high wear resistance and hardness, good heat resistance, and sufficient strength to work the materials. In some cases, dimensional stability may be very important. Tool steels also must be economical to use and be capable of being formed or machined into the desired shape for the tool.

Since the property requirements are so special, tool steels are usually melted in electric furnaces using careful metallurgical quality control. A great effort is made to keep porosity, segregation, impurities, and nonmetallic inclusions to as low a level as possible. Tool steels are subjected to careful macroscopic and microscopic inspections to ensure that they meet strict "tool steel" specifications.

Although tool steels are a relatively small percentage of total steel production,[1] they have a strategic position in that they are used in the production of other steel products and engineering materials. Some applications of tool steels include drills, deep-drawing dies, shear blades, punches, extrusion dies, and cutting tools.

For some applications, especially where extremely high-speed cutting is important, other *tool materials* such as sintered carbide products are a more

---

[1] In 1978, the United States shipment of tool steels was 92,800 tons, or 0.1 percent of the total shipment of steel products. (*Source:* "Annual Statistical Report," American Iron and Steel Institute, Washington, D.C., 1978.)

economical alternative to tool steels. The exceptional tool performance of sintered carbides results from their very high hardness and high compressive strength. Other tool materials are being used more and more often industrially.

9-1 CLASSIFICATION OF TOOL STEELS

The most commonly used classification system of tool steels is that established by the American Iron and Steel Institute (Ref. 9). The AISI system of classifying tool steels is based on quenching method, application method, special characteristics, and composition. In it, tool steels are classified into the groups and subgroups listed in Table 9-1.

9-2 WATER-HARDENING TOOL STEELS

Chemical Compositions and Typical Applications

Water-hardening tool steels are usually plain-carbon steels with 0.6 to 1.4% C, but most often with 0.8 to 1.1% C content. In one modification of these alloys 0.25% V is added, and in another 0.50% Cr. Table 9-2 lists the chemical compositions and typical applications of the water-hardening, W-type tool steels.

Table 9-1 Classification of tool steels†

Group		Letter symbol	Reference table in chapter for different types
1.	Water-hardening tool steels	W	Table 9-2
2.	Shock-resistant tool steels	S	Table 9-3
3.	Cold-work tool steels		
	Oil-hardening	O	Table 9-4
	Medium alloy, air-hardening	A	Table 9-5
	High carbon, high chromium	D	Table 9-6
4.	Hot-work tool steels	H	Table 9-7
	Chromium type	H1 to H19	
	Tungsten type	H20 to H39	
	Molybdenum type	H40 to H59	
5.	High-speed tool steels		Table 9-8
	Tungsten type	T	
	Molybdenum type	M	
6.	Special-purpose tool steels		
	Low alloy	L	
7.	Mold tool steels	P	

† After Ref. 9.

Table 9-2 Chemical compositions and typical applications of water-hardening (W-type) tool steels†

AISI type	% C	% W	% Mo	% Cr	% V	Typical applications
W1	0.60–1.40	. . .	. . .	. . .	. . .	Low carbon: blacksmith tools, blanking
W2	0.60–1.40	. . .	. . .	. . .	0.25	tools, caulking tools, cold chisels, forging dies, rammers, rivet sets, shear blades, punches, sledges. Medium carbon: arbors, beading tools, blanking dies, reamers, bushings, cold heading dies, chisels, coining dies, countersinks, drills, forming dies, jeweler dies, mandrels, punches, shear blades, woodworking tools. High carbon: glass cutters, jeweler dies, lathe tools, reamers, taps and dies, twist drills, woodworking tools. Vanadium content of W2 imparts finer grain, greater toughness, and shallow hardenability.
W5	1.10	. . .	. . .	0.50	. . .	Heavy stamping and draw dies, tube-drawing mandrels, large punches, reamers, razor blades, cold-forming rolls and dies, wear plates.

† After "ASM Databook," published in *Metal Progress*, mid-June 1977.

Heat Treatment and Microstructures

Water-hardening tool steels are the least expensive of all tool steels but, because of their relatively simple composition, they have in general the lowest wear resistance. Water-quenched tool steels also have very low hardenability, as shown in the IT diagram of a W1 tool steel in Fig. 9-1a. These steels do not harden adequately unless they are drastically water-quenched. Even with the drastic quench, except for very thin samples, only the outer part of the steel or *case* is transformed to a martensitic structure while the interior or *core* has a softer pearlitic structure (Fig. 9-1b). The water-hardening tool steels, because of their simple composition, serve as a basis for comparison for other such steels.

Figure 9-2a shows the microstructure of a W1 (1.10% C) tool steel after normalizing and air cooling to produce a pearlitic structure. Figure 9-2b shows the microstructure of a W1 (0.94% C) tool steel after water quenching to produce martensite, and Fig. 9-2c shows the microstructure of this steel after tempering. If the W1 steel is overheated to too high an austenitizing temperature, coarse tempered martensite along with retained austenite is formed, as shown in Fig. 9-2d.

Vanadium (0.25 percent) is added to the W2 water-hardening tool steel to inhibit grain growth during austenitizing. Vanadium dissolves in the carbide

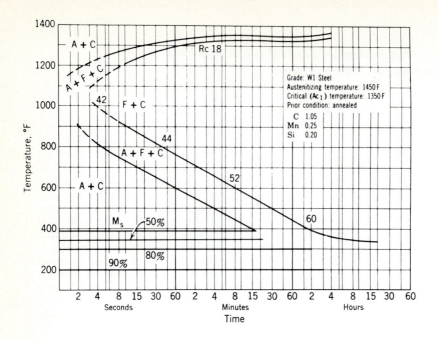

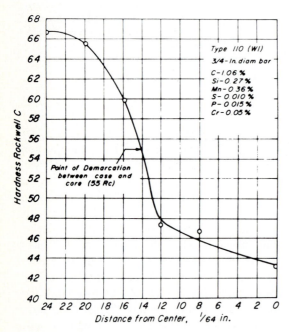

Figure 9-1 (*a*) Isothermal transformation (I-T) diagram for a type W1 water-hardening tool steel. Open part of curve around 1000°F indicates lack of good data for precise position of lines. (*After P. Payson, "Metallurgy of Tool Steels," 1962, p. 63. Reproduced by permission of John Wiley & Sons, Inc.*) (*b*) Hardness penetration on 3/4-in bar of W1 tool steel brine quenched from 815°C. Visual measurement of case depth is 10/64 in from surface. At this point the hardness is R_c55. (*After G. A. Roberts, J. C. Hamaker, and A. R. Johnson, "Tool Steels," 3d ed., American Society for Metals, Metals Park, OH, 1962.*)

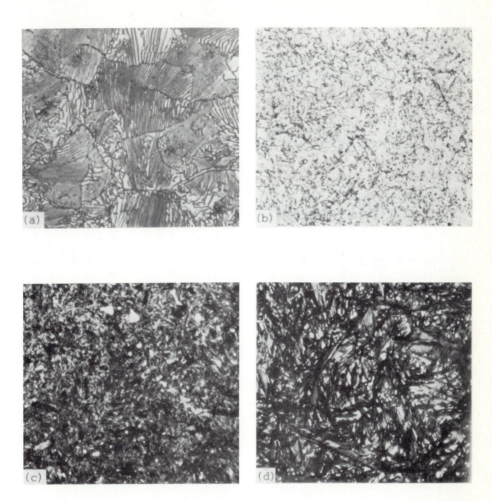

Figure 9-2 Microstructures of water-hardening (W-type) tool steels. (*a*) W1 water hardening (1.10% C, 0.31% Mn) tool steel; normalized by austenitizing at 927°C and air cooling. Bhn 227. Structure consists of lamellar pearlite with thin cementite precipitate at grain boundaries. (4% picral; ×1000.) (*b*) W1 water hardening (0.94% C, 0.21% Mn) tool steel; austenitized at 788°C and quenched in brine. Rockwell C 65. Structure is largely untempered martensite with some undissolved carbide particles. (3% nital; ×1000.) (*c*) W1 water hardening (0.94% C, 0.21% Mn) tool steel; austenitized at 788°C, brine-quenched and tempered to R_c58. Structure is tempered martensite; white spots are carbide precipitates. (3% nital; ×1000.) (*d*) W1 water hardening (0.94% C, 0.21% Mn) tool steel; austenitized at 857°C, brine-quenched and tempered at 163°C. Structure is coarse-tempered martensite and retained austenite (white), which is result of *overheating*. (3% nital; ×1000.) (*After Metals Handbook, 8th ed., vol. 7, American Society for Metals, Metals Park, OH, 1972, p. 102.*)

Table 9-3 Chemical compositions and typical applications of shock-resistant (S-type) tool steels†

AISI type	% C	% W	% Mo	% Cr	% V	% Other	Typical applications
S1	0.50	2.50	...	1.50	...	...	Bolt header dies, chipping and caulking chisels, pipe cutters, concrete drills, expander rolls, forging dies, forming dies, grippers, mandrels, punches, pneumatic tools, scarfing tools, swaging dies, shear blades, track tools, master hobs.
S2	0.50	...	0.50	...	...	1.00 Si	Hand and pneumatic chisels, drift pins, forming tools, knockout pins, mandrels, nail sets, pipe cutters, rivet sets, screw driver bits, shear blades, spindles, stamps, tool shanks, track tools.
S5	0.55	...	0.40	...	...	0.80 Mn, 2.00 Si	Hand and pneumatic chisels, drift pins, forming tools, knockout pins, mandrels, nail sets, pipe cutters, rivet sets and busters, screw driver bits, shear blades, spindles, stamps, tool shanks, track tools, lathe and screw-machine collets, bending dies, punches, rotary shears.
S6	0.45	...	0.40	1.50	...	1.40 Mn, 2.25 Si	Shear blades, aluminum impact extrusion dies and punches, rivet sets, cold-coining dies, cold header punches, knockout punches.
S7	0.50	...	1.40	3.25	...	...	Shear blades, punches, slitters, chisels, forming dies, hot header dies, blanking dies, rivet sets, gripper dies, engraving dies, plastic molds, die-casting dies, master hobs, beading tools, caulking tools, chuck jaws, clutches, pipe cutters, swaging dies.

† After "ASM Databook," published in *Metal Progress*, mid-June 1977.

M_3C (where the letter "M" stands for a metal) and lowers its solubility. Upon austenitizing the steel, grain growth will be inhibited as a result of the stable carbides. However, since the carbides are nucleating agents for pearlite, the hardenability of the steel will be decreased.

Chromium (0.50 percent) is added to the W5 water-hardening tool steel to increase hardenability. Chromium enters the carbide as $(Fe,Cr)_3C$, but has little effect on its solubility. Thus, when the carbide dissolves during austenitizing, the austenite is enriched in chromium. Since chromium inhibits the pearlitic reaction, it increases the hardenability of the W5 tool steel.

9-3 SHOCK-RESISTANT TOOL STEELS (S TYPE)

Chemical Compositions and Typical Applications

Shock-resistant tool steels are used for applications where repetitive impact stresses are encountered, such as in shear blades, chisels, and rivet sets. In these steels, the most important property is toughness, with hardness being secondary. Therefore, these steels have a lower carbon content, i.e., about 0.50 percent, than most other tool steels and are used at a slightly lower hardness, i.e., R_C 56 to 60. Table 9-3 lists the chemical compositions and typical properties of currently used shock-resistant S-type tool steels.

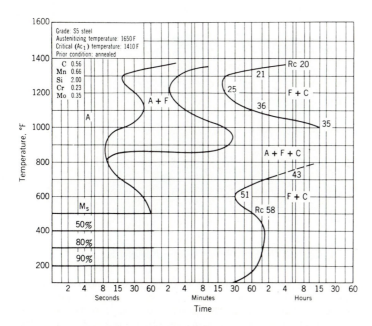

Figure 9-3 I-T diagram of type S5 shock-resistant tool steel. (*After P. Payson, "Metallurgy of Tool Steels," Wiley, 1962, p. 60. Reproduced by permission of John Wiley & Sons, Inc.*)

Heat Treatment and Microstructures

One of the most important shock-resistant tool steels is S5, which is a low-price, general-purpose tool steel. The S5 alloy has a high silicon and relatively low carbon content, and to ensure complete conversion of ferrite to austenite it is austenitized at the high temperature of 927°C. It has medium hardenability, as is shown by its IT diagram in Fig. 9-3. Figure 9-4 shows the microstructure of the S5 tool steel after (1) normalizing, (2) oil quenching, and (3) oil quenching and tempering at 400°C.

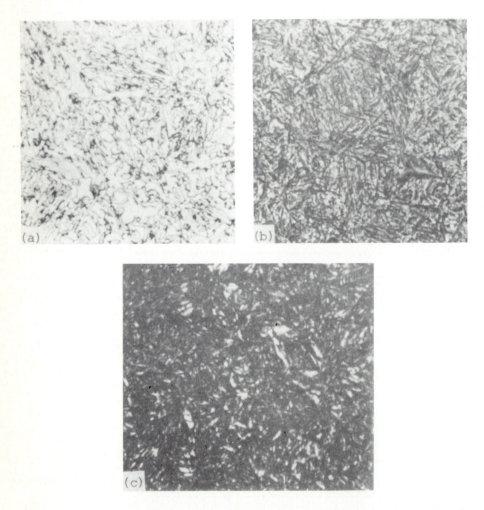

Figure 9-4 Microstructures of shock-resisting (S-type) tool steels. (*a*) S5 shock-resisting tool steel, normalized by austenitizing at 927°C for 1 h and air cooling. Structure consists of a mixture of martensite and coarse pearlite. (*b*) S5 shock-resisting tool steel, austenitized at 899°C and oil-quenched. The structure consists of fine untempered martensite. (*c*) S5 shock-resisting tool steel, austenitized at 899°C, oil-quenched and tempered at 399°C. The structure consists of fine tempered martensite. (Etchant: 2% nital; X1000.) (*After Metals Handbook, 8th ed., vol. 7, American Society for Metals, Metals Park, OH, 1972, p. 104.*)

For increased depth of hardening, the S7 tool steel, which contains 1.4% Mo and 3.25% Cr, has been developed. The low carbon level of 0.50 percent allows for high toughness, and the high amounts of Mo and Cr increase hardenability. The old S1 tool steel has decreased in popularity because it contains 2.5% W, which makes it relatively expensive without any qualitative advantages over the less expensive tool steels.

9-4 COLD-WORK (OIL-HARDENING) TOOL STEELS (O TYPE)

Cold-work tool steels are widely used for cold-work tool and die applications where resistance to wear and toughness are important. The principal groups of

Table 9-4 Chemical compositions and typical applications of cold-work oil-hardening (O-type) tool steels†

AISI type	% C	% W	% Mo	% Cr	% V	% Other	Typical applications
O1	0.90	0.50	...	0.50	...	1.00 Mn	Blanking dies, plastic mold dies, drawing dies, trim dies, paper knives, shear blades, taps, reamers, tools, gauges, bending and forming dies, bushings, punches.
O2	0.90	...	...	...	...	1.60 Mn	Blanking, stamping, trimming, cold-forming dies and punches, cold-forming rolls, threading dies and taps, reamers, gauges, plugs and master tools, broaches, circular cutters and saws, thread roller dies, bushings, plastic-molding dies.
O6	1.45	...	0.25	...	...	0.80 Mn, 1.00 Si	Blanking dies, forming dies, mandrels, punches, cams, brake dies, deep-drawing dies, cold-forming rollers, bushings, gauges, blanking and forming punches, piercing and perforating dies, taps, paper-cutting dies, wear plates, tool shanks, jigs, machine spindles, arbors, guides in grinders and straighteners.
O7	1.20	1.75	...	0.75	...	...	Mandrels, slitters, skiving knives, taps, reamers, drills, blanking and forming dies, gauges, chasers, brass finishing tools, dental burrs, paper knives, roll turning tools, burnishing dies, pipe-threading dies, rubber-cutting knives, woodworking tools, hand reamers, scrapers, spinning tools, broaches, blanking and cold-forming punches.

† After "ASM Databook," published in *Metal Progress*, mid-June 1977.

cold-work tool steels are (1) oil-hardening, (2) air-hardening, and (3) high-carbon, high-chromium types. In this section, only cold-work (oil-hardening, O-type) tool steels will be dealt with. Sections 9-5 and 9-6 will treat air-hardening and high-chromium types.

Chemical Compositions and Typical Applications

Oil-hardening cold-work tool steels are among the most widely used tool steels. Their properties include a high as-quenched hardness, high hardenability from low quenching temperatures, freedom from cracking on quenching intricate sections, and the maintenance of a sharp edge for cutting purposes. However, they cannot be used for cutting at high speed or for hot working. Table 9-4 lists the chemical compositions and typical applications for the currently used oil-hardening O-type tool steels.

Heat Treatment and Microstructures

One of the most widely used of the tool steels is the oil-hardening type O1. Its high manganese content, along with 0.50% Cr and 0.50% W, increases the hardenability of the steel so that drastic water quenching can be avoided. Figure 9-5 shows how the IT diagram of this alloy is modified so that slower oil-quenching rates may be used to produce a martensitic structure than the rates that would be achieved by water quenching. By using the slow oil quench, there

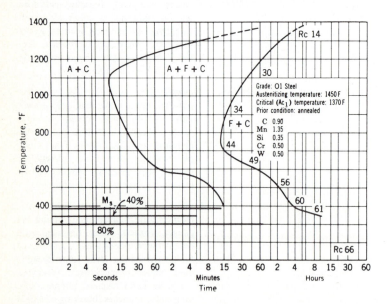

Figure 9-5 IT diagram of type O1 oil-hardening tool steel. (*After P. Payson, "Metallurgy of Tool Steels," Wiley, 1962, p. 64. Reproduced by permission of John Wiley & Sons, Inc.*)

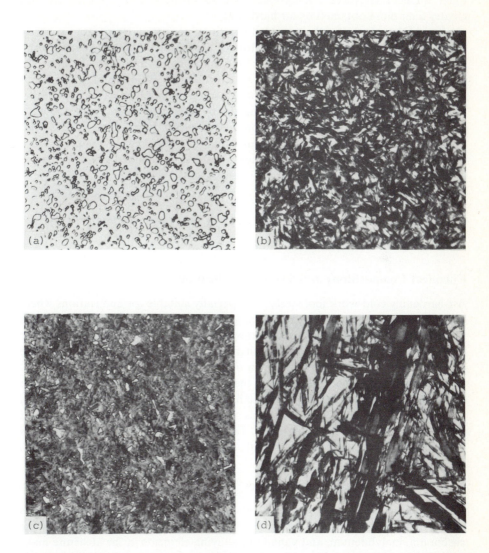

Figure 9-6 Microstructures of O1 (oil-hardening) tool steel (C, 0.94; Si, 0.30; Mn, 1.20; W, 0.50; Cr, 0.50). (*a*) Fully annealed condition. Structure consists of spheroidal carbide particles in a matrix of tempered martensite. (*b*) Normal oil-quenched and tempered condition; austenitized at 815°C, oil-quenched to room temperature; tempered at 150°C for 2 h. Structure consists of carbide particles in a matrix of tempered martensite. (*c*) Oil-quenched condition; austenitized at 815°C, oil-quenched to room temperature. Structure consists of carbide particles, untempered martensite, probably some bainite and retained austenite (white). (*d*) Overheated structure; austenitized at 927°C or higher, oil-quenched to room temperature, tempered at 150°C for 2 h. Structure consists of only a few carbide particles due to high solution temperature; also contains untempered martensite; probably some bainite and retained austenite. (Etchant: 4% nital; ×1000.) (*Courtesy of J. Stepanic, Latrobe Steel Co.*)

is much less dimension change, distortion, and cracking hazard than with the water quench. However, dimensional changes using the oil quench are still greater than those obtained by air hardening.

Type O1 tool steel in the annealed condition consists of ferrite and spheroidized carbide particles (Fig. 9-6a). Most of the carbides are dissolved during austenitizing at 815°C, but a small amount remains undissolved. After oil quenching to room temperature, the structure consists mainly of untempered martensite (Fig. 9-6b), but also present are some undissolved carbides, bainite, and retained austenite. After tempering the oil-quenched steel 2 h at 150°C, a matrix of tempered martensite is produced, along with some undissolved carbides (Fig. 9-6c). If this alloy is austenitized at too high a temperature, a coarse structure is produced after quenching and is retained after tempering (Fig. 9-6d).

9-5 COLD-WORK (MEDIUM-ALLOY, AIR-HARDENING) TOOL STEELS (A TYPE)

Chemical Compositions and Typical Applications

Air-hardening cold-work tool steels are especially suitable for applications where exceptional toughness and reasonably good abrasion resistance are required such as for blanking, forming, and drawing dies. These steels can be used for intricate dies since their dimensional changes after hardening and tempering are only about one-quarter that of the Mn oil-hardening tool steels (i.e., O1 alloy).

The principal alloying elements in the air-hardening cold-work tool steels, in addition to the 1 to 2% C, are chromium, manganese, molybdenum, vanadium, and nickel. Two important alloys of this series are the 5% Cr type (A2 alloy) and the 1% Cr, 2 to 3% Mn type (A4 alloy). Table 9-5 lists the chemical compositions and typical applications for the currently used air-hardening (A-type) tool steels.

Heat Treatment and Microstructures

The air-hardening cold-work tool steel type A2 is used for tool applications where toughness is more important than wear resistance. The solution of the chromium, molybdenum, and vanadium alloying elements in the austenite make this alloy highly hardenable, as can be seen by its IT diagram in Fig. 9-7. Fairly large sections of this alloy may be air-cooled after austenitizing at 968°C. Very slow cooling of large sections could cause precipitation of carbides and the formation of bainite, which would lead to more retained austenite. In order to avoid this problem, large sections can be quenched in a salt bath at 540°C and then air-cooled to room temperature.

The microstructural changes that occur during the heat treatment of air-hardening cold-work tool steels are exemplified by the changes which occur in

Table 9-5 Chemical compositions and typical applications of cold-work (medium-alloy) air-hardening (A-type) tool steels†

AISI type	% C	% W	% Mo	% Cr	% V	% Other	Typical applications
A2	1.00	...	...	1.00	5.00	...	...	Thread-rolling dies, extrusion dies, trimming dies, blanking dies, coining dies, mandrels, shear blades, slitters, spinning rolls, forming rolls, gauges, beading dies, burnishing tools, ceramic tools, embossing dies, plastic molds, stamping dies, bushings, punches, liners for brick molds.
A3	1.25			1.00	5.00	1.00	...	
A4	1.00	...		1.00	1.00	...	2.00 Mn	Blanking dies, forming dies, trimming dies, punches, shear blades, mandrels, bending dies, forming rolls, broaches, knurling tools, gauges, arbors, bushings, slitting cutters, cold-treading rollers, drill bushing, master hobs, cloth-cutting knives, pilot pins, punches, engraver rolls.
A6	0.70	...		1.25	1.00	...	2.00 Mn	Blanking dies, forming dies, coining dies, trimming dies, punches, shear blades, spindles, master hobs, retaining rings, mandrel, plastic dies.
A7	2.25	1.00‡		1.00	5.25	4.75	...	Brick mold liners, drawing dies, briquetting dies, liners for shot-blasting equipment and sand slingers, burnishing tools, gauges, forming dies.
A8	0.55	1.25		1.25	5.00	...	...	Cold slitters, shear blades, hot-pressing dies, blanking dies, beading tools, cold-forming dies, punches, coining dies, trimming dies, master hobs, rolls, forging die inserts, compression molds, notching dies, slitter knives.
A9	0.50	...		1.40	5.00	1.00	1.50 Ni	Solid cold-heading dies, die inserts, heading hammers, coining dies, forming dies and rolls, die casings, gripper dies. Hot-work applications: punches, piercing tools, mandrels, extrusion tooling, forging dies, gripper dies, die casings, heading dies, hammers, coining and forming dies.
A10	1.35	...		1.50	...	...	1.80 Mn, 1.25 Si, 1.80 Ni	Blanking dies, forming dies, gauges, trimming shears, punches, forming rolls, wear plates, spindle arbors, master cams and shafts, stripper plates, retaining rings.

† After "ASM Databook," published in *Metal Progress*, vol. 112, no. 1, mid-June 1977.
‡ Optional.

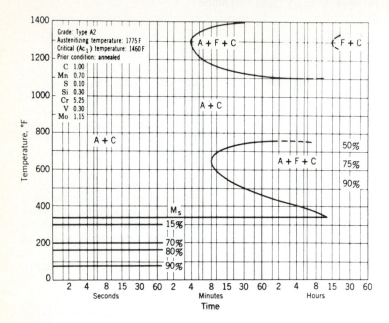

Figure 9-7 IT diagram of A2 type cold-work medium-alloy air-hardening tool steel. (*After P. Payson, "Metallurgy of Tool Steels," Wiley, 1962, p. 267. Reproduced by permission of John Wiley & Sons, Inc.*)

the A2 tool steel. In the annealed condition, the microstructure of the A2 tool steel consists of low-alloy ferrite and about 15 percent by weight of carbides, which are principally of the M_7C_3 and $M_{23}C_6$ types (the letter "M" standing for a metal) (Fig. 9-8a). When this steel is austenitized, it must be heated above about 970°C since most of the alloy carbides do not dissolve very rapidly until the temperature is above about 927°C. After austenitizing, most of the carbon, chromium, molybdenum, and vanadium are dissolved in the austenite. However, about 5 percent residual carbides still remain, as can be seen in the austenitized, air-cooled, and tempered microstructure of Fig. 9-8b. The residual carbides are mainly of the M_7C_3 type, but some $M_{23}C_6$ carbides may also be present. Double tempering of the alloy is necessary to reduce the amount of retained austenite and thus prevent dimensional changes due to the transformation of the austenite to martensite at room temperature.

If the A2 tool steel is *overheated* during austenitizing (for example, at 1010°C), a coarse grain structure that contains fused carbides is produced (Fig. 9-8c). This type of structure is undesirable, since full hardness cannot be developed in the alloy in this condition by subsequent heat treatment. If the alloy is tempered before being allowed to air-cool to room temperature, a "hot-tempered" condition is produced in which large amounts of retained

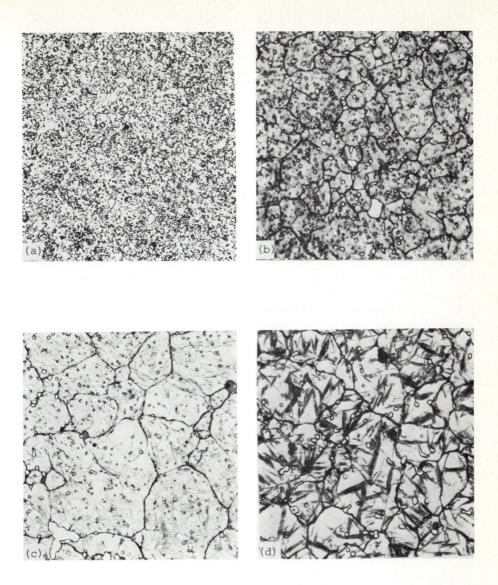

Figure 9-8 Microstructures of A2 (cold-work, air-hardening) tool steel (C, 1.0; Si, 0.30; Mn, 0.90; Cr, 5.25; V, 0.25; Mo, 1.10). (*a*) Fully annealed condition. Structure consists of spheroidal carbide particles in a matrix of ferrite. (*b*) Austenitized at 955°C, air-cooled to room temperature, double-tempered at 232°C for 2 h each. Structure consists of carbide particles in a matrix of tempered martensite. This structure is characteristic of A2 in the normal heat treated condition. (*c*) Austenitized at 1010°C, air-cooled to room temperature, no tempering; structure consists of carbide particles untempered martensite and retained austenite. Note coarse grain structure and fused carbides indicative of overheating. (*d*) Austenitized at 982°C, air-cooled to 204°C, double-tempered at 510°C for 2 h each. Structure consists of carbide particles, tempered and untempered martensite, and retained austenite. This structure is characteristic of "hot tempering," which is caused by tempering before air cooling to room temperature is complete. (Etchant: 4% nital; ×1000.) (*Courtesy of J. Stepanic, Latrobe Steel Co.*)

377

austenite are present after tempering (Fig. 9-8d). This condition is undesirable, since full hardening will not be attained and the material will be subject to possible dimensional changes later from the presence of the retained austenite.

9-6 COLD-WORK (HIGH-CARBON, HIGH-CHROMIUM) TOOL STEELS (D-TYPE)

Chemical Compositions and Typical Applications

The high-carbon, high-chromium tool steels were introduced into the United States about 1915 and were originally developed as a possible substitute for high-speed cutting tool steels. Since these steels did not have sufficient hardness at high cutting speeds and were also too brittle, they had limited use for this purpose. However, it was discovered that their high wear resistance and exceptional nondeforming properties made them very useful for cold-work die steels. The chemical compositions and typical applications of the principal D-type tool steels in use today are listed in Table 9-6.

The excellent wear resistance of the D-type cold-work tool steels is the result of their high chromium (12 percent) and carbon (1.50 to 2.35 percent) contents. Differences in the wear resistance among the high-chromium, high-carbon tool steels is mainly the result of their carbon contents. For example, the 1.50% C, 12% Cr alloy (D2) contains about 30 to 40 percent fewer carbides under equilibrium conditions than does the 2.25% C, 12% Cr alloy (D3), as is indicated

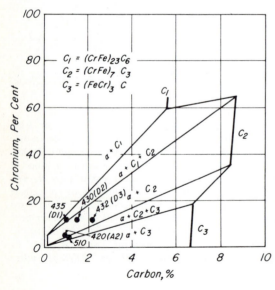

Figure 9-9 Isothermal section of the iron-chromium-carbon system at 700°C. Indicated on the diagram are the approximate compositions of the high-carbon high-chromium tool steels D2 and D3. The air-hardening cold-work tool steel type A2 is also indicated. (*After G. A. Roberts, J. C. Hamaker, and A. R. Johnson, "Tool Steels," 3d ed., American Society for Metals, Metals Park, OH, 1962, p. 499.*)

Table 9-6 Chemical compositions and typical applications of cold-work, high-carbon, high-chromium (D-type) tool steels†

AISI type	% C	% W	% Mo	% Cr	% V	% Other	Typical applications
D2	1.50	...	1.00	12.00	1.00	...	Blanking dies, cold-forming dies, drawing dies, lamination dies, thread-rolling dies, shear blades, slitter knives, forming rolls, burnishing tools, punches, gauges, knurling tools, lathe centers, broaches, cold-extrusion dies, mandrels, swaging dies, cutlery.
D3	2.25	...	...	12.00	...	...	Blanking dies, cold-forming dies, drawing dies, lamination dies, thread-rolling dies, shear blades, slitter knives, forming rolls, seaming rolls, burnishing tools, punches, gauges, crimping dies, swaging dies.
D4	2.25	...	1.00	12.00	...	...	Blanking dies, brick molds, burnishing tools, thread-rolling dies, hot-swaging dies, wiredrawing dies, forming tools and rolls, gauges, punches, trimmer dies, dies for deep drawing.
D5	1.50	...	1.00	12.00	...	3.00 Co	Cold-forming dies, thread-rolling dies, blanking dies, coining dies, trimming dies, draw dies, shear blades, punches, quality cutlery, rolls.
D7	2.35	...	1.00	12.00	4.00	...	Brick mold liners and die plates, briquetting dies, grinding wheel molds, dies for deep drawing, flattening rolls, shot and sandblasting liners, slitter knives, wear plates, wiredrawing dies, Sendzimir mill rolls, ceramic tools and dies, lamination dies.

† After "ASM Databook," published in *Metal Progress*, vol. 112, no. 1, mid-June 1977.

by the relative position of the D2 and D3 points on the isothermal section of the iron-chromium-carbon system at 700°C (Fig. 9-9).

The high chromium content of the D-type tool steels provides them with resistance to oxidation at high temperatures and good resistance to staining when hardened and polished.

Small amounts of molybdenum, vanadium, cobalt, and tungsten are added to these steels to form different types. Molybdenum increases hardenability and toughness, but hardly affects the austenitic grain size or amount of retained austenite. Vanadium refines the grain size, but decreases hardenability when more than 0.8 percent is added. Vanadium also decreases the retained austenite and in amounts up to 1 percent increases toughness.

Heat Treatment and Microstructures

In order to obtain the least amount of dimensional change with these alloys, they must be heated slowly and uniformly to the austenitizing temperature. Salt baths or controlled-atmosphere furnaces are commonly used for hardening high-carbon, high-chromium tool steels.

In the D2 tool steel, the addition of 0.8% Mo suppresses the formation of pearlite and allows full hardness to be obtained by air cooling. The small molybdenum addition greatly increases this steel's hardenability, as is indicated by its IT diagram (Fig. 9-10).

The D2 tool steel is usually air-cooled from austenitizing temperatures of about 1010 to 1038°C. If heated to too high austenitizing temperatures, its hardness upon tempering will be lower up to about 450°C (Fig. 9-11). The reason for this hardness lowering is that, after austenitizing above about 1090°C, more carbon and chromium are dissolved in the austenite so that the M_s temperature is lowered and consequently more retained austenite is formed. When the tempering temperature exceeds about 500°C, much of the retained

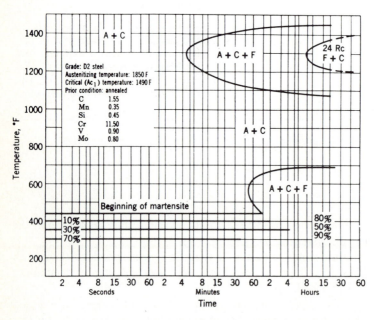

Figure 9-10 IT diagram of D2 (high-carbon, high-chromium) tool steel. (*After P. Payson, "Metallurgy of Tool Steels," Wiley, 1962, p. 151. Reproduced by permission of John Wiley & Sons, Inc.*)

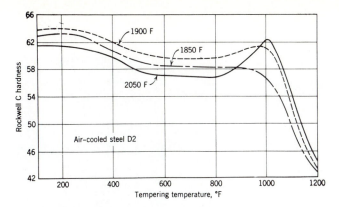

Figure 9-11 Effect of austenitizing temperature on the hardness of type D2 tool steel after temper-ing. A large amount of retained austenite in the 1120°C (2050°F) sample causes low hardness in the tempered condition. Above about 450°C (842°F) in the 1120°C (2050°F) sample, transformation of the austenite to martensite along with precipitation of chromium carbides causes an increase in hardness. (*After P. Payson, "Metallurgy of Tool Steels," Wiley, 1962, p. 263. Reproduced by permission of John Wiley & Sons, Inc.*)

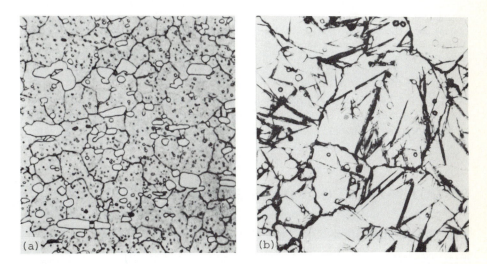

Figure 9-12 Microstructures of D2 (high-carbon, high-chromium cold-work) tool steels (C, 1.50; Si, 0.30; Mn, 0.50; Cr, 12.00; V, 0.90; Mo, 0.75). (*a*) Austenitized at 1025°C, air-cooled to room temperature; tempered at 288°C for 2 h. The structure consists of carbide particles in a matrix of about 60–70% tempered martensite and about 30–40% retained austenite. This structure is the normal D2 heat-treated condition when tempered in the low-temperature range. (*b*) Austenitized at 1150°C, air-cooled to room temperature; double-tempered at 510°C for 2 h each. The structure consists of carbide particles in a matrix of retained austenite with some tempered and untempered martensite also present. This structure is characteristic of D2 in the overheated condition.

Figure 9-12 *Continued* (*c*) Austenitized at 982°C, air-cooled to room temperature, double-tempered at 510°C for 2 h each. The structure consists of carbide particles in a matrix of tempered martensite. This structure is an underheated condition. (*d*) Austenitized at 1023°C, air-cooled to room temperature, double-tempered at 510°C for 2 h each. The structure consists of carbide particles in a matrix of tempered martensite. This structure is characteristic of D2 in the normal heat-treated condition. (Tempering in the secondary range is employed for semihot work applications.) (Etchant: 4% nital; X1000.) (*Courtesy of J. Stepanic, Latrobe Steel Co.*)

austenite is transformed to martensite, and this transformation is partly responsible for the hardness peak observed. Precipitation of chromium carbide may also contribute to the increase in hardness at this high temperature. The microstructures of the D2 tool steel after various heat treatments are shown in Fig. 9-12.

9-7 HOT-WORK TOOL STEELS (H TYPE)

Chemical Compositions and Typical Applications

For tool steels used in hot-work applications such as hot extrusion, hot forging, and die-casting dies, the following characteristics are important:

1. Resistance to deformation at the hot-working temperatures. Carbon steels become soft and weak at such temperatures and therefore cannot be used for hot work.
2. Relative resistance to both mechanical and thermal shock (especially if water cooling is used). To increase the shock resistance of these tool steels, the carbon content must be kept to a low level.

3. Resistance to erosion and wear at elevated temperatures.
4. Resistance to heat-treating deformation. Intricate dies should not warp during heat treatment; this is solved using a highly hardenable steel.
5. Resistance to heat checking (development of fine shallow cracks on the surface of a tool).

Hot-work tool steels are of three principal types: (1) chromium base, types H1 to H19; (2) tungsten base, types H20 to H39; and (3) molybdenum base, types H40 to H59. The chemical compositions and typical applications of some currently used hot-work tool steels are listed in Table 9-7.

The 5% Cr hot-work tool steels such as H11, H12, and H13 have high hardenability and can be hardened in relatively large sections by air cooling

Table 9-7 Nominal chemical compositions and typical applications of hot-work tool steels (H-type)†

AISI type	% C	% W	% Mo	% Cr	% V	% Other	Typical applications
				Hot work (chromium)			
H10	0.40	...	2.50	3.25	0.40	...	Mandrels, extrusion and forging dies, die holders, bolsters and dummy blocks, punches, die inserts, gripper and header dies, hot shears, aluminum die-casting dies, inserts for forging dies and up-setters, shell-piercing tools.
H11	0.35	...	1.50	5.00	0.40	...	die-casting dies, punches, piercing tools, mandrels, extrusion tooling, forging dies, high-strength structural components.
H12	0.35	1.50	1.50	5.00	0.40	...	Extrusion dies, dummy blocks, holders, gripper and header dies, forging-die inserts, punches, mandrels, sleeves for cold-heading dies.
H13	0.35	...	1.50	5.00	1.00	...	Die-casting dies and inserts, dummy blocks, cores, ejector pins, plungers, sleeves, slides, extrusion dies, forging dies and inserts.
H14	0.40	5.00	...	5.00	...	...	Backer blocks, die holders, aluminum and brass extrusion dies, press liners, dummy blocks, forging dies and inserts, gripper dies, shell-forging points and mandrels, hot punches, pushout rings, dies and inserts for brass forging.
H19	0.40	4.25	...	4.25	2.00	4.25 Co	Extrusion dies and die inserts, dummy blocks, punches, forging dies and die inserts, mandrels, hot-punch tools.

Table 9-7-Continued

AISI type	% C	% W	% Mo	% Cr	% V	% Other	Typical applications
							Hot work (tungsten)
H21	0.35	9.00	. . .	3.50	. . .	. . .	Mandrels, hot-blanking dies, hot punches, blades for flying shear, hot trimming dies, extrusion and die-casting dies for brass, dummy blocks, piercer points, gripper dies, hot-nut tools (crowners, cutoffs, side dies, piercers), hot headers.
H22	0.35	11.00	. . .	2.00	. . .	. . .	Mandrels, hot-blanking dies, hot punches, blades for flying shear, hot-trim dies, extrusion dies, dummy blocks, piercer points, gripper dies.
H23	0.30	12.00	. . .	12.00	. . .	. . .	Extrusion and die-casting dies for brass, brass and bronze permanent molds.
H24	0.45	15.00	. . .	3.00	. . .	. . .	Punches and shear blades for brass, hot-blanking and drawing dies, trimming dies, dummy blocks, hot-press dies, hot-punches, gripper dies, hot-forming rolls, hot-shear blades, swaging dies, hot-heading dies, extrusion dies.
H25	0.25	15.00	. . .	4.00	. . .	. . .	Hot-forming dies, die-casting and forging dies, die inserts, extrusion dies and liners, shear blades, blanking dies, gripper dies, punches, hot-swaging dies, nut piercers, piercer points, mandrels, high-temperature springs.
H26	0.50	18.00	. . .	4.00	1.00	. . .	Mandrels, hot-blanking dies, hot punches, blades for flying shear, hot-trimming dies, extrusion dies, dummy blocks, piercer points, gripper dies, pipe-threadings dies, nut chisels, forging-press inserts, extrusion dies for brass and copper.
							Hot work (molybdenum)
H42	0.60	6.00	5.00	4.00	2.00	. . .	Cold-trimming dies, hot-upsetting dies, dummy blocks, header dies, hot-extrusion dies, cold-header and extrusion dies and die inserts, hot-forming and swaging dies, nut piercers, hot punches, mandrels, chipping chisels.

† After "ASM Databook," published in *Metal Progress*, vol. 112, no. 1, mid-June 1977.

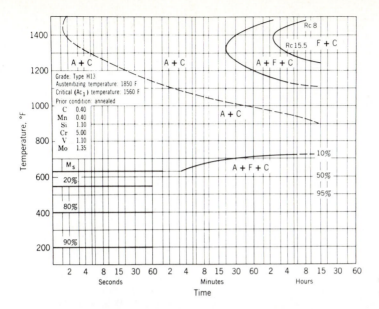

Figure 9-13 IT diagram for type H13 hot work tool steel. Since the steel contains residual carbides as austenitized at 1010°C, the constituents, austenite and carbide, are present to the left and below the carbide precipitation line as well as to the right and above. (*After P. Payson, "Metallurgy of Tool Steels," Wiley, 1962, p. 227. Reproduced by permission of John Wiley & Sons, Inc.*)

with a minimum amount of dimensional change. The relatively high silicon content of the H13 steel is for improving the oxidation resistance at the austenitizing temperature of 1010°C. The IT diagram of the H13 hot-work tool steel is shown in Fig. 9-13. Since this alloy contains 5% Cr and 1.5% Mo in solid solution, the reaction times are quite long except that required for the carbide precipitation.

Heat Treatment

When the H13 tool steel is austenitized at 1010°C for about 1 h, the molybdenum and chromium carbides are dissolved in solid solution. Only the vanadium carbide (MC) remains undissolved. Air cooling to room temperature produces a structure containing martensite, retained austenite, and probably some bainite. By tempering this structure twice at about 578°C, the retained austenite can be converted to tempered martensite. During the first tempering, the retained austenite from air cooling after austenitizing is converted to martensite (this process is called *conditioning*). During the second tempering, the new martensite is transformed to tempered martensite. Double tempering of these hot-work tool steels is therefore necessary to obtain maximum dimensional stability.

The hardness versus tempering temperature curves for double tempering (2 + 2 h at temperature) is shown for types H12 and H13 tool steels (Fig. 9-14).

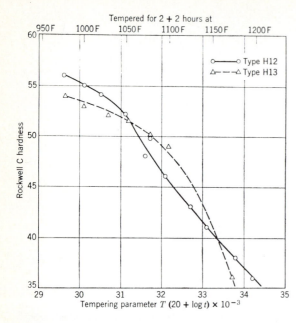

Tempered for 2 + 2 hours at

Figure 9-14 Tempering curves for H13 and H12 hot-work tool steels. Note that H13 has better resistance to softening in the temperature range of most importance for hot-work steels. (*After P. Payson, "Metallurgy of Tool Steels," Wiley, 1962, p. 229. Reproduced by permission of John Wiley & Sons, Inc.*)

Note that the H13 steel has quite good resistance to softening since a Rockwell C hardness of 45 is maintained after 4 h of tempering at 1130°F (610°C).

Microstructures

In the annealed condition, H13 tool steel consists of about 3.5 percent by weight of alloy carbides (M_6C containing principally molybdenum, M_7C_3 containing principally chromium, and MC containing principally vanadium) finely dispersed in a relatively low-alloy ferrite (Fig. 9-15a). The structure after air cooling and tempering is shown in Fig. 9-15b and consists of fine carbide particles in a matrix of tempered martensite.

9-8 SECONDARY HARDENING OF MOLYBDENUM AND TUNGSTEN STEELS[1]

Secondary Hardening in General

When plain-carbon steels are tempered, a progressive decrease in strength and a corresponding increase in ductility occur as the tempering temperature is increased in the range of 100 to 700°C. The formation of cementite and its gradual coarsening in the ferritic matrix are the principal causes for the changes in mechanical properties.

[1] Ref. 4.

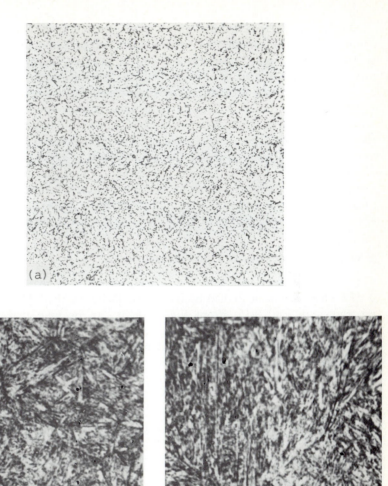

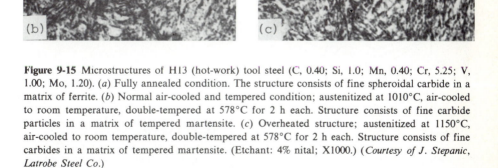

Figure 9-15 Microstructures of H13 (hot-work) tool steel (C, 0.40; Si, 1.0; Mn, 0.40; Cr, 5.25; V, 1.00; Mo, 1.20). (*a*) Fully annealed condition. The structure consists of fine spheroidal carbide in a matrix of ferrite. (*b*) Normal air-cooled and tempered condition; austenitized at 1010°C, air-cooled to room temperature, double-tempered at 578°C for 2 h each. Structure consists of fine carbide particles in a matrix of tempered martensite. (*c*) Overheated structure; austenitized at 1150°C, air-cooled to room temperature, double-tempered at 578°C for 2 h each. Structure consists of fine carbides in a matrix of tempered martensite. (Etchant: 4% nital; X1000.) (*Courtesy of J. Stepanic, Latrobe Steel Co.*)

By replacing the cementite with a more stable alloy carbide, e.g., molybdenum or tungsten carbides, the softening observed in plain-carbon steels can be greatly reduced; and if sufficient amounts of alloying elements are added, an increase in hardening in the 500 to 650°C range will occur. This rehardening effect upon tempering is called *secondary hardening*.

The alloy carbides of molybdenum and tungsten are more stable than cementite and form in its place if sufficient activation energy is provided. The rate of growth of these alloy carbides in tempered martensite is chiefly determined by the activation energy for the diffusivity of these elements in ferrite. Since the diffusion rates of the alloying elements in ferrite are much slower than those of carbon in plain-carbon steels, the alloy carbides produced are much finer and coarsen at a slower rate. Thus, the strength properties of the tempered alloy steel martensites are much higher than those of the plain-carbon steels.

Secondary Hardening in Molybdenum Steels

The effect of 0.5 to 3.0% Mo in producing secondary hardening in quenched 0.1% C steels has been studied by Irving and Pickering (Ref. 19), and the results are shown in Fig. 9-16. From Fig. 9-16, it can be seen that above about 0.5% Mo, marked secondary hardening occurs and that the amount of secondary hardening increases as the amount of molybdenum increases. The secondary hardening effect in molybdenum steels is caused by a very fine precipitate of Mo_2C, which is shown in an electron micrograph by Raynor et al. (Ref. 18) of a 4% Mo, 0.2% C steel (Fig. 9-17). The Mo_2C strengthens the ferrite and reaches a maximum secondary hardening effect at about 550°C. The main strengthening

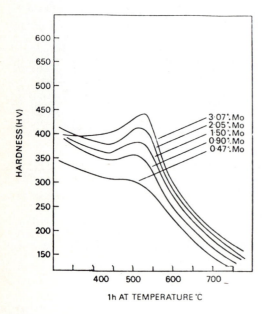

Figure 9-16 Effect of molybdenum on the tempering of quenched 0.1% C steels. [*After K. J. Irvine and F. B. Pickering, J. Iron Steel Inst. "The Tempering Characteristics of Low-Carbon Low-Alloy Steels," 194(1960):137.*]

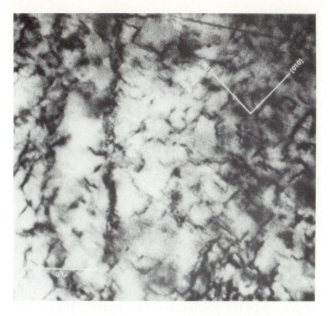

Figure 9-17 Electron micrograph of an Fe-4% Mo-0.2% C alloy tempered 7 h at 550°C; note the Mo$_2$C precipitate which shows evidence of nucleating on dislocations. Sample is in the underaged condition. [*After D. Raynor, J. A. Whiteman, and R. W. K. Honeycombe, "Precipitation of Molybdenum and Vanadium Carbides in High-Purity Iron Alloys," J. Iron Steel Inst. 204(1966):349.*]

precipitation reaction is the nucleation and growth of small Mo$_2$C needles in the dislocation network formed by the quenching to martensite. The needles grow along the three cube directions in the ferrite and form a Widmanstätten structure. The orientation relationship was found to be

$$(0001)_{Mo_2C}\|(001)_\alpha$$

$$[11\bar{2}0]_{Mo_2C}\|[100]_\alpha \quad \text{(Mo}_2\text{C needle growth direction)}$$

At peak hardness (25 h at 550°C), the needles were found to be about 100 to 200 Å long and 10 to 20 Å in diameter. The Mo$_2$C was also found to nucleate at former austenitic and martensitic lath boundaries. The overall process of coarsening of the Mo$_2$C is complex. This process is responsible for the rapid softening which takes place after peak hardness is reached.

Secondary Hardening in Tungsten Steels

The secondary hardening reaction in tungsten steels parallels that in molybdenum steels since in both the separate nucleation of acicular hexagonal M$_2$C carbides (where "M" stands for a metal) occurs during tempering. However, the extent of secondary hardening in tungsten steels is much less than in molybdenum steels, as shown in Fig. 9-18.

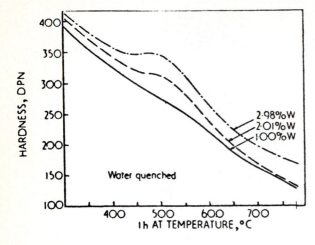

Figure 9-18 Effect of tungsten on the tempered hardness of quenched 0.1% C steels. [*After K. J. Irvine and F. B. Pickering, J. Iron Steel Inst. "The Tempering Characteristics of Low-Carbon Low-Alloy Steels," 194(1960): 137.*]

Electron microscopic examination shows that the precipitation of W_2C is very similar to that of Mo_2C: The morphology of the precipitate in each is the same, although the size and density of the precipitates are different. The particle density is less in the tungsten steels, presumably due to lower diffusivity of the larger tungsten atoms in ferrite. The tungsten steel is softer because the tungsten carbide particles are larger and further apart. It is possible that the dislocation network has had time to coarsen before it is locked in place by the W_2C particles. Figure 9-19 shows a typical dispersion of W_2C after overaging 100 h at 600°C. The W_2C does not coarsen as rapidly as the Mo_2C upon overaging. Again, this difference is attributed to the slower diffusion of the tungsten atoms.

The Fe_3C in 6.3% W, 0.23% C steel has been observed to transform to W_2C in three ways (Ref. 10):

1. By nucleation at the Fe_3C/ferrite interfaces, where W_2C gradually replaces the Fe_3C laths
2. By nucleation of W_2C on dislocations inherited from the martensitic transformation
3. By nucleation of W_2C in the vicinity of grain boundary spheroids of Fe_3C

M_6C is first detected on the prior austenitic grain boundaries and the martensitic laths during overaging. Coarse laths and spheroids of M_6C also form within the ferritic grains. Prolonged overaging produces very coarse particles of M_6C at the grain boundaries, which would probably be detrimental to low-temperature ductility and toughness.

The tempering sequence is as follows:

$$Fe_3C \rightarrow W_2C \rightarrow M_6C$$
$$\searrow M_{23}C_6 \rightarrow M_6C \quad \text{(within the matrix)}$$

$$Fe_3C \rightarrow W_2C + M_{23}C_6 \rightarrow M_6C + M_{23}C_6 \quad \text{(on the grain boundaries)}$$

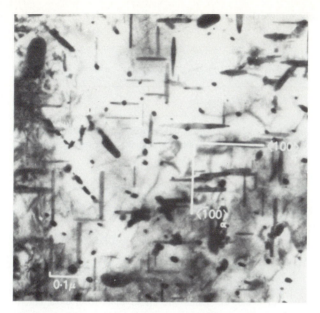

Figure 9-19 Electron micrograph of Fe-6.3% W, 0.23% C steel after tempering for 100 h at 600°C. Structure shows coarse W$_2$C needles. Alloy is in overaged condition. [*After A. T. Davenport and R. W. K. Honeycombe, "The Secondary Hardening of Tungsten Steel," Met. Sci. 9(1975):201.*]

9-9 HIGH-SPEED TOOL STEELS (T AND M TYPES)

Chemical Compositions and Typical Applications

High-speed tool steels are highly alloyed steels that are used for high cutting rates of very hard metals (hence the name "high speed"). Since the cutting speeds involved with these steels often cause high temperatures at the tool tip into the red range, they must be resistant to tempering at these temperatures. The ability of a steel to resist softening in the red-heat range is termed *red hardness* and is an important property of high-speed tool steels. These steels must also have good wear resistance and high hardness to be able to retain a sharp cutting edge for a prolonged period.

High-speed steels have been developed for many different applications, and contain tungsten and/or molybdenum for carbide formation and red hardness, vanadium for increased abrasion resistance, and chromium for reducing oxidation and increased hardness. Sometimes cobalt is added to improve the high-temperature hardening. High-speed steels are divided into two groups: (1) tungsten, type T, and (2) molybdenum, type M. The chemical compositions and typical applications of selected high-speed tool steels are given in Table 9-8.

Table 9-8 Nominal chemical compositions and typical applications for selected high-speed tool steels†

AISI type	% C	% W	% Mo	% Cr	% V	% Other	Typical applications
			High speed (tungsten)				
T1	0.75	18.00	...	4.00	1.00	...	Drills, taps, reamers, hobs, lathe and planer tools, broaches, crowners, burnishing dies, cold-extrusion dies, cold-heading die inserts, lamination dies, chasers, cutters, taps, end mills, milling cutters.
T2	0.80	18.00	...	4.00	2.00	...	Lathe and planer tools, milling cutters, form tools, broaches, reamers, chasers.
T4	0.75	18.00	...	4.00	1.00	5.00 Co	Lathe and planer tools, drills, boring tools, broaches, roll-turning tools, milling cutters, shaper tools, form tools, hobs, single-point cutting tools.
T5	0.80	18.00	...	4.00	2.00	8.00 Co	Lathe and planer tools, form tools, cutoff tools, heavy-duty tools requiring high red hardness.
T6	0.80	20.00	...	4.50	1.50	12.00 Co	Heavy-duty lathe and planer tools, drills, checking tools, cutoff tools, milling cutters, hobs.
T8	0.75	14.00	...	4.00	2.00	5.00 Co	Boring tools, lathe tools, heavy-duty planer tools, tool bits, single-point cutting tools for stainless steel.
T15	1.50	12.00	...	4.00	5.00	5.00 Co	Form tools, lathe and planer tools, broaches, milling cutters, blanking dies, punches, heavy-duty tools requiring good wear resistance.
			High speed (molybdenum)				
M1	0.85	1.50	8.50	4.00	1.00	...	Drills, taps, end mills, reamers, milling cutters, hobs, punches, lathe and planer tools, form tools, saws, chasers, broaches, routers, woodworking tools.
M2	0.85; 1.00	6.00	5.00	4.00	2.00	...	Drills, taps, end mills, reamers, milling cutters, hobs, form tools, saws, lathe and planer tools, chasers, broaches and boring tools.
M3-1	1.05	6.00	5.00	4.00	2.40	...	Drills, taps, end mills, reamers and counterbores, broaches, hobs, form tools, lathe and planer tools, cheeking tools, milling cutters, slitting saws, punches, drawing dies, routers, woodworking tools.

Table 9-8-Continued

AISI type	% C	% W	% Mo	% Cr	% V	% Other	Typical applications
					High speed (molybdenum)		
M3-2	1.20	6.00	5.00	4.00	3.00	. . .	Drills, taps, end mills, reamers and counterbores, broaches, hobs, form tools, lathe and planer tools, cheeking tools, slitting saws, punches, drawing dies, woodworking tools.
M4	1.30	5.50	4.50	4.00	4.00	. . .	Broaches, reamers, milling cutters, chasers, form tools, lathe and planer tools, cheeking tools, blanking dies and punches for abrasive materials, swaging dies.
M6	0.80	4.00	5.00	4.00	1.50	12.00 Co	Lathe tools, boring tools, planer tools, form tools, milling cutters.
M7	1.00	1.75	8.75	4.00	2.00	. . .	Drills, taps, end mills, reamers, routers, saws, milling cutters, lathe and planer tools, chasers, borers, woodworking tools, hobs, form tools, punches.
M10	0.85; 1.00	. . .	8.00	4.00	2.00	. . .	Drills, taps, reamers, chasers, end mills, lathe and planer tools, woodworking tools, routers, saws, milling cutters, hobs, form tools, punches, broaches.
M30	0.80	2.00	8.00	4.00	1.25	5.00 Co	Lathe tools, form tools, milling cutters, chasers.
M33	0.90	1.50	9.50	4.00	1.15	8.00 Co	Drills, taps, end mills, lathe tools, milling cutters, form tools, chasers.
M34	0.90	2.00	8.00	4.00	2.00	8.00 Co	Drills, taps, end mills, lathe tools, milling cutters, form tools, chasers.
M36	0.80	6.00	5.00	4.00	2.00	8.00 Co	Heavy-duty lathe and planer tools, boring tools, milling cutters, drills, cutoff tools, tool-holder bits.
M41	1.10	6.75	3.75	4.25	2.00	5.00 Co	Drills, end mills, reamers, form cutters, lathe tools, hobs, broaches, milling cutters, twist drills, end mills. Hardenable to Rockwell C67 to C70.
M42	1.10	1.50	9.50	3.75	1.15	8.00 Co	
M43	1.20	2.75	8.00	3.75	1.60	8.25 Co	
M44	1.15	5.25	6.25	4.25	2.00	12.00	
M46	1.25	2.00	8.25	4.00	3.20	8.25 Co	
M47	1.10	1.50	9.50	3.75	1.25	5.00 Co	

† After "ASM Databook," published in *Metal Progress*, vol. 112, no. 1, mid-June 1977.

Development of High-Speed Tool Steels

The first high-speed tool steels to be developed were tungsten-based. In 1904 the addition of 1% V to an 18% W, 4% Cr steel led to the development of the 18-4-1 high-speed steel (18% W, 4% Cr, 1% V), which was designated T1. This alloy was the standard high-speed tool steel for many years and is still used extensively today. Cobalt was added to high-speed steel in Germany about 1912, and is still in use today to increase red hardness (see Table 9-8). As time progressed, a whole series of tungsten-based high-speed tool steels were developed with the tungsten level ranging from 12 to 20 percent, along with about 4% Cr and 1 to 5% V. For some alloys, 5 to 12% Co was added to improve high-temperature hardness.

Until about 1930, the price of molybdenum was approximately the same as that of tungsten, and so molybdenum was not used extensively in high-speed steels. However, with the discovery of large deposits of molybdenum in Colorado, many molybdenum-type high-speed steels were developed. Type M1 was developed first and contained 9% Mo and 1.5% W. Later, M2 with 6% W, 5% Mo, and 2% V was developed. In the United States today, about 80 percent of the high-speed steels used are of the molybdenum types. However, in England the tungsten-base high-speed steels are still preferred.

Molybdenum high-speed tool steels are less costly, and for this reason they dominate the U.S. market. However, molybdenum steels are more susceptible to decarburization and require better temperature control during heat treatment. The general-purpose high-speed tool steel in the United States today is the M2 type, while in England the T1 alloy is still dominant.

Tungsten-Type High-Speed Tool Steels[1]

For these high-speed tool steels, type T1 will be used to illustrate the changes in structure that occur during solidification, hot working, and heat treatment.

Solidification and the cast structure The cast structure of the 18-4-1 high-speed steel can best be considered by referring to an approximate phase diagram which is a "pseudo" binary section of the Fe–18% W–4% Cr versus carbon system (Fig. 9-20).

When liquid of an 18% W–4% Cr tool steel containing approximately 0.75% C is solidified, δ-ferrite dendrites form first from the melt at about 1475°C. Since the solubility of carbon and the alloying elements is very restricted in the δ ferrite, the liquid becomes enriched in them and upon further cooling precipitates austenite, which is rich in alloying elements around the ferrite dendrites.

As the temperature is decreased to about 1350°C, a four-phase region is passed through where carbide and austenite are precipitated from the liquid and

---

[1] Ref. 12.

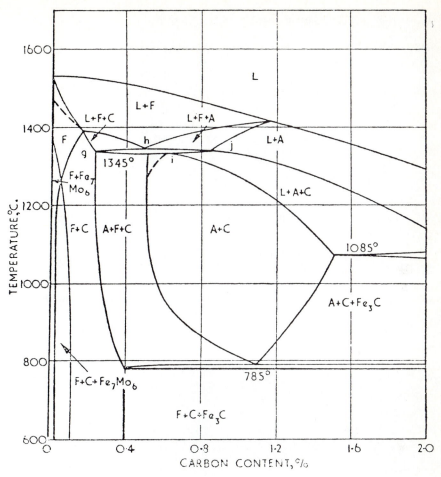

L:liquid; A:austenite; F:ferrite; C:Fe$_3$W$_3$C

Figure 9-20 Phase diagram of the Fe-W-Cr-C systems. Section is at 18% W and 4% Cr. (After Murakami and Hatta, with alterations indicated by dotted lines.) [*After K. Kuo, J. Iron Steel Inst. 181(1955):128.*]

δ ferrite transforms to austenite. The overall reaction is as follows:

$$\text{Liquid} + \delta \rightarrow \gamma + M_6C \qquad (M_6C \simeq Fe_3W_3C)$$

Since the reaction takes place slowly, equilibrium solidification is not complete and the actual solidification takes place by two reactions:

$$\text{Liquid} \rightarrow \gamma + M_6C \qquad \text{(ferrite eutectic)}$$

and $\qquad\qquad\qquad \delta \rightarrow \gamma + M_6C \qquad$ (ledeburite eutectoid)

As a result of the uneven cooling in a highly alloyed tool steel, the products of the reactions and their distribution vary greatly throughout the ingot section

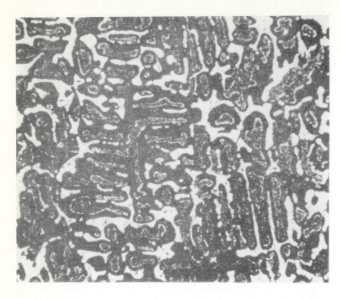

Figure 9-21 Microstructure of as-cast T1 high-speed steel showing heavily cored regions. (Etch: 5% nital; ×100.) (*After G. Hobson and D. S. Tyas, "High Speed Steels," Metals and Materials, May 1968, p. 147.*)

Figure 9-22 Microstructure of T1 high-speed tool steel showing lightly banded carbide regions. Longitudinal direction. Structure after hot working. (Etch: 5% nital; ×100.) (*After G. Hobson and D. S. Tyas, "High Speed Steels," Metals and Materials, May 1968, p. 147.*)

and heavy coring results in the as-cast structure (Fig. 9-21). Since the distribution of carbides is an important factor in the performance of a high-speed steel, the as-cast structure must be hot-worked extensively to disperse the carbides evenly through the structure.

Hot working After casting, the structure must be mechanically hot-worked by plastic deformation to eliminate the inhomogeneities. The cellular as-cast structure is disintegrated by plastic deformation and the austenite and carbide grains are refined. Elongation of the networks, partial networks, and carbide segregates occur in the direction of forging; but with high reductions in cross-sectional areas, carbide band thickness is reduced to narrow limits, as shown in Fig. 9-22. A reduction in cross-sectional area of 90 to 95 percent is considered necessary to achieve a satisfactory breakdown of the carbide structure.

Heat treatment of high-speed tool steels The most important factor affecting the ultimate performance of a cutting tool is its heat treatment. If the heat treatment is not carried out correctly, an inferior performance will be the result.

Structure of T1 high-speed steel in the annealed condition In the annealed condition, the structure of T1 high-speed steel consists of about 30% complex carbides in a matrix of ferrite (Fig. 9-23a). These carbides have been identified as belonging to three groups: (1) M_6C, (2) $M_{23}C_6$, and (3) MC.

1. The M_6C double-carbide composition varies from Fe_4W_2C to Fe_3W_3C. It dissolves moderate amounts of chromium, vanadium, and cobalt and is of importance in the secondary hardening reactions which produce the red-hardness property. (See Sec. 9-8.)
2. The $M_{23}C_6$ carbide is essentially a chromium carbide, but can dissolve large amounts of iron, vanadium, and molybdenum. This carbide itself dissolves to a large extent in the austenite, and hence is important in the formation of martensite in the heat-treated structure.
3. The MC carbide, or vanadium carbide, ranges in composition from VC to V_4C_3 and has some solubility for iron, chromium, tungsten, and molybdenum. It provides the wear resistance because of its high hardness and good abrasion resistance.

Structure of T1 high-speed steel in the quenched and tempered condition

Austenitizing After one or two preheat treatments to minimize thermal stresses and to serve as a solutionizing treatment for alloy carbides, the T1 high-speed steel is austenitized at 1250 to 1290°C. This temperature range is just below the liquidus, and heating must be carefully controlled to avoid melting the eutectics.

The temperature of austenitizing must be as high as possible to make sure that as many of the alloy carbides as possible are taken into solution so that after quenching and tempering the maximum hardness will be attained. The

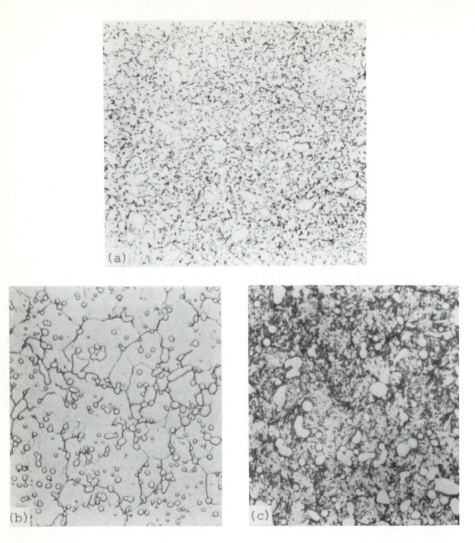

Figure 9-23 Microstructures of T1 high-speed tool steel (nominal composition: C, 0.75; W, 18.0; Cr, 4.0; V, 1.0). (*a*) Fully annealed condition; structure consists of large and small spheroidal carbide particles in a matrix of ferrite. (2% nital; X1000.) (*b*) Quenched condition; austenitized at 1279°C 3 to 4 min; salt-quenched to 607°C, air-cooled. Structure consists of undissolved carbide particles in untempered martensite. (10% nital; X1000.) (*c*) Normal quenched and tempered condition; austenitized at 1279°C 3 to 4 min; salt quenched to 607°C; air-cooled and double-tempered at 538°C. Structure consists of undissolved carbide particles in matrix of tempered martensite. (4% nital; X1000.) (*After Metals Handbook, 8th ed., vol. 7, American Society for Metals, Metals Park, OH, 1972, p. 118.*)

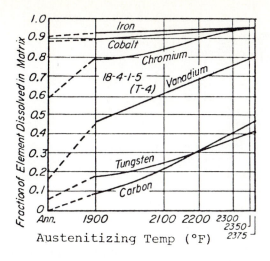

Figure 9-24 Partition of elements between carbides and matrix in T4 high tungsten tool steel. (*After F. Kayser and M. Cohen, Metal Progress, June 1952, p. 79.*)

partition of the alloy elements between carbides and matrix in a tungsten T4 high-speed steel is shown in Fig. 9-24 and is similar to that for the T1 alloy, which has the same composition as the T4 alloy less the 5% Co. It should be noted that, in the annealed condition, over half the chromium is dissolved in the matrix and that this partition is increased to about 90 percent in the commercial austenitized condition. Most of the increase in the solutionizing of the $M_{23}C_6$ carbide takes place below 650°C. Tungsten increases in the matrix gradually with tempering so that about half of it is dissolved after commercial austenitization. In the T1 steel most of the available vanadium dissolves in the M_6C and $M_{23}C_6$ carbides. The steel must not be held too long at austenitizing temperatures since grain growth and decarburization will occur.

Quenching Quenching is carried out to produce an austenitic structure. A hot quench to 560°C followed by air cooling to room temperature is used to minimize distortion and cracking. Also, the hot quench in the 540 to 650°C range minimizes grain boundary precipitation since this can occur at high temperatures, as noted in the IT diagram of the T1 tool steel (Fig. 9-25).

The quenched microstructure of T1 tool steel consists of 60 to 80% highly alloyed tetragonal martensite, 15 to 30% retained austenite, and about 5 to 10% undissolved M_6C and VC carbides (Fig. 9-23b).

Tempering The changes in microstructure of the T1 high-speed steel upon tempering can be divided into the following four stages:

1. *First stage* (room temperature to 400°C). From room temperature to 400°C, the martensite loses its tetragonality, decomposes into the cubic structure, and forms heaxagonal ε carbide. This carbide, which is precipitated at about 270°C, dissolves upon further heating and is replaced by cementite in the 300

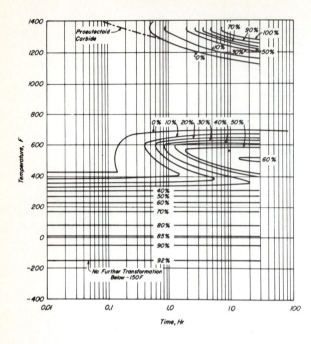

Figure 9-25 I-T diagram for T1 type high-speed tool steel (austenitizing temperature, 1290°C). [*After P. Gordon, M. Cohen, and R. S. Rose, Trans. ASM 31(1943):161.*]

to 400°C range. In the room temperature to 400°C range the Rockwell hardness decreases from 2 to 6 points, as shown in Fig. 9-26.

2. *Second stage* (470 to 570°C). Some cementite dissolves and precipitates of M_2C carbide begin at about 500°C, leading to secondary hardening (Fig. 9-26).

3. *Third stage* (500 to 620°C). On cooling from tempering temperature, transformation of retained austenite occurs, probably preceded by precipitation of

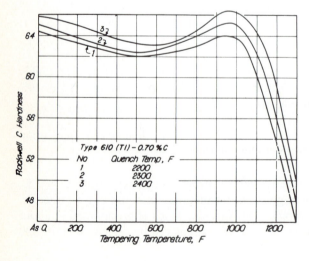

Figure 9-26 Effect of tempering temperature on the hardness of a 0.70% C type T1 high-speed tool steel after quenching from 1204°C (2200°F), 1260°C (2300°F) and 1315°C (2400°F). For this steel the normal austenitizing temperature is 1290°C. Tempering time, .5 h. (*After G. A. Roberts, J. C. Hamaker, and A. R. Johnson, "Tool Steels," 3d ed., American Society for Metals, Metals Park, OH, 1962, p. 647.*)

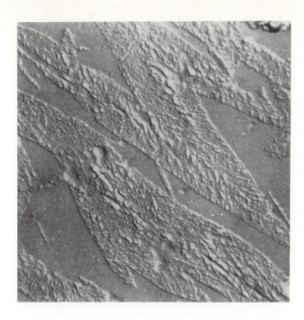

Figure 9-27 Electron micrograph of T1 high-speed tool steel (18-4-1) quenched to 130°C and tempered 24 h at 565°C. Structure shows smooth background of untempered martensite contrasted to heavy carbide precipitation within tempered martensite. [*After R. J. Beltz and R. W. Lindsay, Trans. ASM 61(1968):790.*]

the alloy carbide from the austenite. Figure 9-27 shows an electron micrograph which has regions of untempered martensite produced by the transformation of the retained austenite to martensite upon cooling after tempering at 565°C. In order to eliminate the newly formed martensite, the T1 steel is double-tempered. Figure 9-23c shows the microstructure of this alloy after double-tempering at 538°C.

4. *Fourth stage* (above 620°C). Above 620°C, the M_2C and Fe_3C dissolve while at the same time the alloy carbides M_6C and $M_{23}C_6$ precipitate and coalesce, resulting in a rapid decrease in hardness in the 620 to 650°C range.

Molybdenum-Type High-Speed Tool Steels

Molybdenum can be used to replace all or part of the tungsten in high-speed tool steels and, because of the lower cost of molybdenum, about 80 percent of the high-speed tool steel used in the United States today is of the molybdenum type. In the M2 grade, which is the most popular in the United States, the tungsten content is reduced to 6 percent while the molybdenum content is 5 percent. The vanadium content in this alloy has been increased to 2 percent as compared to the T1 (18-4-1) alloy, which has 1 percent (Table 9-8). In the M1 grade, the tungsten content is reduced to 1.5 percent, and the molybdenum content increased to 8 percent. This alloy has proven satisfactory for drills and similar applications at a cost reduction over the M2 alloy.

Molybdenum tool steels are more susceptible to decarburizing and require better temperature control during heat treatment than the tungsten-base high-speed steels. By the use of salt baths and sometimes surface coatings, decarburization during manufacture and heat treatment can be kept to a minimum.

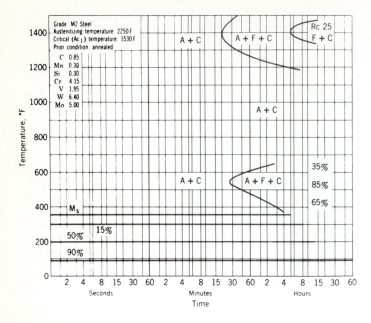

Figure 9-28 I-T diagram of type M2 high-speed tool steel. (*After P. Payson, "Metallurgy of Tool Steels," Wiley, 1962, p. 66. Reproduced by permission of John Wiley & Sons, Inc.*)

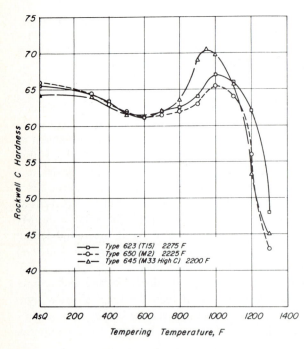

Figure 9-29 Effect of tempering temperature on the hardness of M2 high-speed tool steel. (*After G. A. Roberts, J. C. Hamaker and A. R. Johnson, "Tool Steels," 3d ed., American Society for Metals, Metals Park, OH, 1962, p. 648.*)

Molybdenum high-speed steels have lower peritectic reactions and must therefore be austenitized at lower temperatures. For example, M2 alloy is austenitized in the 1190 to 1230°C range, whereas the tungsten T1 alloy is austenitized between 1260 and 1300°C.

The IT diagram for the M2 high-speed steel is shown in Fig. 9-28. As in the case of the T1 alloy, long times are necessary for transformations to occur. However, to avoid transformation at the higher temperatures, i.e., above 700°C, the M2 alloy, like the T1 alloy, is hot-quenched to 565°C. The effect of tempering temperature on the hardness of the M2 tool steel is shown in Fig. 9-29. This graph is similar to that for the T1 steel in that a secondary hardness peak is reached after tempering at about 550°C.

The microstructures of the M2 tool steel in various conditions (Fig. 9-30) also closely resemble those of the T1 steel, as can be seen by comparing those structures. This is understandable since the molybdenum atoms can replace the tungsten atom and form the same type of carbides. For example, in the annealed condition there is 28 vol % carbides in the M2 steel, which is close to the 29.2 vol % carbides in the T1 steel. Since the atomic weight of molybdenum is about half that of tungsten, the weight percent carbide in the M2 steel is 20.8 compared to 28.1 wt % for the T1 steel.

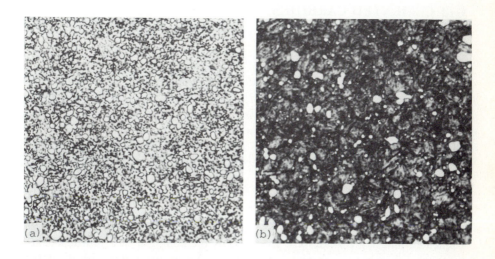

Figure 9-30 Microstructures of M2 high-speed tool steel (C, 0.85; W, 6.30; Cr, 4.15; V, 1.85; Mo, 5.05). (*a*) Fully annealed condition; structure consists of carbide particles in a matrix of ferrite. (*b*) Normal quenched and tempered condition; austenitized at 1200°C quenched to 565°C; air-cooled to room temperature; double-tempered at 550°C for 2 h each. Structure consists of carbide particles in a matrix of tempered martensite.

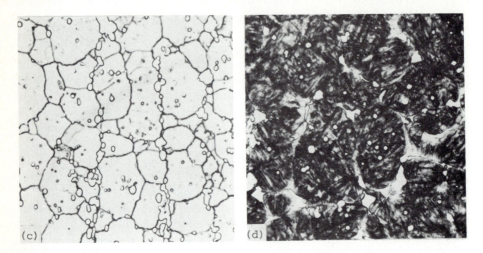

Figure 9-30 *Continued* (*c*) Quenched condition; austenitized at 1200°C, quenched to 565°C; air-cooled to room temperature; no tempering; longitudinal section. Structure consists of carbide particles, untempered martensite, and retained austenite. (*d*) Overheated condition; austenitized at 1245°C; quenched to 565°C; air-cooled to room temperature; double-tempered at 550°C for 2 h each; structure consists of carbide particles (separate and fused together) and tempered martensite. (Etchant: 4% nital; X1000.) (*Courtesy of J. Stepanic, Latrobe Steel Co.*)

9-10 CEMENTED CARBIDES

Definition and Application

Cemented carbides consist of finely divided hard particles of carbides of refractory metals (e.g., W, Ti, Ta) sintered together by a film of cobalt metal. Cobalt is used as a binder for the carbides since it wets their surface and slightly dissolves the solid carbide particles. Extremely hard and wear-resistant cutting tools are made with the cemented carbides. Tool bits so made, for example, allow machining cutting speeds up to five times that of high-speed steels and thus provide great cost savings for repetitive machining operations in spite of their initial high cost of production.

Production

Cemented tungsten carbides are produced by blending fine WC powders (1 to 3 μm) with cobalt metal powder so that the WC grains are coated with the cobalt. The mixture is then sintered in hydrogen above the melting point of the cobalt. The WC particles are wetted by the liquid cobalt metal, with a small amount of the carbides (about 1 percent) being dissolved by the cobalt. It is the low solubility of WC in cobalt compared with its solubility in iron or nickel that is the reason why cobalt is used almost exclusively as the binder metal. Cobalt also has a superior ability to wet the carbides at elevated temperatures, which is important in the sintering operation.

Classification

Cemented carbides can be divided into two broad types: (1) the type made with mainly tungsten carbide, and (2) the type containing large amounts of titanium and tantalum carbides as well as tungsten carbide. Table 9-9 lists the compositions of different carbide groups along with hardness values and typical applications. The "straight" tungsten carbides are used principally for the machining of

Table 9-9 Classification of sintered carbides†

Carbide group	Composition, % (remainder WC)		Rockwell A hardness	Density, g/cm³
	Co	TaC + TiC		
	Straight tungsten carbide			
1	2.5–6.5	0–3	93–91	15.2–14.7
2	6.5–15	0–2	92–88	14.8–13.9
3	15–30	0–5	88–85	13.9–12.5
	Added carbide, predominantly TiC			
4	3–7	20–42	93.5–92.0	11.0–9.0
5	7–10	10–22	92.5–90.0	12.0–11.0
6	10–12	8–15	92.0–89.0	13.0–12.0
	Added carbide, predominantly TaC			
7	4.5–8	16–25	93.0–91.0	12.5–12.0
8	8–10	12–20	92.0–90.0	13.0–11.5
	Added carbide, exclusively TaC			
9	5.5–16	18–30	91.5–84.0	14.8–13.5

Group	Typical uses

1. Finishing to medium roughing cuts on cast iron, nonferrous metals, superalloys, and austenitic alloys; low-impact dies.
2. Rough cuts on cast iron, especially on planers; moderate-impact dies.
3. High-impact die applications.
4. Light high-speed finishing cuts on steel. High crater resistance. Low shock resistance.
5. Medium cuts and speeds on steel. Good crater resistance and moderate shock resistance. Dies with moderate impact involving pickup
6. Roughing cuts on steel. Good shock resistance together with wear and crater resistance. Moderate-impact die applications involving pickup.
7. Light cuts on steel where a combination of edge wear and crater resistance is required.
8. General purpose and heavy cutting of steel requiring resistance to wear and cratering. Also resists abrasive wear caused by scale.
9. Wear-resistant applications particularly involving heat; gauge elements, special machining applications. Special applications involving mechanical shock and heat such as hot trimming of flash.

† After *Metals Handbook*, vol. 1: "Properties and Selection of Metals," American Society for Metals, Metals Park, OH, 1961, p. 660.

cast irons, nonferrous alloys, and nonmetallic materials. The mixed carbide type is used for machining carbon and alloy steels.

Microstructure

The microstructure of a 94% WC and 6% Co cemented carbide tool material is shown in Fig. 9-31a at 1500 ×. The angular grains of the WC can be seen embedded in a matrix of cobalt. Figure 9-31b shows the blocky shapes of the carbide grains at higher magnification. Figure 9-32 shows the microstructure of

(a)

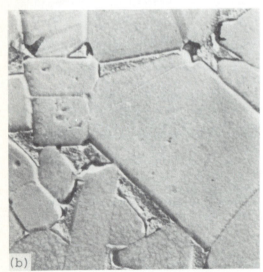

(b)

Figure 9-31 Cemented carbide containing 94% WC, 6% Co; mixed grain size. Structure consists of tungsten carbide in a matrix of cobalt. (*a*) (Murakami's reagent; X1500.) (*b*) Electron replica; blocky shapes are carbide grains. (Murakami's reagent; X15,000.) (*After Metals Handbook, 8th ed., vol. 7, American Society for Metals, Metals Park, OH, 1972, p. 128.*)

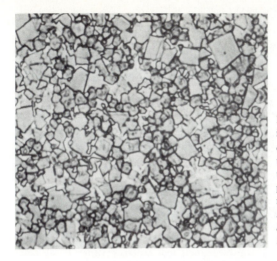

Figure 9-32 Cemented carbide containing 72% WC, 11% TaC, 8% TiC, 9% Co; density, 12.6 g cm³. Light angular particles are WC; dark gray rounded particles are TaC-TiC-WC solid solution phases. The matrix is cobalt. (Murakami's reagent; X1500.) (*After Metals Handbook, 8th ed., vol. 7, American Society for Metals, Metals Park, OH, 1972, p. 130.*)

a mixed WC, TiC, and TaC cemented carbide in a cobalt matrix. The angular particles of WC can be distinguished from the rounded particles of TaC-TiC-WC solid solution phases in the cobalt matrix.

Engineering Properties

The high carbide content (90 to 95 percent by volume) of cemented carbides makes them harder than either high-speed tool steels which contain up to 35% carbides or cast-nickel and cobalt-base tool materials which also contain up to 35% carbides. Figure 9-33 shows how cemented carbides retain their hardness at

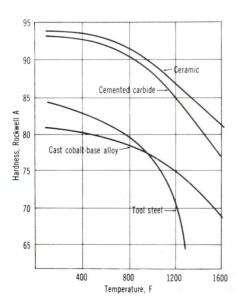

Figure 9-33 Effect of elevated temperatures on the hardness of tool materials. (*After E. W. Gobiler, "Advances in Cemented Carbide Tooling," Metal Progress, Aug. 1968, p. 95.*)

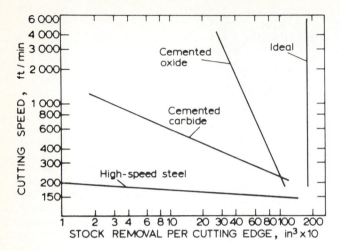

Figure 9-34 Cutting speed versus stock removal for various tool materials. (*After H. C. Child, "Materials for Metal Cutting," The Metals Society, 1970, p. 173.*)

elevated temperatures better than the tool steels and cobalt-base alloys. This property is important for machining since temperatures above 815°C may exist at the cutting edge during high-speed metal machining.

The cemented carbides also have high compressive strengths (up to 800 ksi) and retain much of this strength at elevated temperatures. This is also an important property since compressive forces up to 150 ksi at 538 to 815°C may exist at the cutting edge at high speeds.

In steel machining, cutting speed is the principal factor in selecting a cutting-tool material. Cutting speeds as high as 1000 ft/min are possible with the cemented carbides, whereas the tool steels are limited to speeds of 200 ft/min (Fig. 9-34).

PROBLEMS

1. Define tool steels.

2. What are some of the important applications for tool steels?

3. How are tool steels classified by the AISI system?

4. What are the advantages of using plain-carbon tool steels?

5. What are the limitations in using plain-carbon tool steels?

6. What is the purpose of adding vanadium to the water-hardening tool steels? Why does vanadium lead to a decreased hardenability?

7. What is the purpose of adding chromium to water-hardening tool steels?

8. How is the chemical composition of shock-resistant tool steels modified for shock resistance? For increased hardenability?

9. What are the principal types of cold-work tool steels?

10. What are the advantages of cold-work oil-hardening tool steels over the water-hardening types? Which elements are added to increase hardenability?

11. Describe the microstructure of the O1 tool steel after oil quenching to room temperature.

12. For what applications are the cold-work air-hardening tool steels used?

13. How is the composition of the air-hardening cold-work tool steels modified to increase hardenability above that of the oil-hardening type?

14. Describe the microstructure of the A2 tool steel in the quenched and tempered condition.

15. How is the "hot-tempered" condition produced in an A2 tool steel?

16. What are the principal applications of the cold-work, high-carbon, high-chromium tool steels?

17. What is the origin of the excellent wear resistance of the high-carbon, high-chromium tool steels? Oxidation resistance?

18. Describe the microstructure of the D2 tool steel in the quenched and tempered condition. What type of heat treatment can lead to about 25% or more retained austenite in this alloy?

19. For what typical applications are the hot-work tool steels used? What are the three principal types?

20. What are the important characteristics of hot-work tool steels?

21. What alloying elements are added to obtain the special properties required of hot-work tool steels and what are their functions?

22. Describe the microstructure of the H13 hot-work tool steel in the quenched and tempered condition.

23. What is high-speed (tool) steel? What are the two principal types?

24. What is red hardness? Why is it an important property of high-speed steels?

25. Which elements contribute to red hardness in high-speed steels?

26. What are the advantages of the tungsten-type high-speed steels? Disadvantages?

27. What are the advantages of the molybdenum-type tool steels? Disadvantages?

28. Describe the reactions which take place during the solidification of the T1 (18-4-1) high-speed steel from the liquid state to room temperature.

29. Describe the as-cast microstructure of the T1 high-speed steel.

30. Why must high-speed steels be considerably hot-worked before heat treating?

31. Describe the heat treatments necessary to produce the final quenched and tempered T1 high-speed steel.

32. What is the purpose of double tempering and what changes in structure does it cause?

33. Describe the mechanism of secondary hardening in an Fe–6% W–0.23% C alloy.

34. How do molybdenum high-speed steels differ from the tungsten ones?

35. Describe the mechanism of secondary hardening in an Fe–4% Mo–0.2% C alloy.

36. Describe the two principal types of cemented carbide tool materials.

37. Why is cobalt used as the matrix metal for cemented carbides?

38. Describe how tungsten cemented carbides are produced.

39. What are the advantages of mixed cemented carbide tool materials?

40. For what applications are the two groups of cemented carbides used?

41. How do the cutting properties of cemented carbides and high-speed steels compare?

REFERENCES

1. G. A. Roberts, J. C. Hamaker, and A. R. Johnson: "Tool Steels," American Society for Metals, Metals Park, OH, 1962.
2. P. Payson: "Metallurgy of Tool Steels," Wiley, New York, 1962.
3. "Materials for Metal Cutting," Iron and Steel Institute Publication 126, London, 1970.

4. R. W. K. Honeycombe: "Structure and Strength of Alloy Steels," Climax Molybdenum Co., London, 1974.
5. E. R. Petty: "Physical Metallurgy of Engineering Materials," American Elsevier, New York, 1968.
6. *Metals Handbook*, vol. 7: "Atlas of Microstructures of Industrial Alloys," American Society for Metals, Metals Park, OH, 1972.
7. J. T. Berry: "Recent Developments in the Processing of High-Speed Steels," Climax Molybdenum Co., Greenwich, CO, 1970.
8. J. T. Berry: "High Performance High Hardness High Speed Steels," Climax Molybdenum Co., Greenwich, CO, 1970.
9. "Tool Steels." In *Steel Products Manual*, American Iron and Steel Institute, Washington, D.C., 1976.
10. A. T. Davenport and R. W. K. Honeycombe: "The Secondary Hardening of Tungsten Steels," *Met. Sci.* 9(1975)201.
11. R. Schatter and J. Stepanic: "Silicon Additions Improve High-Speed Steels," *Metal Progress*, June 1976, p. 56.
12. G. Hobson and D. S. Tyas: "High Speed Steels," *Metals and Materials*, May 1968, p. 144.
13. R. J. Beltz and R. W. Lindsay: "The Conditioning of Retained Austenite in High-Speed Steel," *Trans. ASM* 61(1968):790.
14. G. A. Roberts: "Vanadium in High-Speed Steel," *Trans. AIME* 236(1966):950.
15. E. W. Globier: "Advances in Cemented Carbide Tooling," *Metal Progress*, Aug. 1966, p. 95.
16. B. O. Haglund: "Understanding the Sintering of Cemented Carbide," *Metal Progress*, Dec. 1976.
17. B. Fredriksson: "Optimizing Tempering Time for High-Speed Steels," *Metal Progress*, Aug. 1977.
18. D. Raynor, J. A. Whiteman, and R. W. K. Honeycombe: "Precipitation of Molybdenum and Vanadium Carbides in High-Purity Iron Alloys," *J. Iron Steel Inst.* 204(1966):349.
19. K. J. Irvine and F. B. Pickering: "The Tempering Characteristics of Low-Carbon Low-Alloy Steels," *J. Iron Steel Inst.* 194(1960):137.

TITANIUM AND ITS ALLOYS

Titanium and its alloys are relatively new engineering metals since they have been in use as structural materials only since 1952. Titanium alloys are attractive since they have a high strength-to-weight ratio, high elevated-temperature properties to about 550°C, and excellent corrosion resistance, particularly in oxidizing acids and chloride media and in most natural environments.

Unfortunately, titanium and its alloys cost somewhat more than common metals because they are difficult to extract from their ores and sophisticated melting and fabricating techniques must be used in their manufacture. The higher cost of titanium alloy fabrication is principally the result of the metal's high reactivity and affinity for interstitial elements such as oxygen, nitrogen, hydrogen, and carbon. Nevertheless, titanium and its alloys do compete effectively in many areas where their special properties can be used to advantage. For example, high strength-to-weight ratio and high elevated-temperature properties of titanium alloys are of prime importance in the aerospace industry. The excellent corrosion resistance of titanium makes it particularly useful for the chemical and food industries. New uses for titanium and its alloys are being constantly sought and discovered.

10-1 PRODUCTION OF TITANIUM

Extraction of Titanium Sponge[1]

Titanium metal is obtained from the mineral rutile, which consists of approximately 97 to 98% titanium dioxide (TiO_2). The titanium oxide in the rutile is first chemically converted to pure titanium tetrachloride ($TiCl_4$).

---

[1] M. Hoch, "Winning and Refining Titanium," in vol. 1 of Ref. 1.

In the *Kroll process*, $TiCl_4$ is reacted with liquid *magnesium* at about 773 to 873°C in a closed stainless-steel vessel. The end products of this high-temperature reaction are titanium sponge, magnesium chloride ($MgCl_2$), and some excess magnesium. The chemical reaction is

$$TiCl_4 \text{ (gas)} + 2Mg \text{ (liquid)} \rightarrow Ti \text{ (solid)} + 2MgCl_2 \text{ (liquid)}$$

In the *Hunter process*, $TiCl_4$ is reacted with *sodium* instead of magnesium. A main advantage of the Hunter process is that the reduction of the $TiCl_4$ can be carried out in two steps. In the first, $TiCl_4$ is reduced by sodium to $TiCl_2$ and NaCl in a continuous reducer (Fig. 10-1). The chemical reaction for this step is

$$TiCl_4 + 2Na \rightarrow TiCl_2 + 2NaCl$$

The reaction vessel has an agitator and is maintained under positive pressure by argon gas at a temperature of 232°C.

In the second step of the Hunter process, the melt containing the $TiCl_2$ and NaCl is reacted with additional sodium in a sinter pot which has an upper atmosphere of argon gas (Fig. 10-1). The $TiCl_2$ and sodium react in the sinter pot according to the reaction

$$TiCl_2 + 2Na \rightarrow Ti + 2NaCl$$

This reaction is carried out at a temperature below 1037°C. Since a large amount of heat is released in the first step of the Hunter process, closer temperature control can be maintained during the second step. This closer

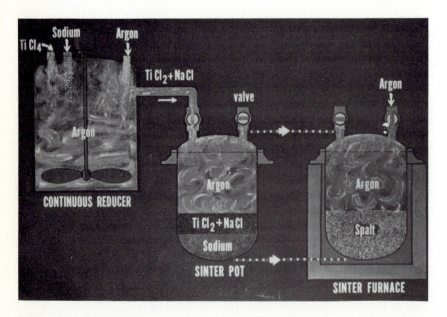

Figure 10-1 Processing steps in the production of titanium sponge from $TiCl_4$ and Na by the Hunter process. (*Courtesy of RMI Company.*)

control enables large crystals of titanium to be obtained, which can be up to 150 mm in length. Finally, for both the Kroll and Hunter processes the titanium sponge is separated from the salts and excess unreacted metal by acid leaching or vacuum distillation.

Preparation of Titanium Ingots

Since molten titanium reacts with oxygen and nitrogen in the air, special processes must be used to produce titanium ingots from titanium sponge. In conventional practice, the titanium sponge is crushed and compacted into electrode compacts which are welded together to form a long consumable electrode for vacuum arc melting (Fig. 10-2). Vacuum arc melting is necessary since it prevents the molten titanium from reacting with the oxygen and nitrogen in the air. The consumable electrode becomes the anode in the vacuum arc furnace and a water-cooled copper crucible serves as the cathode. An arc is struck between the compacted electrode and the copper crucible, and the molten metal collects and solidifies in the copper crucible. Using this process, ingots of up to 10 tons with diameters of 36 in can be produced (Fig. 10-3).

For alloy ingots the alloying materials are mixed with the crushed titanium sponge before compacting. Care must be taken that the alloying additions are uniformly distributed throughout the long consumable electrode. For alloy ingots, double melting is used to insure homogeneity of the ingots. In this procedure, the ingot from the first melting serves as the electrode for the second melting.

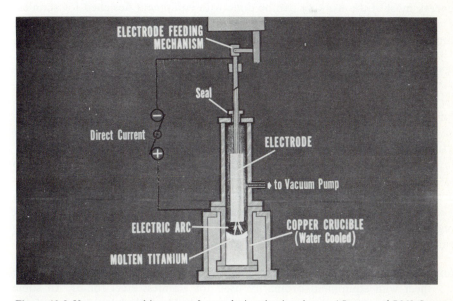

Figure 10-2 Vacuum arc melting set up for producing titanium ingots. (*Courtesy of RMI Company.*)

Figure 10-3 Removing copper crucible from a homogeneous titanium ingot. This ingot is 36 in in diameter and weighs 15,000 lb. (*Courtesy of RMI Company.*)

Primary Working

In spite of the reactivity of titanium with hydrogen, oxygen, and nitrogen, working procedures for producing high-quality titanium bar, plate, sheet, strip, extrusions, wire, tubing, and other mill products have been developed. In general, most of the working can be done with equipment designed for stainless or other specialty steels. However, some of it must be carried out with close temperature and atmospheric control.

Ingot breakdown The surface of commercially vacuum-arc-melted titanium ingots are first conditioned after casting to provide defect-free surfaces for forging. *Conditioning*, which is the process of removing irregularities and imperfections from the ingot surface, is usually done by grinding. Most commonly the ingots are hot-forged by a heavy press with open flat dies. By using forge pressing, the deformation rate can be closely controlled to avoid cracking the ingot.

Highly alloyed titanium ingots are first preheated at about 700 to 760°C to minimize undesirable thermal gradients and then are heated to forging temperature. Commercially pure titanium ingots are forged in the 1040 to 980°C range as the ingot breakdown progresses. If appreciable surface cracking occurs during forging, the ingots may be conditioned by grinding and reheated to forging temperature.

Rolling of plate and sheet Slabs from the forging operation can be rolled to plate and sheet with equipment ordinarily used for stainless steel rolling. In some cases, specially designed equipment is required. The initial breakdown of the forged slabs is usually done by hot rolling using two-high or three-high rolling mills. The hot-rolling temperature range depends on the alloy being rolled. For example, commercially pure titanium is hot-rolled to plate in the 760 to 788°C range and to sheet in the 730 to 700°C range. Care must be taken to use an oxidizing atmosphere to prevent hydrogen contamination. Intermediate annealing is done in furnaces which have oxidizing atmospheres.

With the exception of commercially pure titanium and a few alloys, most titanium alloys are hot-rolled to finished plate and sheet thicknesses. For thin gauges (i.e., 0.020 in), sheet is usually *pack-rolled*. In this process, about four or five sheets are placed between two sheets of steel and the whole pack is welded together. Parting agents are used to prevent the sheets from sticking together. Pack rolling is necessary to maintain the high temperatures needed for hot rolling and also to prevent surface contamination. Vacuum annealing is generally used for commercially pure titanium strip product. All alloy sheet is air-annealed, descaled, and pickled.

10-2 PURE TITANIUM

Important Physical Properties

Titanium is a relatively light metal having a density of 4.54 g/cm^3, which is intermediate between that of aluminum (2.71 g/cm^3) and iron (7.87 g/cm^3). Titanium has a high melting point of 1668°C, which is higher than that of iron (1536°C), and a modulus of elasticity of 16.8×10^6 lb/in^2, which is intermediate between the values for aluminum and iron. The density, melting point, and modulus of elasticity of titanium are compared with those properties of aluminum and iron in Table 10-1.

Titanium exists in two allotropic crystal forms. These are α, which has the hexagonal close-packed (HCP) structure, and β, which has the BCC crystal structure. In pure titanium, the α phase is stable up to 883°C. Above 883°C, the β *transus temperature*, the hexagonal α phase is transformed on heating to the BCC β phase (Fig. 10-4).

Table 10-1 Selected physical properties of titanium as compared to those of aluminum and iron

	Titanium	Aluminum	Iron
Density, gm/cm^3	4.54	2.70	7.87
Modulus of elasticity, $\times 10^6$ lb/in^2	16.8	9.0	28.5
Melting point, °C	1668	660	1536
Crystal structure at room temperature	HCP	FCC	BCC

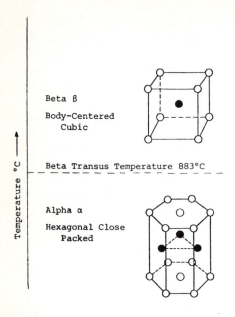

Beta β
Body-Centered Cubic

Temperature °C →

Beta Transus Temperature 883°C

Alpha α
Hexagonal Close Packed

Figure 10-4 Allotropic crystalline forms of pure titanium.

Deformation Properties

Pure titanium can be cold-rolled at room temperature to above 90 percent reduction in thickness without serious cracking[1]. Such extensive deformability is unusual for HCP metals, and is most probably related to the low c/a ratio of titanium. This ratio for titanium is 1.587, which is -2.81 percent from the ideal of 1.633 (Table 10-2). In contrast, pure magnesium, which has a c/a ratio of 1.624 (-0.55 percent from the ideal), cannot be cold-rolled more than 50 percent even in its highest-purity form.

The relatively high ductility of HCP titanium is attributed to the many operative slip systems and available twinning planes in the titanium crystal lattice. In titanium, because of its low c/a ratio, slip occurs on the $\{10\bar{1}0\}$ prism planes and the $\{10\bar{1}1\}$ pyramidal planes as well as on the basal planes (Fig. 10-5). The contribution of twinning to plastic deformation is much more important in titanium than in other HCP metals such as magnesium, zinc, and cadmium. The principal twinning planes in titanium are of the $\{10\bar{1}2\}$, $\{11\bar{2}1\}$, and $\{11\bar{2}2\}$ types (Fig. 10-5).

The type of slip in titanium is also very dependent on the concentration of interstitial impurity atoms such as oxygen and nitrogen. In crystals with an impurity content of 0.01 wt%, deformation occurs at room temperature predominantly on prism $\{10\bar{1}0\}$ planes, but some basal slip also occurs. With higher impurity contents (0.1 wt% O and N), slip occurs mainly on the $\{10\bar{1}1\}$ pyramidal planes (see Ref. 18 on this topic).

[1] F. D. Rosi, C. A. Dube, and B. H. Alexander, *Trans. AIME* 197(1953):257.

Table 10-2 Grouping of important HCP metals according to their c/a ratios†

Metal	c/a Ratio	% Deviation from ideal	Group
Cd	1.886	+ 15.5	I
Zn	1.856	+ 13.6	I
Ideal c.p.h.	1.633	0	
Mg	1.624	− 0.55	II
Co	1.624	− 0.55	II
Zr	1.589	− 2.69	III
Ti	1.587	− 2.81	III
Be	1.568	− 3.98	III

† After F. D. Rosi, C. A. Dube, and B. H. Alexander, *Trans. AIME* 197(1953):257.

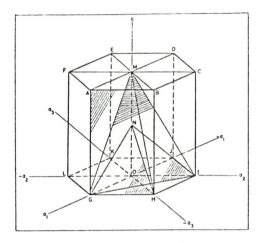

ABHG	prism planes	{10ī0}
GHM	pyramidal planes	{10ī1}
GHN	pyramidal planes	{10ī2}
GIM	pyramidal planes	{11ī1}
GIN	pyramidal planes	{11ī2}

Slip direction: OG—digonal axis, of the form $\langle 11\bar{2}0 \rangle$.

Figure 10-5 Hexagonal lattice showing the position of the operative slip and twinning planes and slip direction in the plastic deformation of titanium at room temperature. [*After F. D. Rosi, C. A. Dube, and B. H. Alexander, Trans. AIME (1953):257.*]

10-3 TITANIUM ALLOY SYSTEMS AND PHASE DIAGRAMS

In order to interpret the various microstructures that are observed in titanium alloys, it is necessary to have some knowledge of the different titanium alloy stabilized systems and binary phase diagrams. However, it must be remembered that binary phase diagrams are for conditions approaching equilibrium and that most commercial alloys are cooled at faster rates. Also, most titanium alloys are ternary or quaternary types and are not binary alloys.

Binary titanium alloys are divided into two stabilized systems: α and β. In the α stabilized system, the α phase field is enlarged with the addition of α-stabilizing elements. In the β-stabilized system, the β phase field is enlarged with the addition of β-stabilizing elements.

α-Stabilized System

In the binary α-stabilized system, the alloying element is more soluble in the α phase and the β transus is raised, as is indicated in Fig. 10-6.

Some of the substitutional elements which stabilize the titanium α phase are aluminum, gallium, and germanium. Of the three, *aluminum* is by far the most important. In fact, almost all titanium alloys contain aluminum since it adds to the ductility and lightness of titanium. The strong effect of aluminum in raising the β transus of titanium is shown in the Ti-Al phase diagram of Fig. 10-7.

Some interstitial alloying elements also stabilize the α phase. Oxygen, nitrogen, and carbon are all α-stabilizing elements. Since oxygen is an impurity found in all commercial titanium alloys, it is an important α-stabilizing element. The oxygen level is used in some cases to establish strength levels.

β-Stabilized Systems

In the β-stabilized systems, alloying elements stabilize the β phase in titanium. There are two β-stabilized systems: the β isomorphous and the β eutectoid.

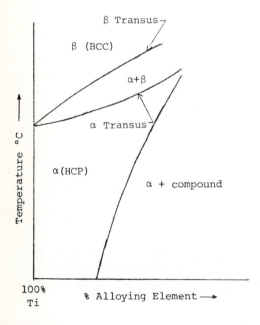

Figure 10-6 Alpha stabilized system. Note how the alpha-phase field is extended and the beta transus is raised.

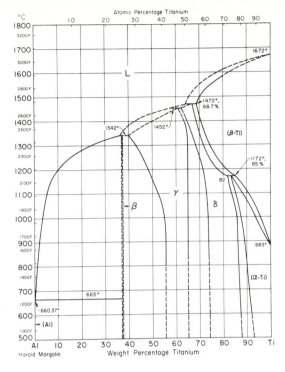

Figure 10-7 Phase diagram for the titanium-aluminum system. (*After Metals Handbook, 8th ed., vol. 8, American Society for Metals, 1973, p. 264.*)

β Isomorphous system In the β isomorphous system, the alloying element is completely miscible in the β phase and decomposition of the β phase to α plus another phase or compound does not occur. The β transus temperature decreases as the amount of alloying element is increased, as shown in Fig. 10-8. Alloying elements which are of the β isomorphous type are vanadium, molybdenum, tantalum, and columbium. The two most important of these are *vanadium* and *molybdenum*, whose phase diagrams are shown in Figs. 10-9 and 10-10.

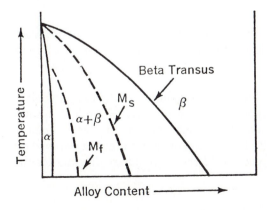

Figure 10-8 Beta isomorphous system. The most important titanium alloying elements of this system are vanadium and molybdenum.

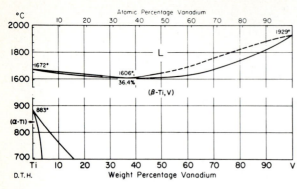

Figure 10-9 Phase diagram for the titanium-vanadium system. (*After Metals Handbook, 8th ed., vol. 8, American Society for Metals, Metals Park, OH, 1973, p. 337.*)

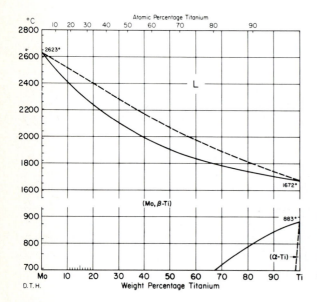

Figure 10-10 Phase diagram for the titanium-molybdenum system. (*After Metals Handbook, 8th ed., vol. 8, American Society for Metals, Metals Park, OH, 1973, p. 321.*)

β Eutectoid system In this system, the alloying element stabilizes the β phase, but under very slow cooling conditions the β phase can transform to α plus another phase or compound (Fig. 10-11). The β eutectoid alloying elements are of two types: (1) rapid or active eutectoid formers and (2) slow or sluggish eutectoid formers. Rapid eutectoid formers in titanium are silicon and copper. These elements cause rapid decomposition of the β phase to produce a compound and the α phase. Slow eutectoid formers are elements such as chromium, manganese, iron, nickel, and cobalt. These elements are more sluggish in their rate of eutectoid decomposition. Figure 10-12 shows the phase diagram for the Ti-Cr system.

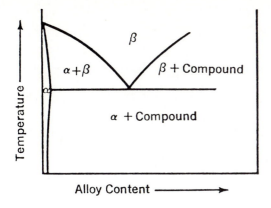

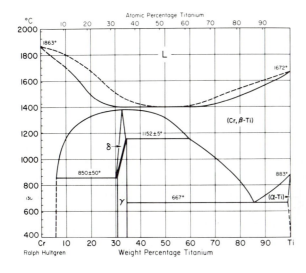

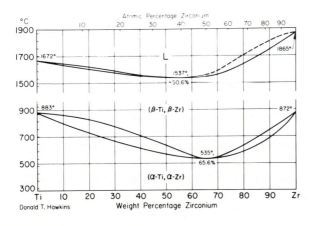

Figure 10-11 Beta eutectoid system. The most important titanium alloying elements of this system are chromium, iron and silicon.

Figure 10-12 Phase diagram for the titanium-chromium system. (*After Metals Handbook, 8th ed., vol. 8, American Society for Metals, Metals Park, OH, 1973, p. 292.*)

Figure 10-13 Phase diagram for the titanium-zirconium system. (*After Metals Handbook, 8th ed., vol. 8, American Society for Metals, Metals Park, OH, 1973, p. 338.*)

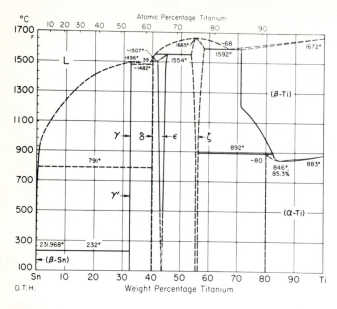

Figure 10-14 Phase diagram for the titanium-tin system. (*After Metals Handbook, 8th ed., vol. 8, American Society for Metals, Metals Park, OH, 1973, p. 335.*)

Other alloying elements Tin and zirconium are added to many titanium alloys. These elements have extensive solid solubility in both α and β phases, but do not strongly promote phase stability. They are useful alloying elements since they contribute to solid-solution strengthening and to slowing down some adverse transformations in titanium alloys such as the formation of the ω phase. Figure 10-13 shows the Ti-Zr phase diagram and Fig. 10-14 the Ti-Sn diagram.

10-4 CLASSIFICATION OF TITANIUM ALLOYS

Titanium alloys are classified according to the phases present in their structure. Alloys that consist mainly of the α phase are called α *alloys*, whereas those that contain principally the α phase along with small amounts of β-stabilizing elements are termed *near-α titanium alloys*. Alloys that consist of mixtures of α and β phases are classified as α-β *alloys*. Finally, titanium alloys in which the β phase is stabilized at room temperature after cooling from a solution heat treatment are classified as β *alloys*. Each group has certain distinguishing characteristics, which are briefly described below. Subsequent parts of this chapter deal with commercially pure titanium (Sec. 10-5), α titanium alloys (Sec. 10-6), near-α titanium alloys (Sec. 10-7), α-β titanium alloys (Sec. 10-8), and β titanium alloys (Sec. 10-9).

α **Titanium alloys** *α* and near-*α* alloys are generally non-heat-treatable and weldable. They have medium strength, good notch toughness, and good creep resistance at elevated temperatures.

α-β **Titanium alloys** Most *α-β* alloys are heat-treatable to a moderate increase in strength. Their strength levels are medium to high. They also have good forming properties, but do not have as good creep resistance at elevated temperatures as the *α* and near-*α* alloys.

β **Alloys** The *β*-rich alloys are heat-treatable to very high strengths and are readily formable. However, these alloys have relatively high density and in the high-strength condition have low ductility. Because of these disadvantages, they are not used much at present.

10-5 COMMERCIALLY PURE TITANIUM

Chemical Compositions and Typical Applications

Commercially pure titanium which is unalloyed ranges in purity from 99.5 to 99.0% Ti. The main elements in unalloyed titanium are iron and the interstitial elements carbon, oxygen, nitrogen, and hydrogen. The chemical compositions and applications of the principal grades of unalloyed titanium are listed in Table 10-3.

Commercially pure titanium can be considered an *α*-phase alloy in which the oxygen content determines the grade and strength. Oxygen is present to a certain level in all titanium sponge, but its amount can be adjusted to modify the strength of commercially pure titanium. In this respect, oxygen is an important "alloying element." Carbon, nitrogen, and hydrogen are present as impurities in titanium. The sources of carbon contamination in the sponge are the $TiCl_4$, which is continuously introduced into the extraction equipment, the walls of the reaction vessel from which iron and carbon are absorbed, and oil vapor from the vacuum system (Ref. 1, vol. 3, p. 212). Iron, which is also an impurity, is a *β* stabilizer. The principal source of the iron is the reaction vessel. Nitrogen within low limits can be useful as an interstitial strengthening element. Even though hydrogen has very low solubility in *α* titanium, its presence is always undesirable since it has an embrittling effect on titanium. Commercially pure titanium is lower in strength but more corrosion-resistant and less expensive than titanium alloys. It is used primarily when strength is not the main requirement.

Commercially pure titanium has excellent resistance to many chemical environments. It is resistant to nitric acid, moist chlorine, solutions of chlorine, chlorinated organic compounds and inorganic chloride solutions, and especially to hot chloride solutions. One area where titanium is finding increasing use is in the petroleum-processing industry, especially for heat exchangers. Titanium is

Table 10-3 Chemical compositions (maximum values) and typical applications of unalloyed titanium†

% Ti	Grade	ASTM No.	% C	% Fe	% N	% O	% H	Typical applications
99.5	1	B265	0.08	0.20	0.03	0.18	0.015	Airframes; chemical, desalination, and marine parts; plate-type heat exchangers; cold-spun or pressed parts; platinized anodes; high formability.
99.2	2	B265	0.08	0.25	0.03	0.20	0.015	Airframes; aircraft engines; marine chemical parts; heat exchangers; condenser and evaporator tubing; formability.
99.1	3	B265	0.08	0.25	0.05	0.30	0.015	Chemical, marine, airframe, and aircraft engine parts which require formability strength, weldability, and corrosion resistance.
99.0	4	B265	0.08	0.50	0.05	0.40	0.015	Chemical, marine, airframe, and aircraft engine parts; surgical implants; high-speed fans; gas compressors; good formability and corrosion resistance, high strength.

† After "ASM Databook," *Metal Progress*, mid-June, vol. 114, no. 1, 1978.

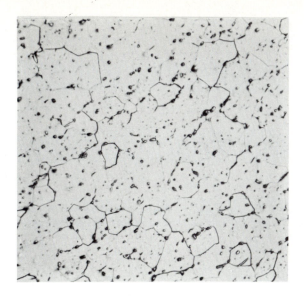

Figure 10-15 Unalloyed titanium sheet annealed 1 h at 700°C and air-cooled. Structure consists of equiaxed alpha grains and beta spheroids stabilized by the presence of 0.3% Fe in the alloy. (Etchant: 10% HF–5% HNO₃; X250.) (*Courtesy of RMI Company*.)

used in refineries since it is resistant to sulfides, chlorides, and many other chemicals encountered in petroleum refining.[1]

The addition of about 0.2% Pd to commercially pure titanium (99.2 to 99.5 percent) improves its corrosion resistance in reducing media. This extends the range of titanium's application to hydrochloric, phosphoric, and sulfuric acid solutions and other service areas where operating conditions vary between oxidizing and mildly reducing.[2]

Microstructures

Equiaxed grain structures are principally developed by cold work followed by annealing above the recrystallization temperature. Figure 10-15 shows an equiaxed α structure in unalloyed titanium, which was produced by annealing at 700°C. Small particles of spheroidal β were stabilized in this alloy by its 0.3% Fe content. Figure 10-16 shows the microstructure of elongated α produced by unidirectional hot rolling and represents a typical heavily worked unalloyed titanium structure.

Mechanical Properties

The tensile strength of unalloyed titanium is determined principally by the levels of nitrogen and oxygen and, to a lesser extent, by the carbon content. Figure 10-17 shows the effects of oxygen, nitrogen, and carbon additions on the tensile

[1] L. C. Covington, *Metal Progress*, Feb. 1977, p. 38.
[2] "Basic Facts about Titanium," Reactive Metals, Inc., Niles, OH, 1970, p. 8.

Figure 10-16 Unalloyed titanium sheet as hot-rolled structure shows elongated alpha produced by the deformation. (Etchant: 10% HF–5% HNO_3; X500.) (*Courtesy of RMI Company.*)

properties of pure (iodide) titanium. Since oxygen is the main element controlling the strength of unalloyed titanium, the strengthening effect of the interstitial elements oxygen, nitrogen, and carbon is expressed in terms of an oxygen equivalent as

$$\% \, O_{equiv} = \% \, O + 2.0(\% \, N) + 0.67(\% \, C)$$

Each 0.1% O equivalent of interstitial elements in pure titanium increases the

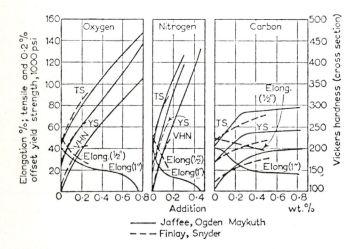

Figure 10-17 Effect of oxygen, nitrogen, and carbon additions on the mechanical properties of iodide titanium. (*After R. I. Jaffee, "The Physical Metallurgy of Titanium Alloys," Progress in Metal Physics, vol. 7, 1958, p. 109. Pergamon, Elmsford, NY. By permission.*)

Table 10-4 Mechanical properties of commercially pure titanium and low-alloyed titanium†

Commercially pure or low-alloyed titanium	Condition	Room temperature				Average mechanical properties — Extreme temperatures						
		Tensile strength, psi	Yield strength, psi	Elongation, %	Reduction in area, %	Test temp., °F	Tensile strength, psi	Yield strength, psi	Elongation, %	Reduction in area, %	Charpy impact strength ft · lb	Hardness, Bhn
99.5	Annealed	48,000	35,000	30	55	600	22,000	14,000	32	80	· · ·	120
99.2	Annealed	63,000	50,000	28	50	600	28,000	17,000	35	75	32	200
99.1	Annealed	75,000	65,000	25	45	600	34,000	20,000	34	75	28	225
99.0	Annealed	96,000	85,000	20	40	600	45,000	25,000	25	70	15	265
99.2, 0.15 Pd	Annealed	63,000	50,000	28	50	600	27,000	16,000	37	75	32	200
98.9, 0.8 Ni, 0.3 Mo	Annealed	75,000	65,000	25	42	400	50,000	36,000	37			200
					600	47,000	30,000	32				

† After "ASM Databook," *Metal Progress*, vol. 114, no. 1, mid-June 1978.

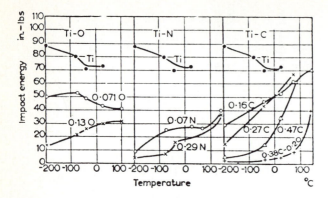

Figure 10-18 Effect of interstitial content on the notch-bend impact toughness of high-purity titanium in the fine-grained (0.01–0.05 mm) equiaxed condition. (*After R. I. Jaffee, "The Physical Metallurgy of Titanium Alloys," Progress in Metal Physics, Pergamon, Elmsford, NY, vol. 7, 1958, p. 111. Used by permission*)

strength of pure titanium by roughly 17.5 ksi. Table 10-4 lists the average mechanical properties for the various grades of unalloyed titanium.

Although the interstitial elements increase the strength of titanium, they are detrimental to toughness as measured by the notched impact test. Therefore, when high toughness is desired for certain applications, the alloy will be produced with extra-low interstitials. Commercially, these alloys are referred to as *extra-low interstitial* (ELI) alloys. Figure 10-18 shows the effects of nitrogen, oxygen, and carbon additions on lowering the impact resistance or toughness of pure titanium.

10-6 α TITANIUM ALLOYS

Chemical Compositions and Typical Applications

There is only one important all-α titanium alloy in commercial use today, and it has the nominal composition of Ti–5% Al–2.5% Sn (Table 10-5). It is an all-α alloy because aluminum and tin both stabilize the α phase in titanium.

Aluminum is one of the most important alloying elements for titanium since it strengthens the latter by solid-solution strengthening and also reduces its density. Tin is added to Ti–5% Al–2.5% Sn since it also contributes solid-solution strengthening. Oxygen, which is present to a certain degree in all titanium alloys, is also a strong α stabilizer like aluminum and strengthens titanium. However, like all interstitial elements in titanium, oxygen lowers its ductility, and thus a special low-oxygen Ti–5% Al–2.5% Sn is produced for applications requiring good low-temperature ductility. The Ti–5% Al–2.5% Sn alloy is weldable and has good stability and oxidation resistance at elevated temperatures. Its strength, however, is only moderate.

Table 10-5 Chemical compositions and typical applications of α titanium alloys†

α Alloys	Condition	Typical applications
5% Al, 2.5% Sn	Annealed	Weldable alloy for forgings and sheet-metal parts such as aircraft engine compressor blades and ducting; steam turbine blades; good oxidation resistance and strength at 600 to 1100°F; good stability at elevated temperatures.
5% Al, 2.5% Sn (low O$_2$)	Annealed	Special grade for high-pressure cryogenic vessels operating down to −423°F.

† After "ASM Databook," *Metal Progress*, mid-June, vol. 114, no. 1, 1978.

Microstructure

All-α titanium alloys have the HCP crystal structure of titanium. Small amounts of β phase may be present due to β-stabilizer impurities such as iron. For example the recrystallized microstructure of Ti–5% Al–2.5% Sn alloy (Fig. 10-19) shows small particles of β phase in an otherwise all-α structure. The 0.3% Fe impurity content of this alloy is responsible for the precipitation of the small β-phase particles.

Aluminum is the most important substitutional alloying element in α titanium alloys since it greatly stabilizes the α phase while increasing the strength and lowering the density of titanium. However, the amount of aluminum that is alloyed with titanium is usually limited to about 5 to 6 wt% since coherent ordered α$_2$ phase (Ti$_3$Al) forms, which embrittles the Ti-Al alloys.

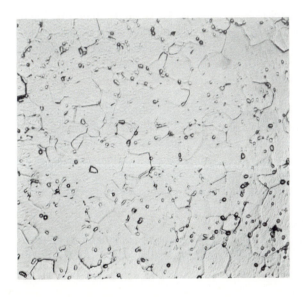

Figure 10-19 Ti–5% Al–2.5% Sn alloy in sheet form. Alloy was heated at 815°C for 30 min; air-cooled. Structure shows spheroidal beta in equiaxed alpha. This material contains 0.3% iron which acts as a beta stabilizer. (Etchant: 10% HF–5% HNO$_3$; X250.) (*Courtesy of RMI Company.*)

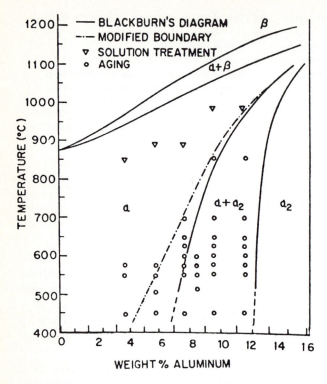

Figure 10-20 Titanium rich end of Ti-Al phase diagram. [*After T. K. Namboodhiri, C. J. McMahon, and H. Herman, "Decomposition of the α-Phase in Titanium-Rich Ti-Al Alloys," Met. Trans. 4(1973):1331.*]

Figure 10-20 shows the aluminum-rich end of the Ti-Al phase diagram. There has been some disagreement where the $\alpha/\alpha + \alpha_2$ phase boundary exists. More recently it has been shown that the $\alpha/\alpha + \alpha_2$ phase boundary can exist with an aluminum content as low as 5 wt% at temperatures below 500°C (Fig. 10-20). In a Ti-8% Al alloy aged 200 h at 695°C, homogeneous α_2 precipitation occurs as shown in Fig. 10-21.

The presence of the coherent α_2 phase is associated with an embrittlement of the Ti-Al alloys. When the titanium content of Ti-Al alloys reaches 6 wt%, during deformation, definite coplanar arrays of dislocations are created (Fig. 10-22a) which produce regions highly susceptible to early fatigue cracking. In pure titanium, a cellular distribution of dislocations is produced during deformation which is characteristic of ductile metals (e.g., aluminum and copper) (Fig. 10-22b).

Additions of tin, zirconium, and oxygen (often present as an impurity) also stabilize the α phase in titanium and increase the metal's strength. According to Rosenberg (Ref. 2, p. 851), the maximum aluminum equivalent of these alloying

Figure 10-21 Homogeneous α_2 precipitation with some precipitation on dislocations in aged Ti-8 wt% Al alloy with 1780 ppm oxygen. [*After J. Y. Lim, C. J. McMahon, D. P. Pope, and J. C. Williams, "The Effect of Oxygen on the Structure and Mechanical Behavior of Aged Ti-8 wt% Al," Met. Trans. 7A(1976):139.*]

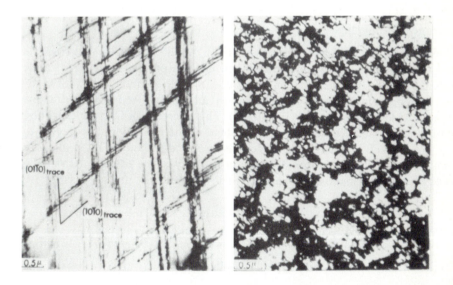

Figure 10-22 Dislocation arrangements in (*a*) Ti-6 wt% Al and (*b*) commercially pure titanium after ≈ 4 percent deformation. [*After M. J. Blackburn and J. C. Williams, "Strength, Deformation Molds and Fracture in Titanium-Aluminum Alloys," Trans. ASM 62(1969):398.*]

431

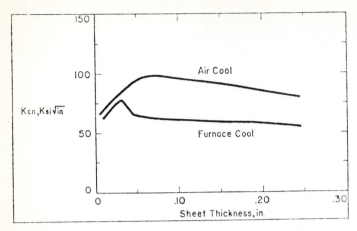

Figure 10-23 Effect of cooling rate after annealing on the fracture toughness of Ti–5% Al–2.5% Sn alloy sheet at −253°C. (*After S. Seagle, and L. J. Bartlo, "Physical Metallurgy and Metallography of Titanium Alloys," Met. Eng. Quart., Aug. 1968, p. 2.*)

elements that may be added to titanium to avoid excessive α_2 phase is

$$Al_{equiv} = Al + \frac{Sn}{3} + \frac{Zr}{6} + 10(O) \leq 9 \text{ wt\%}$$

Hence for applications requiring good ductility at low temperatures, a low-oxygen-type Ti–5% Al–2.5% Sn alloy is produced.

Mechanical Properties

The mechanical properties of the all-α Ti–5% Al–2.5% Sn alloy are listed in Table 10-6. Note that lowering the oxygen content of this alloy significantly lowers its tensile strength by 8 ksi. Slow cooling the Ti–5% Al–2.5% Sn alloy from elevated temperatures as opposed to air cooling reduces its fracture toughness at low temperatures (Fig. 10-23). This decrease in fracture toughness is attributed to the presence of an early stage in the formation of coherent α_2 phase. The effect of aluminum in decreasing the ductility of Ti-Al alloys is shown in Fig. 10-24. The rapid decrease in ductility in the 6 to 8% Al range is attributed to the formation of α_2 phase.

10-7 NEAR-α TITANIUM ALLOYS

Chemical Compositions and Typical Applications

Near-α titanium alloys are those which contain some β phase dispersed in an otherwise all-α-phase structure. Small amounts of molybdenum and vanadium (about 1 to 2 percent), which are β-stabilizing elements, are added to these alloys to retain some β phase at room temperature.

Table 10-6 Mechanical properties of α titanium alloys†

| | Room temperature | | | | Average mechanical properties | | | | | | | |
| | | | | | Extreme temperatures | | | | | | Charpy | |
α Alloys	Tensile strength, psi	Yield strength, psi	Elon- gation, %	Reduction in area, %	Test temp., °F	Tensile strength, psi	Yield strength, psi	Elon- gation, %	Reduction in area, %	impact strength, ft · lb	Hardness, R_C
5% Al, 2.5% Sn*	125,000	117,000	16	40	600	82,000	65,000	18	45	19	36
5% Al, 2.5% Sn (low O₂)*	117,000	108,000	16	...	−320	180,000	168,000	16	...	20	35
					−423	229,000	206,000	15	...		

† After "ASM Databook," *Metal Progress*, vol. 114, no. 1, mid-June 1978.
*Annealed condition.

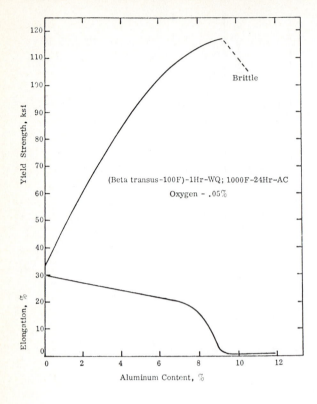

Oxygen – .05%

(Beta transus–100F)–1Hr–WQ; 1000F–24Hr–AC

Brittle

Figure 10-24 Effect of aluminum content on embrittlement of Ti–Al alloys. (*After RMI Company Data.*)

Table 10-7 Chemical compositions and typical applications of near-α titanium alloys†

Composition	Condition	Typical applications
8% Al–1% Mo–1% V	Duplex-annealed	Airframe and jet-engine parts requiring high strength to 850°F (455°C); good creep and toughness properties; good weldability.
6% Al–2% Sn–4% Zr–2% Mo		Parts and cases for jet-engine compressors; airframe skin components.
5% Al–5% Sn–2% Zr–2% Mo–0.25% Si	975°C ($\frac{1}{2}$ h) air-cooled, +600°C (2h) air-cooled	Jet engine parts; high creep strength to 1000°F (538°C).
6% Al–1% Mo–2% Cb–1% Ta	As-rolled, 1-in plate	High toughness; moderate strength; good resistance to sea water and hot-salt stress corrosion; good weldability.

† After "ASM Databook," *Metal Progress*, vol. 114, no. 1, mid-June 1978.

Table 10-7 lists the chemical compositions and typical applications of most of the commercial near-α titanium alloys. Tin and zirconium are added to several of these alloys (e.g., Ti–6% Al–2% Sn–4% Zr–2% Mo) so that their aluminum contents can be reduced while still maintaining their strength. Of the near-α titanium alloys, Ti–8% Al–1% Mo–1% V and Ti–6% Al–2% Sn–4% Zr–2% Mo alloys are the most commonly used and in 1978 accounted for about 3 percent of the titanium market.

The Ti–8% Al–1% Mo–1% V alloy was originally developed for moderately high-temperature applications in the compressor section of jet engines and has been used for aircraft skin components. It has desirable properties such as good weldability, good creep resistance and toughness, high strength, low ductility, and high modulus. A disadvantage of this alloy is that it is susceptible to stress-corrosion cracking in a salt environment.

Microstructures

The Ti–8% Al–1% Mo–1% V alloy is one of the two most commonly used near-α titanium alloys and, since its microstructural variations are quite well established (Refs. 12 and 13), it will be used as an example for this group of alloys. This alloy is normally used in the annealed condition and is not heat treated. Two heat treatments are in common usage for this alloy, termed *mill annealing* and *duplex annealing*. Mill annealing consists of heating the alloy at 790°C for 8 h and then furnace cooling. More commonly, this alloy is duplex-annealed by reheating the mill-annealed material at 790°C for 0.25 h and air cooling. The microstructure of duplex-annealed Ti–8% Al–1% Mo–1% V alloy shows β particles in an α matrix (Fig. 10-25). Electron micrographs show essentially the same type of annealed microstructure as the optical micrographs.

The phases present in the Ti–8% Al–1% Mo–1% V alloy after cooling from 790°C can best be considered by referring to the pseudo-binary phase diagram

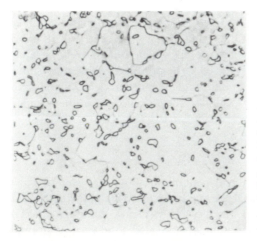

Figure 10-25 Duplex annealed Ti–8% Al–1% Mo–1% V alloy. Alloy was mill annealed at 790°C for 8 h and furnace-cooled, then reheated to 790°C, held for 0.25 h, and air-cooled. Structure consists of equiaxed alpha grains and outlined intergranular beta. (Etchant: 2% HF, 8% HNO₃, 90% H₂O; X700.) [*After M. J. Blackburn, "Relationship of Microstructure to Some Mechanical Properties of Ti–8 Al–1 V–1 Mo," Trans. ASM 59(1966):694.*]

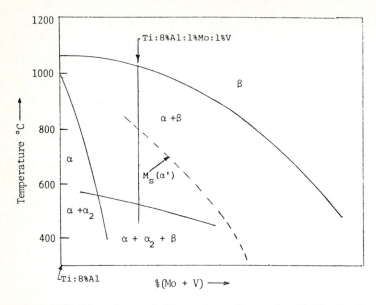

Figure 10-26 Schematic pseudo-binary phase diagram for Ti–8% Al alloy with additions of molybdenum and vanadium. α: hexagonal; α_2: ordered hexagonal; β: body-centered cubic; M_s: martensite start temperature. [*After M. J. Blackburn, "Relationship of Microstructure to some Mechanical Properties of Ti–8 Al–1 V–1 Mo," Trans. ASM 59(1966):876.*]

for the combined addition of molybdenum and vanadium to a Ti–8 wt% Al base alloy (Fig. 10-26). Mill annealing at 790°C and slow furnace cooling produces a structure containing α phase, α_2 ordered phase, and β phase. By reheating at 790°C for the duplex annealing, followed by air cooling, more disordered α phase is produced than with slow furnace cooling. The duplex-annealed structure is more desirable since the disordered α-phase imparts more ductility and impact resistance to the alloy than the ordered α_2-phase, which tends to embrittle it.

The duplex-annealed alloy when strained about 5 percent shows a similar structure to a binary Ti–6% Al alloy deformed about 4 percent (Fig. 10-22a). That is, coplanar arrays of dislocations are present which are relatively far apart (Fig. 10-27). This type of structure is believed to be produced when cross slip of dislocations is difficult due to the presence of the ordered α_2 phase. Pure titanium strained about 4 percent has a cellular network of dislocations which is associated with high ductility in metals (Fig. 10-22b).

Mechanical Properties

The mechanical properties of selected near-α titanium alloys are listed in Table 10-8. The tensile strengths of the first three are moderately high (142 to 152 ksi) at room temperature and show good ductility (15 percent). Although the Ti–8% Al–1% Mo–1% V alloy is used in the duplex-annealed condition, its strength

Table 10-8 Mechanical properties of near-α titanium alloys†

Composition	Condition	Room temperature				Average mechanical properties					Charpy impact strength, ft·lb	Hardness, R_C
							Extreme temperatures					
		Tensile strength, psi	Yield strength, psi	Elongation, %	Reduction in area, %	Test temp., °F	Tensile strength, psi	Yield strength, psi	Elongation, %	Reduction in area, %		
8% Al–1% Mo–1% V	Duplex-annealed	145,000	138,000	15	28	600	115,000	90,000	20	38	24	35
						800	107,000	82,000	20	44		
						1000	90,000	75,000	25	55		
6% Al–2% Sn–4% Zr–2% Mo		142,000	130,000	15	35	600	112,000	85,000	16	42	…	32
						800	102,000	75,000	21	55		
						1000	94,000	71,000	26	60		
5% Al–5% Sn–2% Zr–2% Mo–0.25% Si	975°C ($\frac{1}{2}$ hr) air-cooled, +600°C (2 hr) air-cooled	152,000	140,000	13	…	600	115,000	82,000	15	…		
						800	113,000	77,000	17	…		
						1000	100,000	73,000	19	…		
6% Al–1% Mo–2% Cb–1% Ta	As-rolled 1-in plate	124,000	110,000	13	34	600	85,000	67,000	20	…	23 (−80°F)	30
						800	75,000	60,000	20	…		
						1000	70,000	55,000	20	…		

† After "ASM Databook," *Metal Progress*, vol. 114, no. 1, mid-June, 1978.

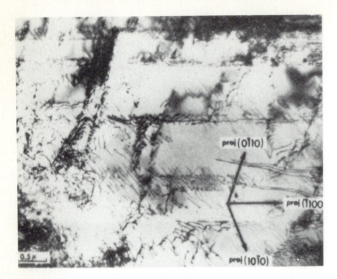

Figure 10-27 Duplex annealed Ti–8% Al–1% Mo–1% V alloy showing dislocation structure within the α-phase after straining 5 percent. Electron transmission micrograph. Note the coplanar arrays of dislocations. [*After M. J. Blackburn, "Relationship of Microstructure to some Mechanical Properties of Ti–8 Al–1 V–1 Mo," Trans. ASM 59(1966):694.*]

can be increased about 25 percent by solution heat treatment and aging. However, these processes are reported to make it susceptible to stress-corrosion cracking in salt-water environments.

10-8 α-β TITANIUM ALLOYS

Chemical Compositions and Typical Applications

This class of titanium alloys contains one or more β-stabilizing elements in sufficient quantity to permit the retention of appreciable amounts of β phase at room temperature, resulting in an $\alpha + \beta$ structure. α-β titanium alloys can be solution-heat-treated, quenched, and aged for increased strength. The chemical compositions and typical applications of the most important of these alloys are listed in Table 10-9

Ti–6% Al–4% V is by far the most important and widely used titanium alloy, accounting for 55 percent of the titanium market in 1978. It can be readily welded, forged, and machined, and is available in a wide variety of mill product forms such as sheet, extrusions, wire, and rod. Ti–6% Al–4% V is also used extensively for ordnance forgings. It is heat-treatable to an ultimate tensile strength of 165 ksi and has good metallurgical stability to 482°C. One of its disadvantages is that, since it is a "lean" α-β alloy, it has low hardenability, so that sections of only up to about 1 in can be hardened all the way through.

The Ti–6% Al–6% V–2% Sn alloy was developed for forgings and extrusions

Table 10-9 Chemical compositions and typical applications of α-β titanium alloys†

Alloy composition	Condition	Typical applications
6% Al, 4% V	Annealed; solution + age	Rocket motor cases; blades and disks for aircraft turbines and compressors; structural forgings and fasteners; pressure vessels; gas and chemical pumps; cryogenic parts; ordnance equipment; marine components; steam-turbine blades.
6% Al, 4% V (low O_2)	Annealed	High-pressure cryogenic vessels operating down to $-320°$F
6% Al, 6% V, 2% Sn	Annealed; solution + age	Rocket motor cases; ordnance components; structural aircraft parts and landing gears; responds well to heat treatments; good hardenability.
7% Al, 4% Mo	Solution + age	Airframes and jet engine parts for operation at up to 800°F; missile forgings; ordnance equipment.
6% Al, 2% Sn, 4% Zr, 6% Mo	Solution + age	Components for advanced jet engines.
6% Al, 2% Sn, 2% Zr, 2% Mo, 2% Cr, 0.25% Si	Solution + age	Strength, fracture toughness in heavy sections; landing-gear wheels.
10% V, 2% Fe, 3% Al	Solution + age	Heavy airframe structural components requiring toughness at high strengths.
8% Mn	Annealed	Aircraft sheet components, structural sections, and skins; good formability, moderate strength.
3% Al, 2.5% V	Annealed	Aircraft hydraulic tubing, foil; combines strength, weldability, and formability.

† After "ASM Databook," *Metal Progress*, vol. 114, no. 1, mid-June 1978.

in applications where higher strengths are required than can be attained using Ti–6% Al–4% V. Additions of about 0.5% Cu and 0.5% Fe are usually added to this alloy to further enhance its strength properties. In 1978, the Ti–6% Al–6% V–2% Sn alloy accounted for about 10 percent of the titanium market.

For special applications requiring higher strengths at elevated temperatures, such as components for advanced jet engines, the Ti–6% Al–2% Sn–4% Zr–6% Mo and Ti–6% Al–2% Sn–2% Zr–2% Mo–2% Cr–0.25% Si alloys have been developed. They are more hardenable and can be used in heavier sections and as well as at higher temperatures.

Microstructures

The microstructures of the α-β titanium alloys depend primarily on the following: (1) chemical composition, (2) processing history, and (3) thermal treatment.

The microstructures of these alloys are often duplex, consisting of constituents of different microstructural scales. Since these alloys are so complex, only the microstructural changes in the most important and studied α-β alloy, Ti–6% Al–4% V, are described and discussed here. Such changes in Ti–6% Al–6% V–2% Sn and Ti–6% Al–2% Sn–4% Zr–6% Mo alloys are even more complicated and are much less understood.

Microstructural changes in Ti–6% Al–4% V **alloy due to thermal treatments** The microstructural changes caused by thermal treatment of the Ti–6% Al–4% V alloy can best be understood by considering the pseudo-binary Ti–6% Al phase diagram shown in Fig. 10-28.

Cooling from above the β transus (1066°C) The structure produced depends upon the type of cooling method used:

1. *Water quenching from 1066°C.* Heating a bar of Ti–6% Al–4% V alloy to 1066°C, which is above the β transus (Fig. 10-28), and holding for 1 h, produces an all-β-phase structure. Upon water quenching from 1066°C, a structure consisting of all α' (titanium martensite) is produced (Fig. 10-29). The microstructure of α' martensite consists of individual platelets which are heavily twinned and have an HCP crystal structure (Fig. 10-30). This titanium martensite is strengthened mainly by grain refinement associated

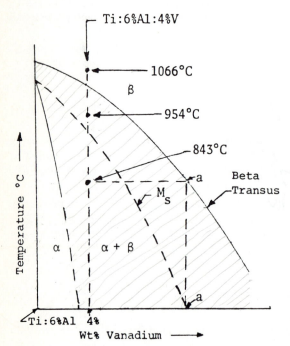

Ti:6%Al:4%V

1066°C

β

954°C

843°C

a   Beta
    -Transus

M_s

α   α + β

a

Temperature °C

Ti:6%Al   4%
Wt% Vanadium

Figure 10-28 Schematic pseudo-binary phase diagram for Ti–6% Al alloy with additions of vanadium. (M_s = martensite start temperature.)

Figure 10-29 Titanium–6% Al–4% V bar solution heat-treated at 1066°C for 30 min and water-quenched. Structure consists of alpha prime formed by martensitic-type shear process. Prior beta grain boundaries are evident. (Etchant: 10% HF–5% HNO$_3$; X500.) (*Courtesy of RMI Company.*)

with the BCC-to-HCP transformation and by the increased dislocation density due to the rapid transformation. Titanium martensites, though, are relatively soft compared to iron-carbon martensites in steels. They do not develop high hardnesses since the interstitial elements carbon, oxygen, and nitrogen are more soluble in the lower-temperature stable hexagonal phase. Titanium martensites are only supersaturated with respect to β-stabilizing elements such as vanadium and molybdenum. Aging or tempering the titanium martensites, however, does produce some increased strengthening because of the precipitation of the β phase from the unstable α' martensite.

Figure 10-30 Alpha prime titanium martensite formed in Ti–6% Al–4% V alloy quenched from 1200°C. Structure shows plates of martensite which vary in size and have a HCP structure. These plates have a high dislocation density and occasionally contain twins. [*After J. C. Williams and M. J. Blackburn, "A Comparison of Phase Transformations in Three Commercial Titanium Alloys," Trans. ASM 60(1967):373.*]

Figure 10-31 Beta phase precipitates formed during the tempering of α' martensite in Ti–6% Al–4% V alloy for 48 h at 500°C. Nucleation occurs on subboundaries and individual dislocations. [*After J. C. Williams and M. J. Blackburn, "A Comparison of Phase Transformations in Three Commercial Titanium Alloys," Trans. ASM 60(1967):373.*]

Figure 10-31 shows β-phase precipitates formed during the tempering of α' martensite in Ti–6% Al–4% V alloy.

2. *Air cooling from 1066°C.* Air cooling the solution-treated Ti–6% Al–4% V alloy from 1066°C, which is about 50°C above the β transus (Fig. 10-28), produces a structure consisting of acicular α that is transformed from the β phase by nucleation and growth. This type of structure, produced by intermediate cooling rates from high temperatures, is shown in Fig. 10-32.

3. *Furnace cooling from 1066°C.* Slow furnace cooling the solution-treated Ti–6% Al–4% V alloy from 1066°C, which is about 50°C above the β transus, produces a structure that more closely approaches equilibrium. As a consequence, coarse platelike α is formed by nucleation and growth, as shown in

Figure 10-32 Ti–6 Al–4 V alloy solution heated at 1066°C, about 50°C above the beta transus and air-cooled; the structure consists of acicular alpha (transformed beta); prior beta grain boundaries appear. (10% HF, 5% HNO_3, 85% H_2O; X250.) (*After Metals Handbook, 8th ed., vol. 7, American Society for Metals, Metals Park, OH, 1972, p. 328.*)

Figure 10-33 Ti–6 Al–4 V alloy solution heated at 1066°C, about 50°C above the beta transus, and furnace-cooled; the structure consists of platelike alpha (light) and intergranular beta (dark). (10% HF, 5% HNO₃, 85% H₂O; X250.) (*After Metals Handbook, 8th ed., vol. 7, American Society for Metals, Metals Park, OH, 1972, p. 328.*)

Fig. 10-33. Due to the slow cooling, some β phase is retained and occurs intergranularly. The furnace-cooled material has an ultimate tensile strength of 151 ksi as compared to 160 ksi for the water-quenched material, which indicates that the α' martensite structure is associated with about a 10-ksi increase in strength before aging.

Cooling from 954°C (about 50°C below the β transus) Results of the three methods of cooling are:

1. *Water quenching from 954°C.* Upon solution heating a bar of Ti–6% Al–4% V at 954°C, the alloy is about 50°C below the β transus, which is approximately 1010°C. Hence, at 954°C, some primary α will coexist with the

Figure 10-34 Ti–6 Al–4 V alloy solution heat-treated at 954°C; water-quenched. Structure consists of primary alpha (white) and alpha prime (gray). (Etchant: 10% HF–5% HNO₃; X500.) (*Courtesy of RMI Company.*)

Figure 10-35 Ti–6 Al–4 V alloy solution treated at 950°C for 5 h and water-quenched. Structure shows primary alpha and transformed beta (acicular alpha). [*After C. Hammond and J. Nutting, "The Physical Metallurgy of Titanium Alloys," Met. Sci. J. 11(1977):474.*]

Figure 10-36 Ti–6 Al–4 V alloy solution heated at 954°C (about 50°C below the beta transus) and air-cooled. Structure shows grains of primary alpha (light) in a matrix of transformed beta containing acicular alpha. (10% HF, 5% HNO_3, 85% H_2O; X250.) (*After Metals Handbook, 8th ed., vol. 7, American Society for Metals, Metals Park, OH, 1972, p. 328.*)

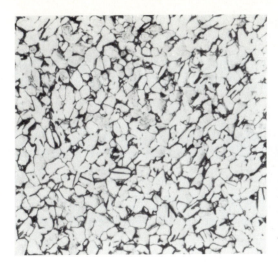

Figure 10-37 Ti–6 Al–4 V alloy solution heated at 954°C (about 50°C below the beta transus) and furnace-cooled. Structure shows equiaxed alpha grains (light) and intergranular beta (dark). (10% HF, 5% HNO_3, 85% H_2O; X250.) (*After Metals Handbook, 8th ed., vol. 7, American Society for Metals, Metals Park, OH, 1972, p. 328.*)

β phase. Upon quenching to room temperature, the β phase is transformed immediately to α' titanium martensite. Thus, a structure consisting of primary α embedded in α' is produced (Fig. 10-34). Essentially the same structure at higher magnification is shown in Fig. 10-35.

2. *Air cooling from 954°C*. Air cooling the solution-heat-treated Ti–6% Al–4% V alloy bar from 954°C produces a structure of primary α in a matrix of transformed β, some of which is acicular α (Fig. 10-36).

3. *Furnace cooling from 954°C*. Slow furnace cooling the solution-treated Ti–6% Al–4% V alloy bar from 954°C produces a structure approaching equilibrium, which consists of equiaxed α and intergranular untransformed β (Fig. 10-37).

Cooling from 843°C (just below the M_s temperature)

Water quenching. At a temperature as low as 843°C, much less β phase is present than at the higher temperatures previously considered. However, the β phase is enriched in the β-stabilizing element vanadium, as indicated by point a in Fig. 10-28. Thus, upon quenching from 843°C, the β phase is retained at room temperature and does not transform. These relationships are shown in the pseudo-binary diagram of Fig. 10-28. Water quenching the solution-treated Ti–6% Al–4% V alloy bar from 843°C, which is below the M_s, produces a structure consisting of primary α and untransformed or retained β (Fig. 10-38). The retained β is metastable, however, and may undergo a subsequent strain-induced transformation.

Microstructural changes in Ti–6% Al–4% V due to processing Another important variable determining the microstructure of α-β titanium alloys is processing, since the temperature of hot working and the extent of hot working can produce

Figure 10-38 Ti–6 Al–4 V solution heat-treated at 843°C for 1 h and water-quenched. Structure consists of retained beta in an alpha matrix. (Etchant: 10% HF–5% HNO₃; X500.) (*Courtesy of RMI Company.*)

great changes. As there are many possible differences in microstructure by processing, however, only a few such variations can be given here.

Forging is an important processing operation in the production of many Ti–6% Al–4% V products. Figure 10-39 shows the microstructure of a Ti–6% Al–4% V bar which was forged with 75 percent reduction at 982°C and reheated 2 h at 732°C followed by air cooling. This worked structure, which was produced at 982°C (just below the β transus), consists of platelike and equiaxed α with a small amount of transformed β. And yet forging a similar bar of this alloy with the same amount of deformation (75 percent), but at a *lower* temperature of 899°C, a structure consisting of fine elongated α-β is produced, as shown in Fig. 10-40. Thus, the lower the temperature of working below the β

Figure 10-39 Ti–6 Al–4 V bar forged 75 percent at 982°C + 2 h at 732°C and air-cooled. Structure shows platelike and equiaxed alpha with a small amount of transformed beta. (Etchant: 10% HF–5% HNO₃; X250.) (*Courtesy of RMI Company.*)

Figure 10-40 Ti–6 Al–4 V bar forged 75 percent at 899°C + 2 h at 732°C and air-cooled. Structure shows fine elongated alpha-beta. (Etchant: 10% HF–5% HNO₃; X250.) (*Courtesy of RMI Company.*)

transus, the more elongated the structure becomes. If extensive hot working is done above the recrystallization temperature but below the β transus, an equiaxed structure will be produced.

Heat Treatment

If the Ti–6% Al–4% V alloy is solution-heat-treated about 40°C below the β transus and subsequently quenched and aged, a maximum tensile strength of about 170 ksi with adequate ductility can be obtained (Fig. 10-41). Solution heat treating *above* the β transus leads to lower ductility and strength. Since vanadium strongly partitions to the β phase, quenching from above the β

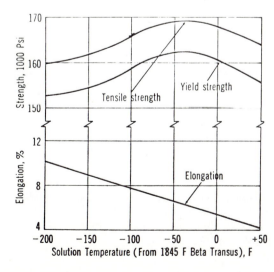

Figure 10-41 Effect of solution heat treatment temperature on the tensile strength and ductility of Ti–6 Al–4 V forgings. This material was aged 8 h at 593°C after solution heat treatment. (*After J. A. Burger and D. K. Hanink, Met. Prog. June 1967, p. 70.*)

transus gives the maximum volume fraction of β phase and also results in the most dilute β phase possible in this alloy. As a result, in β quenched Ti–6% Al–4% V, nucleation and growth of the α phase occurs to a small extent along the prior β-phase boundaries. This continuous grain boundary α phase is believed to reduce the strength and ductility of the Ti–6% Al–4% V alloy by creating an intergranular fracture path.

Quenching the Ti–6% Al–4% V alloy from 954°C produces a structure of primary α and α' titanium martensite, as shown in Fig. 10-34. The α' martensite increases the strength of the annealed Ti–6% Al–4% V from about 144 to 162 ksi, which represents an increase in strength of about 12 percent. This strength increase is believed to be due to grain refinement and to an increase in dislocation density caused by the rapid cooling to form the titanium martensite. Water quenching cannot be delayed more than 10 s after solution heat treatment at 954°C or the alloy will lose strength, especially after aging, as is shown in Fig. 10-42.

Aging the Ti–6% Al–4% V after solution heat treating at 954°C and water quenching increases the tensile strength of this alloy to about 170 ksi, which is approximately a 6 percent increase over that obtained after water quenching. The total increase in strength in the fully heat-treated condition above that of the annealed condition is about 18 percent.

The Ti–6% Al–4% V alloy is aged from 1 to 12 h at 538 to 621°C. During

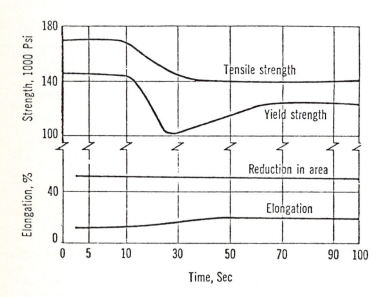

Figure 10-42 Effect of delaying water quenching after solution heat treatment on tensile properties of Ti–6 Al–4 V forgings. In this case samples were solution heat treated at 954°C, water-quenched and aged. Note the rapid decrease in tensile strength if water quench is delayed by more than 10 s. (*After J. A. Burger and D. K. Hanink, Met. Prog. June 1967, p. 70.*)

aging, fine β is precipitated from the α' martensite on subboundaries (Fig. 10-31), dislocations, and twin boundaries. This precipitation is believed to be the cause of the 6 percent increase in strength upon aging (tempering) the α' titanium martensite.

Mechanical Properties

The mechanical properties of the most important commercial α-β titanium alloys are listed in Table 10-10. At room temperature, the tensile strengths of these alloys range from a low of 100 ksi for the annealed Ti–3.5% Al–2.5% V to a high of 185 ksi for the solution-treated and aged Ti–6% Al–6% V–2% Sn.

Higher strengths at elevated temperatures are obtained with Ti–6% Al–6% V–2% Sn, Ti–6% Al–2% Sn–4% Zr–6% Mo, and Ti–6% Al–2% Sn–2% Zr–2% Mo–2% Cr–0.25% Si, since these alloys were designed for high-temperature use. A comparison of the effect of temperature on the tensile strength of heat-treated Ti–6% Al–4% V and Ti–6% Al–6% V–2% Sn is given in Fig. 10-43 along with comparisons to several other titanium alloys.

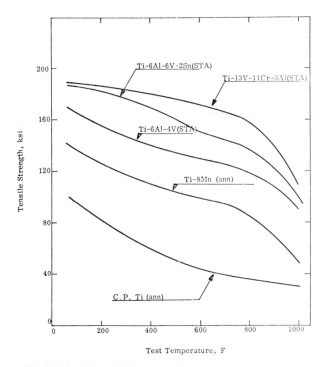

Figure 10-43 Effect of test temperature on the tensile strength of several $\alpha + \beta$ titanium alloys. (*After RMI Company data.*)

Table 10-10 Mechanical properties of α-β titanium alloys†

Alloy composition	Condition	Room temperature				Average mechanical properties — Extreme temperatures						
		Tensile strength, psi	Yield strength, psi	Elongation, %	Reduction in area, %	Test temp., °F	Tensile strength, psi	Yield strength, psi	Elongation, %	Reduction in area, %	Charpy impact strength, ft · lb	Hardness, R_C
6% Al, 4% V	Annealed	144,000	134,000	14	30	600 800 1000	105,000 97,000 77,000	95,000 83,000 62,000	14 18 35	35 40 50	14	36
	Solution + age	170,000	160,000	10	25	600 800 1000	125,000 116,000 95,000	102,000 90,000 70,000	10 12 22	28 35 45	...	41
6% Al, 4% V (low O₂)	Annealed	130,000	120,000	15	35	320	220,000	205,000	14	...	18	35
6% Al, 6% V, 2% Sn	Annealed	155,000	145,000	14	30	600	135,000	117,000	18	42	13	38
	Solution + age	185,000	170,000	10	20	600	142,000	130,000	12	28	...	42
7% Al, 4% Mo	Solution + age	160,000	150,000	16	22	600 800	127,000 123,000	108,000 104,000	18 20	50 55	13 ...	38 42
6% Al, 2% Sn, 4% Zr, 6% Mo	Solution + age	184,000	170,000	10	23	600 800 1000	148,000 138,000 123,000	122,000 110,000 95,000	18 19 19	55 67 70		
6% Al, 2% Sn, 2% Zr, 2% Mo, 2% Cr, 0.25% Si	Solution + age	185,000	165,000	11	33	600	142,000	117,000	14	27		
10% V, 2% Fe, 3% Al	Solution + age	185,000	174,000	10	19	400 600	162,000 160,000	152,000 142,000	13 13	33 42		
8% Mn	Annealed	137,000	125,000	15	32	600	104,000	82,000	18			
3% Al, 2.5% V	Annealed	100,000	85,000	20	...	600	70,000	50,000	25			

† After "ASM Databook," *Metal Progress*, vol. 114, no. 1, mid-June 1978.

10-9 β TITANIUM ALLOYS

Chemical Compositions and Typical Applications

If sufficient amounts of β-stabilizing alloying elements are added to titanium, a structure consisting of all metastable β phase can be obtained at room temperature by quenching or even in some cases by air cooling. The principle alloying elements for the β titanium alloys are vanadium, molybdenum, chromium, and iron. Zirconium is sometimes added since it strengthens both the β and α phases. Aluminum is also added to most of these alloys since it lowers their density, adds some solid-solution hardening, and improves oxidation resistance. The chemical compositions and typical applications of the current β titanium alloys are listed in Table 10-11.

β titanium alloys, because of their BCC crystal structure, are readily cold-worked in the solution-heat-treated and quenched condition and can subsequently be aged to very high strengths. However, they have relatively high densities due to their large percentage of heavy metals such as vanadium and molybdenum. In the high-strength condition, these alloys have low ductilities. In thick sections, they suffer from problems of chemical segregation and large grain size, which result in low-tensile ductility and fatigue performance. As a result, the metastable β titanium alloys are not used much at present.

Microstructure (Ti–13% V–11% Cr–3% Al)

The microstructure of β titanium alloys consists of all metastable β when quenched from above the β transus since the martensitic start temperature for these alloys is below room temperature. At present, there is only one β titanium alloy produced in large quantities, Ti–13% V–11% Cr–3% Al. The microstructure of this engineering alloy will therefore be discussed here.

Table 10-11 Chemical compositions and typical applications of β titanium alloys†

Alloy composition	Typical applications
13% V, 11% Cr, 3% Al	High-strength fasteners, aerospace components, honeycomb panels, (good formability, heat-treatable).
8% Mo, 8% V, 2% Fe, 3% Al	High-strength, tough airframe sheet, plate, fasteners, and forged components.
3% Al, 8% V, 6% Cr, 4% Mo, 4% Zr	High-strength fasteners, torsion bars, aerospace components.
11.5% Mo, 6% Zr, 4.5% Sn	Parts requiring formability and corrosion resistance; high-strength fasteners, high-strength aircraft sheet parts.

† After "ASM Databook," *Metal Progress*, vol. 114, no. 1, mid-June 1978.

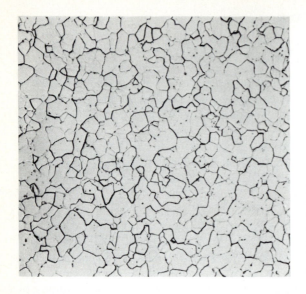

Figure 10-44 Ti–13% V–11% Cr–3% Al bar solution heat-treated at 788°C for 30 min and water-quenched. Structure shows metastable beta. (Etchant: 2% HF–4% HNO₃; X250.) (*Courtesy of RMI Company*.)

Since both vanadium and chromium are β-stabilizing elements, the β transus is lowered to 720°C. (Recall that the β transus for Ti–6% Al–4% V was 1010°C.) In order to produce an all-metastable-β structure at room temperature, this alloy is solution-heat-treated at 788°C, which is 68°C above the β transus. The optical microstructure of Ti–13% V–11% Cr–3% Al after water quenching from 788°C is shown in Fig. 10-44 and consists of all metastable β.

After aging the solution-heat-treated and water-quenched alloy 10 h at 400°C, the metastable β phase decomposes by phase separation into solute rich and solute lean β phases as

$$\beta_{\text{metastable}} \rightarrow \beta_{2\,(\text{rich})} + \beta_{1\,(\text{lean})}$$

The β_2 precipitate, which has a platelike or disk-shaped morphology, is shown in the electron micrograph of Fig. 10-45. If this alloy is aged 10 h at 450°C after solution heat treatment and water quenching, the β_2 phase acts as a nucleation site for the formation of α phase. Figure 10-46 shows the nucleation of the α phase at the β_2-matrix interface. Thus, at this higher temperature, the reaction for the decomposition of the metastable β phase can be written as

$$\beta_{\text{metastable}} \rightarrow \beta_2 + \beta_1 \rightarrow \alpha + \beta_2 + \beta_1 \rightarrow \alpha + \beta_{\text{enriched}}$$

With copious precipitation of the α phase, the β_2 phase disappears, leaving a structure of α-phase precipitates in a matrix of β. Widmanstätten α phase is shown for Ti–13% V–11% Cr–3% Al after aging 300 h at 400°C in Fig. 10-47.

Mechanical Properties

The mechanical properties of β titanium alloys are listed in Table 10-12. These alloys are usually used in the solution-treated and aged condition in order to

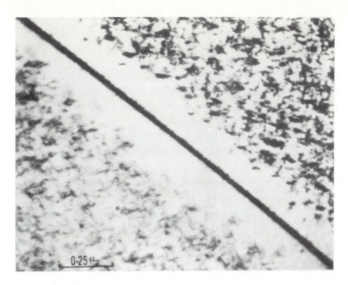

Figure 10-45 Ti–13 V–11 Cr–3 Al alloy solution treated at 800°C, water-quenched, and aged 10 h at 400°C. Structure shows β_2-phase precipitate in β_1 matrix; note the precipitate-free zone adjacent to β-grain boundary. Electron transmission micrograph. [*After G. H. Narayanan and T. F. Archbold, "Decomposition of the Metastable Beta Phase in the All-Beta Alloy Ti–13 V–11 Cr–3 Al," Met. Trans. 1(1970):2281.*]

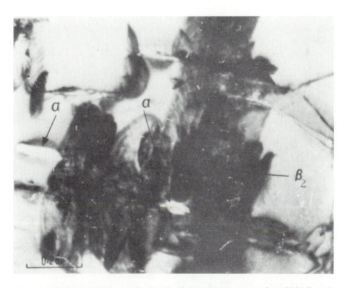

Figure 10-46 Ti–13 V–11 Cr–3 Al alloy solution treated at 800°C and water-quenched and aged at 450°C for 10 h. Structure shows the nucleation of the α phase at the β_2-matrix interface. Electron transmission micrograph. [*After G. H. Narayanan and T. F. Archbold, "Decomposition of the Metastable Beta Phase in the All-Beta Alloy Ti–13 V–11 Cr–3 Al," Met. Trans. 1(1970):2281.*]

Table 10-12 Mechanical properties of β titanium alloys†

Alloy	Condition	Room temperature Tensile strength, psi	Yield strength, psi	Elongation, %	Reduction in area, %	Average mechanical properties / Extreme temperatures Test temp., °F	Tensile strength, psi	Yield strength, psi	Elongation, %	Reduction in area, %	Charpy impact strength, ft·lb	Hardness, R_C
13% V, 11% Cr, 3% Al	Solution + age	177,000	170,000	8	...	600	128,000	115,000	19			
	Solution + age	185,000	175,000	8	...	800	160,000	120,000	12	...	8	40
8% Mo, 8% V, 2% Fe, 3% Al	Solution + age	190,000	180,000	8	...	600	164,000	142,000	15	...	...	40
3% Al, 8% V, 6% Cr, 4% Mo, 4% Zr	Solution + age	210,000	200,000	7	...	600	150,000	130,000	20	...	7.5	42
						800	136,000	110,000	17			
11.5% Mo, 6% Zr, 4.5% Sn	Annealed	128,000	121,000	15	...	600	105,000	95,000	22			
	Solution + age	201,000	191,000	11	...	600	131,000	123,000	16			

† After "ASM Databook," *Metal Progress*, vol. 114, no. 1, mid-June 1978.

Figure 10-47 Ti–13 V–11 Cr–3 Al alloy solution treated at 800°C; water-quenched and aged at 400°C for 350 h. Structure shows Widmanstätten α-phase precipitates in beta matrix. [*After G. H. Narayanan and T. F. Archbold, "Decomposition of the Metastable Beta Phase in the All-Beta Alloy Ti–13 V–11 Cr–3 Al," Met. Trans. 1(1970):2281.*]

obtain their high strengths, and they have the highest strengths of all titanium alloys, reaching up to 210 ksi.

The standard heat treatment for Ti–13% V–11% Cr–3% Al is to solution-heat-treat 0.25 to 1 h at 760 to 815°C, water-quench, and age 2 to 96 h at 482°C. The elevated-temperature properties of Ti–13% V–11% Cr–3% Al are shown graphically in Fig. 10-43, and are compared there with other standard titanium alloys.

PROBLEMS

1. What attractive engineering properties do titanium alloys have? What are their chief disadvantages?

2. Describe (*a*) the Kroll process and (*b*) the Hunter process for producing titanium sponge. Write the chemical equations associated with each.

3. What advantages does the Hunter process have over the Kroll process?

4. How are titanium ingots cast? What special techniques are required and why?

5. How are the cast titanium ingots broken down? What problems are encountered in the process?

6. What special procedures must be used when rolling and annealing titanium alloys?

7. How do the density, melting point, and modulus of elasticity of titanium compare with those properties of aluminum and iron?

8. What are the two allotropic crystalline forms of pure titanium? At what temperature does the transformation from α to β take place?

9. How does the c/a ratio of hexagonal α titanium differ from ideality and from the c/a ratios of magnesium and zinc?

10. Indicate the slip planes which are operative during the deformation of pure titanium using a hexagonal prism drawing. Do the same for the twinning planes.

11. How do nitrogen and oxygen interstitials affect the deformation properties of titanium?

12. What are the three types of alloy-stabilizing systems formed in binary titanium alloys?

13. What are the α-stabilizing elements for titanium? What element is the most important and why?

14. What are the β-isomorphous-stabilizing elements? Which are the most important? What are the β-eutectoid-stabilizing elements?

15. How are titanium alloys classified? What are some of the important engineering properties of each group?

16. What are the main impurities present in commercially pure titanium and what is their origin?

17. What are some of the applications for commercially pure titanium?

18. How do the interstitial elements affect the mechanical properties of commercially pure titanium?

19. For what reason is 0.2% Pd added to commercially pure titanium?

20. What is ELI commercially pure titanium? What are its special applications?

21. What is the most important substitutional α-stabilizing element for titanium alloys? How do additions of this element improve the engineering properties of titanium?

22. Can oxygen be considered an alloying element for titanium? Explain. What is the effect on the mechanical properties of titanium alloys if the oxygen content becomes too high?

23. What are the properties of Ti–5% Al–2.5% Sn that make it an important engineering alloy? What is its chief property disadvantage?

24. Why must the amount of aluminum alloyed with titanium be limited to about 8 percent?

25. What experimental evidence has been obtained to support the belief that an early stage in the formation of coherent α_2 precipitate occurs in slowly cooled Ti–5% Al–2.5% Sn?

26. What are the near-α titanium alloys? What are some of the principal applications of these alloys?

27. What is the advantage of duplex annealing the Ti–8% Al–1% Mo–1% V alloy instead of using it in the mill-annealed condition?

28. What are some of the important engineering properties of the α-β titanium alloys?

29. What is the most important titanium alloy? What are the properties of this alloy that make it so important? What is one of its property disadvantages?

30. What are the chief variables that affect the microstructure of Ti–6% Al–4% V?

31. What other α-β alloys have been developed for strengths higher than those attainable with Ti–6% Al–4% V?

32. Describe the microstructure of α' titanium martensite in Ti–6% Al–4% V. How is α' produced in it?

33. Why does titanium martensite not develop high hardnesses when quenched from high temperatures?

34. What principal microstructural change occurs when α' titanium martensite is tempered at 500°C?

35. How is acicular α produced in Ti–6% Al–4% V? By what mechanism does it form?

36. What type of microstructure is formed when Ti–6% Al–4% V is quenched from 954°C, which is about 50°C below the β transus?

37. What type of structure is produced in Ti–6% Al–4% V by quenching from 843°C, which is just below the martensitic start temperature? Why is some β phase retained untransformed?

REFERENCES

1. R. I. Jaffee and H. M. Burte: "Titanium Science and Technology," vols. 1–4, Plenum, New York, 1973.
2. R. I. Jaffee and N. E. Promisel: "The Science, Technology and Application of Titanium," Pergamon, Elmsford, NY, 1970.
3. C. Hammond and J. Nutting: "The Physical Metallurgy of Titanium Alloys," *Met. Sci. J.* 11(1977):481.
4. P. H. Morton: "Titanium Alloys for Engineering Structures," *Phil. Trans. Roy. Soc. Lond. A* 282(1976):401.
5. S. R. Seagle and L. J. Bartlo: "Physical Metallurgy and Metallography of Titanium Alloys," *Met. Eng. Quart.*, Aug. 8, 1968, p. 1.
6. N. G. Tupper, J. K. Elbaum, and H. M. Burte: "Opportunities for Cost-Affordable Titanium Aerospace Structures," *J. Metals*, Sept. 1978, p. 7.
7. "Facts About the Metallography of Titanium," Reactive Metals, Inc., Niles, OH, 1970.
8. J. Y. Lim et al.: "The Effect of Oxygen on the Structure and Mechanical Behavior of Aged Ti-8 wt% Al," *Met. Trans.* 7*A*(1976):139.
9. T. K. Namboodhiri, C. J. McMahon, and H. Herman, "Decomposition of the α-Phase in Titanium-Rich Ti-Al Alloys," *Met. Trans.* 4(1973):1323.
10. M. J. Blackburn and J. C. Williams: "Strength, Deformation Modes and Fracture in Titanium-Aluminum Alloys," *Trans. ASM* 62(1969):399.
11. J. C. Williams and M. J. Blackburn: "A Comparison of Phase Transformations in Three Commercial Titanium Alloys," *Trans. ASM* 60(1967):373.
12. M. J. Blackburn: "Relationship of Microstructure to some Mechanical Properties of Ti-8Al-1V-1Mo," *Trans. ASM* 59(1966):694.
13. M. J. Blackburn: "Phase Transformations in the Alloy Ti-8Al-1Mo-1V," *Trans. ASM* 59(1966):876.
14. P. J. Fopiano, M. B. Bever, and B. L. Averbach: "Phase Transformations and Strengthening Mechanisms in the Alloy Ti-6Al-4V," *Trans. ASM* 62(1969):324.
15. G. H. Narayanan and T. F. Archbold: "Decomposition of the Metastable Beta Phase in the All-Beta Alloy Ti-13V-11Cr-3Al," *Met. Trans.* 1(1970):2281.
16. L. E. Tanner: "The Isothermal Transformation of Ti-13V-11Cr-3Al," *Trans. ASM* 53(1961):408.
17. F. D. Rosi et al.: "Mechanism of Plastic Flow in Titanium: Determination of Slip and Twinning Elements," *J. Metals*, Feb. 1953, p. 257.
18. A. T. Churchman: "The Slip Modes of Titanium and the Effect of Purity on Their Occurrence during the Tensile Deformation of Single Crystals," *Proc. Roy. Soc.* A226(1954):216.
19. H. Margolin and Y. Mahajan: "Void Formation, Void Growth, and Tensile Fracture in Ti-6Al-4V," *Met. Trans.* 9A(1978):781.
20. J. C. Chesnutt, C. G. Rhodes, and J. C. Williams: "Relationship between Mechanical Properties, Microstrucure, and Fracture Topography in $\alpha + \beta$ Titanium," ASTM Spec. Tech. Publ. 600, 1976.

ELEVEN

NICKEL AND COBALT ALLOYS

Nickel is an excellent structural metal for many engineering applications. It has the desirable FCC crystal structure, so it is tough and ductile. It also has good high- and low-temperature strength as well as high oxidation resistance and good corrosion resistance for most environments. Few metals can match the attractive engineering properties of nickel. Unfortunately, its greatest disadvantage is its relatively high cost, and thus its use as a base metal for alloys is greatly limited. Nickel-base alloys are therefore used when no cheaper types can provide the necessary corrosion- or heat-resisting properties required for special engineering applications.

11-1 PRODUCTION OF NICKEL

In general, there are three major types of nickel deposits: nickel-copper sulfides, nickel silicates, and nickel laterites and serpentines. The sulfide deposits, which are located mainly in Canada, provide most of the Western world's supply of the metal. The second most important source is the nickel silicate ores of New Caledonia. Laterite ores, which have relatively low nickel contents, are located mainly in tropical and subtropical regions of the world. These deposits have not been extensively developed because of the high cost of recovering the nickel.

There are several established processes for the extraction of nickel from its ores with the process used depending mainly on the type of ore being treated. The Canadian Sudbury, Ontario, deposits which are controlled by the Inco

Metals Company are processed in the following manner.[1] After the nickel-copper-iron sulfide ore is crushed and ground, an iron sulfide (pyrrhotite) concentrate is separated magnetically and processed in an iron-ore recovery plant. The remaining ore product is subjected to a froth flotation treatment which produces separate nickel and copper concentrates.

The copper concentrate is sent to the copper smelter for further processing to produce copper products. The nickel concentrate is processed separately and is roasted, smelted in a reverberatory furnace, and converted to a Bessemer matte which consists mainly of nickel and copper sulfides. This copper-nickel matte is cooled under controlled conditions so that discrete crystals of nickel and copper sulfides and a nickel-copper metallic alloy are formed. After the cooled matte is crushed and ground, the metallic alloy is separated magnetically and treated at the Inco Copper Cliff refinery. The remaining copper and nickel sulfides are separated by froth flotation. The copper sulfide is returned to the copper smelter for further processing while the nickel sulfide is roasted to produce various grades of nickel oxides. The purest nickel oxide products are marketed directly and the less pure oxides are processed further at Inco's Port Colborne, Ontario, and Clydach, Wales, nickel refineries to produce commercially pure nickel and other nickel-alloy products.

11-2 COMMERCIALLY PURE NICKEL

Chemical Compositions and Typical Applications

Although high-purity (99.99 percent) nickel is available for special purposes, commercially pure nickel usually contains about 99.5% Ni plus cobalt. Nickel produced from Canadian ores usually contains about 0.5% Co. Table 11-1 lists the nominal chemical compositions and Table 11-2 typical applications of

---

[1] "Smelting and Refining," Inco Metals Company.

Table 11-1 Nominal chemical composition of commercial-purity base nickel alloys†

Nickel alloy	% Ni	% C	% Mn	% Fe	% Si	% Mg	% Cu	% Other
200	99.5‡	0.08	0.2	0.2	0.2		0.13	
201	99.5‡	0.01	0.2	0.2	0.2		0.13	
205	99.5‡	0.08	0.2	0.1	0.08	0.05	0.08	
220	99.5‡	0.04	0.1	0.05	0.08	0.05	0.05	
230	99.5‡	0.05	0.08	0.05	0.02	0.06	0.05	
211	95.0‡	0.1	4.8	0.4	0.08		0.13	
270	99.98	0.01	<0.001	0.003	<0.001	<0.001	<0.001	<0.001 Co

  † After Ref. 4a.
  ‡ Includes cobalt.

Table 11-2 Typical applications of commercial-purity base nickel alloys†

Nickel alloy	Typical applications
200	Food processing equipment; chemical shipping drums; electrical and electronic parts; aerospace and missile components; caustic handling equipment and piping; rocket motor cases; transducers.
201	Caustic evaporators; plater bars; combustion boats.
205	Support wires and rods; lead wires, base pins, getter tabs, anodes; cathode shields and rectifier tubes; magnetostrictive devices.
220	Base material for electronic receiving tubes.
230	Special electron tube applications; this alloy is essentially free of titanium.
270	High-purity nickel; cathode shanks; fluorescent lamps; hydrogen thyratron components; heat exchangers; heat shields.
211	Higher strength and base hardness than nickel 205; sparking electrodes; support wires; grid lateral winding wires; carbon holders for search lights.

† After Ref. 4a.

commercially pure nickel and other slightly alloyed or purified commercial nickel alloys.

Microstructure and Properties

The microstructure of annealed commercially pure nickel (alloy 200) is typical of that of an annealed solid solution, and is shown in Fig. 11-1. Commercially

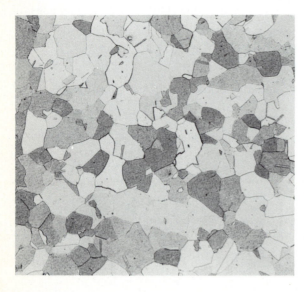

Figure 11-1 Nickel 200, cold drawn, and annealed in a continuous process at 829°C. Structure consists of a solid solution. (NaCN, $(NH_4)_2S_2O_8$; X100.) (*Courtesy of D. J. Tillack, Huntington Alloys, Inc.*)

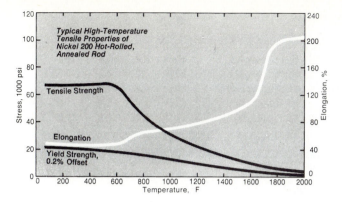

Figure 11-2 Typical high-temperature tensile properties of nickel 200 hot-rolled and annealed rod. (*After "Handbook of Nickel Alloys," Huntington Alloys, Inc., Huntington, WV, 1970.*)

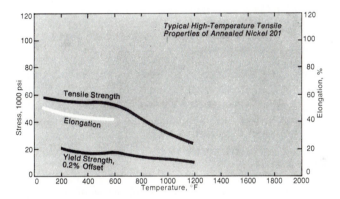

Figure 11-3 Typical high-temperature tensile properties of annealed nickel 201 alloy. (*After "Handbook of Nickel Alloys," Huntington Alloys, Inc., Huntington, WV, 1970.*)

pure nickel has good mechanical properties and excellent resistance to many corrosive environments. This alloy retains much of its strength at elevated temperatures and is tough and ductile at low temperature (Fig. 11-2).

Nickel 201 alloy is similar to nickel 200 except that the carbon content of nickel 201 is limited to 0.02 percent. This low-carbon content lowers the work-hardening rate and increases the ductility of nickel 201 so that it is more adaptable for spinning and cold-working operations (Fig. 11-3).

Nickel 270 is high-purity 99.98% Ni. This metal has excellent thermal conductivity and high ductility, which allows for heavy cold deformations without annealing. Figure 11-4 illustrates typical high-temperature tensile properties of nickel 270 alloy.

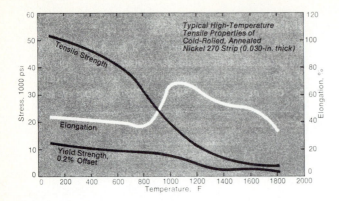

Figure 11-4 Typical high-temperature tensile properties of cold-rolled, annealed nickel 270 strip (0.030-in thick). (*After "Handbook of Nickel Alloys," Huntington Alloys, Inc., Huntington, WV, 1970.*)

11-3 NICKEL-COPPER ALLOYS (MONELS)

Chemical Compositions and Typical Applications

Nickel and copper are completely soluble in each other in all proportions, as shown in the Cu-Ni phase diagram of Fig. 6-42. However, the most important nickel-copper alloys are those containing about 67% Ni and 33% Cu, which are called *Monels*. Table 11-3 lists the nominal chemical compositions of some of the Monels along with typical applications.

Table 11-3 Nominal chemical compositions and typical applications of some Monel nickel-copper alloys†

Monel alloy	% Ni	% Cu	% Al	% Ti	Typical applications
400	66.5*	31.5			Valves and pumps; marine fixtures and fasteners; chemical processing equipment; gasoline and fresh-water tanks; boiler feed-water heaters and other heat exchangers; deaerating heaters.
404	54.5*	44.0			Waveguides; metal to ceramic seals; transistor capsules.
R-405	66.5*	31.5	0.045		Water-meter parts; screw-machine products; valve-seat inserts.
K-500	66.5*	29.5	3.0	0.6	Pump shafts and impellers; doctor blades and scrapers; oil-well drill collars and instruments; springs; valve trim.

† After Ref. 4*a*.
* Plus cobalt.

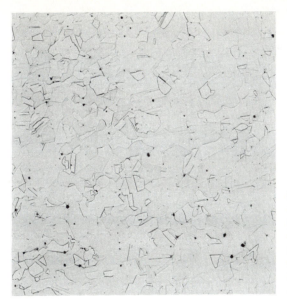

Figure 11-5 Monel alloy 400: cold drawn and annealed in a continuous process at 829°C. Structure consists of a solid solution with a few unidentified nonmetallic inclusions (black). (NaCN, $(NH_4)_2S_2O_8$; X100.) (*Courtesy of D. J. Tillack, Huntington Alloys, Inc.*)

Microstructure and Properties

Monel 400 has high strength, weldability, excellent corrosion resistance, and toughness over a wide range of temperatures. It gives excellent service in sea water under high-velocity conditions where the resistance to the effects of cavitation and erosion are important. Alloy 400 is highly resistant to corrosion by chlorinated solvents, sulfuric acid and many other acids, and practically all alkalis. Monel can be used up to about 538°C in oxidizing atmospheres, and at higher temperatures in a reducing environment. Figure 11-5 shows the annealed

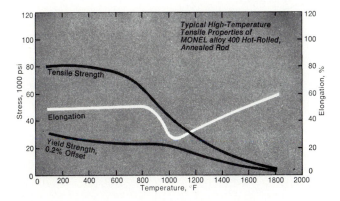

Figure 11-6 Typical high-temperature tensile properties of Monel alloy 400 hot-rolled, annealed rod. (*After "Handbook of Nickel Alloys," Huntington Alloys, Inc., Huntington, WV, 1970.*)

Figure 11-7 Monel alloy R-405: cold drawn and annealed in a continuous process at 829°C. Structure consists of a solid solution of Ni-Cu with sulfide stringers (black). (NaCN, $(NH_4)_2S_2O_8$; X250.) (*Courtesy of D. J. Tillack, Huntington Alloys, Inc.*)

microstructure of Monel 400, which is that of a Ni-Cu solid solution. Figure 11-6 gives typical high-temperature tensile properties of hot-rolled and annealed alloy 400 rod.

Monel R-405 is similar to Monel 400, but sulfur has been added to improve the machining characteristics. The microstructure of this alloy is a solid solution, but sulfide particles are also present (Fig. 11-7).

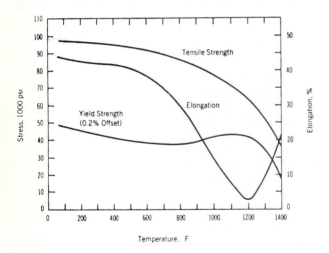

Figure 11-8 High-temperature tensile properties of Monel alloy K-500 rod (hot-rolled, as-rolled). (*After "Monel Nickel-Copper Alloys," Huntington Alloys, Inc., Huntington, WV, 1978.*)

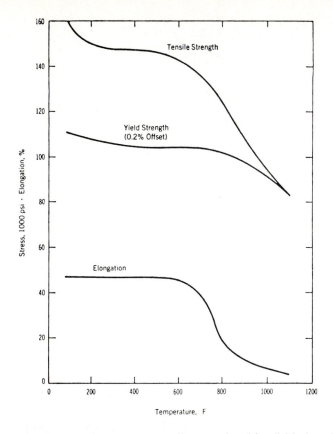

Figure 11-9 High-temperature tensile properties of hot-finished age-hardened Monel alloy K-500. (*After "Monel Nickel-Copper Alloys," Huntington Alloys, Inc., Huntington, WV, 1978.*)

Monel K-500 has the basic Ni-Cu Monel composition, but with the addition of 3.0% Al and 0.6% Ti to form age-hardening precipitates of $Ni_3(Al,Ti)$. Higher strengths are obtained when this alloy is cold-worked before aging. The effects of the precipitation-hardening heat treatment on the high-temperature tensile properties of hot-finished Monel K-500 are shown in Figs. 11-8 and 11-9.

11-4 NICKEL-CHROMIUM ALLOYS

Phase Diagram

Chromium is an important alloying element for many corrosion-resistant and high-temperature-resistant nickel-base alloys. It has a high solid solubility (approximately 30 wt% at room temperature) in nickel, as indicated in the Ni-Cr phase diagram of Fig. 11-10.

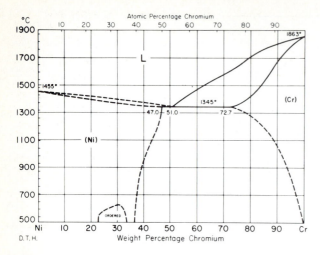

Figure 11-10 Nickel-chromium phase diagram. (*After Metals Handbook, 8th ed., vol. 8, American Society for Metals, Metals Park, OH, 1973, p. 291.*)

Chemical Compositions and Typical Applications

Table 11-4 lists the chemical compositions of some of the important nickel-chromium alloys and their typical applications.

Microstructure and Properties

Inconel 600 is a standard engineering alloy for use in some severely corrosive environments at elevated temperatures. It is essentially a Ni-Cr-Fe ternary alloy containing 15.5% Cr and 8% Fe and has a desirable combination of high strength and workability. This alloy is not heat-treatable but can be strengthened by cold working. Figure 11-11 shows the high-temperature tensile properties of annealed (870°C for 1 h) hot-rolled rod.

Inconel 600 is a stable austenitic solid solution, with the only precipitated phases present being titanium nitrides or carbides (or solutions of the two compounds called cyanonitrides) and chromium carbides. Its microstructure after solution heat treatment at 1200°C and water quenching shows straight grain boundaries relatively free of precipitates since the chromium is kept in solid solution (Fig. 11-12). After solution heat treatment and transferring directly to 870°C for 4 h and water quenching, a globular precipitate of chromium carbides can be seen in the grain boundaries and in the grains (Fig. 11-13).

Inconel 601 has the basic composition of Ni–23% Cr–14% Fe–1.4% Al and is a general-purpose engineering alloy for applications requiring heat and corrosion resistance. The high chromium content of this nickel-base alloy provides good corrosion resistance to many environments and high-temperature oxidation resistance. The aluminum content also enhances its oxidation resistance. The exceptional resistance of Inconel 601 to high-temperature oxidation is

Table 11-4 Chemical compositions and typical applications of some nickel-chromium alloys (wt%)†

Alloy	% Ni	% Cr	% Fe	% Mn	% Si	% Other	Typical applications
Inconel 600	75.0	15.5	8.0	0.5	0.2		Furnace muffles; heat exchanger tubing; chemical and food processing equipment; carburizing baskets; springs.
Inconel 601	60.5	23.0	14.1	0.5	0.2	1.4 Al	Heat-treating baskets; radiant furnace tubes; furnace muffles and retorts; thermocouple protection tubes.
Inconel 625	61.0	21.5	2.5	0.2	0.2	9.0 Mo, 3.6 Cb	Ducting systems; combustion systems; thrust reverser assemblies; fuel nozzles; after-burners; spray bars

† After "ASM Databook," *Metal Progress*, vol. 114, no. 1, mid-June 1978.

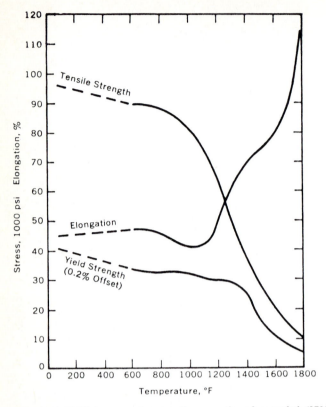

Figure 11-11 High-temperature tensile properties of annealed (870°C for 1 h) hot-rolled rod of inconel 600 alloy. (*After "Inconel Alloy 600," Huntington Alloys, Inc., Huntington, WV, 1978.*)

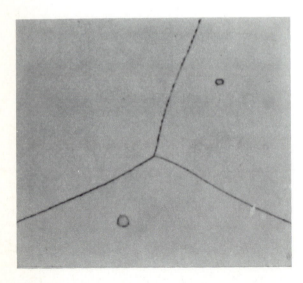

Figure 11-12 Inconel 600, solution heat-treated 1 h at 1200°C and water-quenched. Structure is a solid solution with polyhedral titanium nitride particles in grains. (Etchant: 5% nital, electrolytic; X1500.) (*After "Inconel Alloy 600," Huntington Alloys, Inc., Huntington, WV, 1978.*)

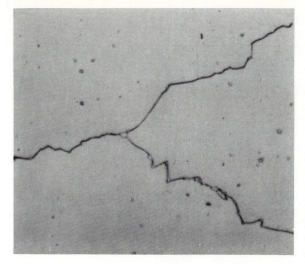

Figure 11-13 Inconel 600, solution heat-treated 1 h at 1200°C, transferred directly to 870°C for 4 h, and water-quenched. Structure shows solid solution with chromium carbide precipitates in grain boundaries and within grains; some titanium nitrides and carbides are also present. (Etchant: 5% nital, electrolytic; X1500.) (*After "Inconel Alloy 600," Huntington Alloys, Inc., Huntington, WV, 1978.*)

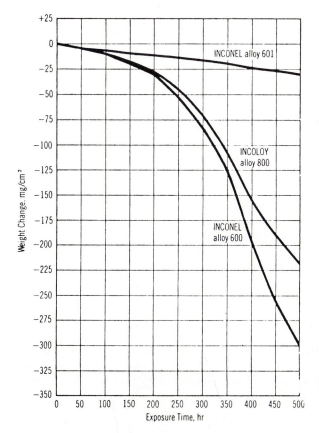

Figure 11-14 Results of oxidation tests at 1205°C. Test cycles consisted of 50 h at exposure temperature followed by air-cooling to room temperature. (*After "Inconel Alloy 601," Huntington Alloys, Inc., Huntington, WV, 1978.*)

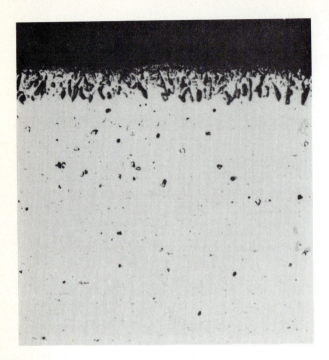

Figure 11-15 Oxide layer of Inconel 601 specimen exposed to 1205°C for 500 h. (Unetched; X75.) (*After "Inconel Alloy 601," Huntington Alloys, Inc., Huntington, WV, 1978.*)

shown in Fig. 11-14. During high-temperature exposure the aluminum, nickel, and chromium oxides form an extremely protective and adherent oxide film on the metal surface, as illustrated in Fig. 11-15.

Inconel 625 has a nominal composition of Ni–22% Cr–5% Fe–9% Mo–3.6% Cb. Its increased strength over that of alloys 600 and 601 is attributed to the solid-solution strengthening effect of the molybdenum and columbium since alloy 625 is not precipitation-hardened (Table 11-4). Its high corrosion resistance to pitting in the presence of chloride ions (e.g., sea water) is due to its high chromium content along with the 9% Mo addition. The columbium addition to the alloy makes it weldable since the carbon it contains will be combined with columbium as columbium carbide. Hence, in a welding operation, chromium carbides will not precipitate at the grain boundaries and make the alloy susceptible to intergranular corrosion.

11-5 NICKEL-BASE SUPERALLOYS

The superalloys are high-temperature heat-resistant alloys that are able to retain high strengths at high temperatures. These complex alloys also have good corrosion and oxidation resistance, and superior resistance to creep and rupture

Table 11-5 Nominal compositions and typical applications of some wrought and cast nickel-base superalloys†

Wrought alloys

Alloy	% Ni	% Cr	% Co	% Mo	% Al	% Ti	% Cb	% C	% B	% Zr	% Other	Typical applications
Inconel X-750	73	15	...	...	0.8	2.5	0.9	0.04	...		6.8 Fe	Gas turbine parts; bolts.
Udimet 500	53.6	18	18.5	4.0	2.9	2.9		0.08	0.006	0.05		Gas turbine parts; sheets; bolts.
Udimet 700	53.4	15	18.5	5.2	4.3	3.5		0.08	0.03	...		Jet engine parts.
Waspaloy	58.3	19.5	13.5	4.3	1.3	3.0		0.08	0.006	0.06		Jet engine blades.
Astroloy	55.1	15.0	17.0	5.2	4.0	3.5		0.06	0.03			Forgings for high temperatures.
René 41	55.3	19.0	11.0	10.0	1.5	3.1		0.09	0.005			Jet engine blades and parts.
Nimonic 80A	74.7	19.5	1.1	...	1.3	2.5		0.06	...			Jet engine parts.
Nimonic 90	57.4	19.5	18.0	...	1.4	2.4		0.07	...			Jet engine parts.
Nimonic 105	53.3	14.5	20.0	5.0	1.2	4.5		0.20	...			Jet engine parts.
Nimonic 115	57.3	15.0	15.0	3.5	5.0	4.0		0.15	...			Jet engine parts.

Cast alloys

Alloy	% Ni	% Cr	% Co	% Mo	% Al	% Ti	% Cb	% C	% B	% Zr	% Other	Typical applications
B-1900	64	8.0	10.0	6.0	6.0	1.0	...	0.10	0.015	0.1	4.0 Ta	Jet engine blades.
MAR-M200	60	9.0	10.0	...	5.0	2.0	1.0	0.13	0.015	0.05	12 W	Jet engine blades.
Inconel 738	61	16.0	8.5	1.7	3.4	3.4	0.9	0.12	0.01	0.10	1.7 Ta, 2.6 W	
René 77	58	14.6	15.0	4.2	4.3	3.3	...	0.07	0.016	0.04	...	Jet engine parts.
René 80	60	14.0	9.5	4.0	3.0	5.0	...	0.17	0.015	0.03	4.0 W	Turbine blade alloy.

† After "ASM Databook," *Metal Progress*, vol. 114, no. 1, mid-June 1978.

at elevated temperatures. In general, there are three main classes of superalloys: nickel base, nickel-iron base, and cobalt base. In this section, only the nickel-base superalloys will be discussed.

Chemical Compositions and Typical Applications

The earliest precipitation-hardenable nickel-base superalloy, Nimonic 80, was developed in Great Britain in 1941. Essentially this alloy is a Ni–20% Cr solid solution, with 2.25% Ti and 1% Al for forming the $Ni_3(Al,Ti)$ precipitates. Over the years, improvements in the performance of these alloys have been made possible by additions of molybdenum, cobalt, columbium, zirconium, boron, iron, and other elements. Today there are about 100 different types of wrought and cast nickel-base superalloys. Table 11-5 lists the chemical compositions and typical applications of some of them. The largest application of the superalloys is in materials for aircraft and industrial gas turbines. However, they are also used in space vehicles, rocket engines, experimental aircraft, nuclear reactors, submarines, steam power plants, petrochemical equipment, and other high-temperature applications.

Microstructure

Figure 11-16 illustrates in a general manner how the microstructure of nickel-base superalloys has developed with time. From 1944 to 1966, the stress to produce failure after 10,000 h at 870°C was raised from 5 ksi to above 30 ksi. This improvement was achieved by increasing solid-solution strengthening and precipitation hardening and by creating an optimum distribution of carbides.

The major phases present in the nickel-base superalloys are

1. γ (gamma) phase—the continuous matrix of FCC austenite
2. γ' (gamma prime) phase—the major precipitate phase
3. Carbides—various types, mainly $M_{23}C_6$ and MC, where "M" stands for a metal

Through the development years of the nickel-base superalloys, (1940 to 1970), the following trends in the change of their microstructure have been observed:

1. The volume fraction of γ' is increased.
2. The size of γ' first increased and then remained constant at about 1 μm
3. γ' became more "cubic."
4. A secondary precipitate of finely divided γ' appeared.

In the progressive development of nickel-based superalloys, some of them generated "problem" structures. Figure 11-17 illustrates and identifies cellular $M_{23}C_6$ carbides and the σ (sigma) phase as problem structures. The cellular

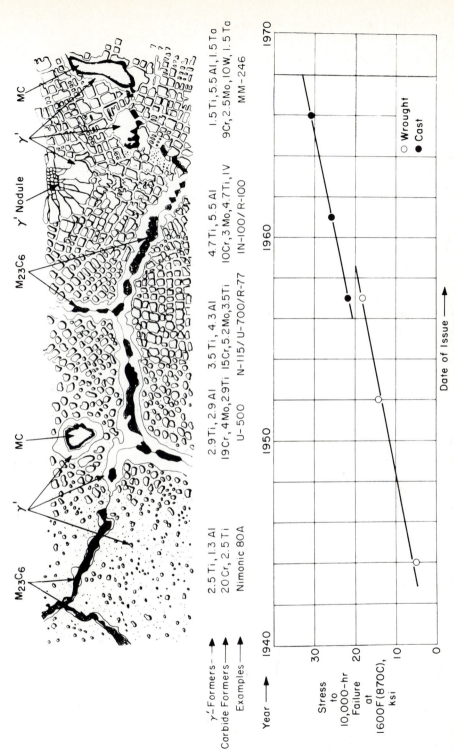

Figure 11-16 Genesis of nickel alloy microstructure, 1940 to 1970. Plot shows stress capability of the alloys as a function of approximate date of issue. Structure shown is as heat-treated for best rupture properties; major features only. Compositions are generalized and typical. (X10,000.) (*After C. T. Sims and W. C. Hagel, "The Superalloys," Wiley, New York, 1972, p. 37. Used by permission of John Wiley & Sons, Inc.*)

carbides lead to shortened rupture life, and the σ phase to low-temperature brittleness as well as shortened rupture life. Improved heat treatments have eliminated the cellular carbide problem, whereas changing the alloy chemistry has led to the elimination of the σ phase in newly developed nickel-base superalloys.

γ **Phase** The γ phase, which is a continuous matrix of nickel-base austenite, is strengthened by the addition of solid-solution elements such as chromium, molybdenum, tungsten, cobalt, iron, titanium, and aluminum. These elements differ from nickel by 1 to 13 percent in atomic diameter, as indicated in Table 11-6.

Aluminum, in addition to being a precipitation strengthener, is a potent solid-solution strengthener. Tungsten, molybdenum, and chromium also are strong solid-solution-strengthening elements. In addition to atomic-size factor, it appears that the position of the element in the periodic table affects solid-solution strengthening. An increase in the electron hole number N_v appears to reduce stacking-fault energy, and thereby makes cross slip more difficult.

At temperatures above $0.6 T_m$, which is the range of high-temperature creep, strengthening is diffusion-dependent. The slow-diffusing elements molybdenum and tungsten are the most beneficial for reducing high-temperature creep in these alloys. Cobalt, by decreasing the stacking-fault energy between partial dislocations, makes cross slip more difficult and thereby increases the high-temperature stability of these alloys.

γ' **Phase** The γ' phase can be precipitated in austenitic nickel superalloys by precipitation-hardening heat treatments. The γ' precipitate in high-nickel

Table 11-6 Difference in atomic diameter between alloying elements and nickel in nickel-base superalloys and electron hole numbers N_v†

	Difference from nickel in atomic diameter, %	Electron hole number N_v
Nickel	‡	0.66
Chromium	+3	4.66
Molybdenum	+12	4.66
Tungsten	+13	4.66
Cobalt	+1	1.71
Iron	+3	2.66
Aluminum	+6	2.66
Titanium	+9	6.66
Niobium	+18	5.66

† Electron *hole* number or electron *vacancy* number represents the average number of electron vacancies in the third subshell of the first long period.

‡ Atomic diameter of nickel = 2.491 Å.

Year →	1940	1950	1960	1970
Undesirable phase or structure	Cellular $M_{23}C_6$	None	Sigma phase	None yet
Characteristics	$M_{23}C_6$ grown in cells with γ interlayers	—	Hard TCP phase usually occurs as plates similar to μ or Laves	—
Effect on alloy	Shortened rupture life	—	Shortened rupture life; low-temperature brittleness	—
Cause	Poor control of microstructure	—	Poor control of alloy chemistry	—
Cure	Improve heat treatments	—	Control chemistry through "Phacomp"	—

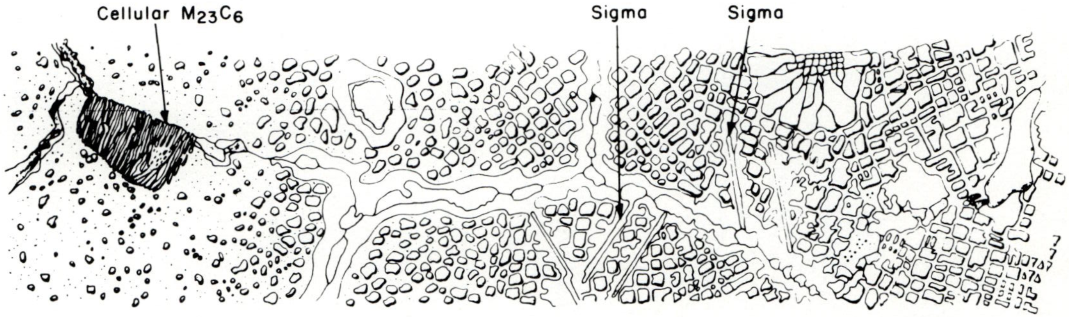

Cellular $M_{23}C_6$

Sigma   Sigma   Sigma

γ'-Formers →	2.5 Ti, 1.3 Al	2.9 Ti, 2.9 Al	3.5 Ti, 4.3 Al	4.7 Ti, 5.5 Al	1.5 Ti, 5.5 Al, 1.5 Ta
Carbide Formers →	20 Cr, 2.5 Ti	19 Cr, 4 Mo, 2.9 Ti	15 Cr, 5.2 Mo, 3.5 Ti	10 Cr, 3 Mo, 4.7 Ti, 1 V	9 Cr, 2.5 Mo, 10 W, 1.5 Ta
Examples →	Nimonic 80A	U-500	N-115/U-700/R-77	N-100/R-100	MM-246

Figure 11-17 Common unwanted phases or structures in nickel alloys, 1940 to 1970. (X10,000.) (*After C. T. Sims and W. C. Hagel, "The Superalloys," Wiley, New York, 1972, p. 39. Used by permission of John Wiley & Sons, Inc.*)

Figure 11-18 Astroloy forging, solution heat-treated 4 h at 1150°C, air-cooled, aged at 1079°C for 4 h oil-quenched, aged at 843°C for 4 h, air-cooled, aged at 760°C for 16 h, air-cooled. Intergranular gamma prime precipitated at 1079°C, fine gamma prime at 843 and 760°C. Carbide particles are also at grain boundaries. Matrix is gamma. (Electrolytic: H_2SO_4, H_3PO_4, HNO_3; X10,000.) (*After Metals Handbook, 8th ed., vol. 7, American Society for Metals, Metals Park, OH, 1972, p. 171.*)

matrices is of the FCC A_3B-type compound. The "A" is composed of relatively electronegative elements such as Ni, Co, and Fe, and the "B" of electropositive elements such as Al, Ti, or Cb. Typically, in a nickel-base superalloy, γ' is $Ni_3(Al,Ti)$, but if cobalt is added it can substitute for some nickel as $(Ni,Co)_3$ (Al,Ti).

Since the nickel atom is relatively incompressible owing to its $3d$ electron state, a high nickel matrix favors the precipitation of γ', which has only about 0.1 percent mismatch with γ. Thus, γ' can nucleate homogeneously with low surface energy and have extraordinary long-term stability. The coherency between γ' and γ is maintained by a tetragonal distortion.

Since γ' (Ni_3Al,Ti) shows long-range order, both superlattice and antiphase boundary (APB) faults occur as the result of shear.[1] Thus, by dislocation interaction APB strengthening occurs in $\gamma - \gamma'$ alloys. Since the degree of order in $Ni_3(Al,Ti)$ increases with temperature, alloys with a high-volume fraction of γ' show a remarkable *increase* in strength with increasing temperature up to about 800°C.

The γ/γ' mismatch determines the γ' particle morphology. With small mismatches (~ 0.05 percent), γ' occurs as spheres. As the mismatch increases, γ' occurs in the cube form with {100} interfaces (Fig. 11-18). Above 1.25 percent mismatch, the γ' occurs as semicoherent plates.

Carbides

Role of carbides in nickel-base heat-resistant alloys The carbon content of nickel-base superalloys varies from 0.02 to about 0.2 percent for wrought alloys and up to about 0.6 percent for cast alloys. Metallic carbides form in the grain

---

[1] B. H. Kear, G. R. Leverant, and J. M. Oblak, *Trans. ASM* 62(1969):639.

boundaries and within the grains. Since carbides are harder and more brittle than the alloy matrix, their distribution along the grain boundaries will affect the high-temperature strength, ductility, and creep properties of the nickel-base heat-resistant alloys. Thus, there is an optimum amount and distribution of carbides along grain boundaries.

If there are no carbides along the grain boundaries, voids will coalesce along them during high-temperature deformation and excess grain boundary sliding will take place. On the other hand, if continuous chains of carbides extend along the grain boundaries, continuous fracture paths will be formed, with resulting low-impact properties. Grain boundary sliding will thus be inhibited and, as a result, excessive stresses will build up and lead to premature fracture. A discontinuous chain of carbides along the grain boundaries is the optimum condition since carbides in this form will hinder grain boundary cracking and at the same time will not restrict ductility due to deformation in the grain boundary region.

Types of carbides The common types of carbides which are formed in the nickel-base superalloys are MC, $M_{23}C_6$, and M_6C.

MC *carbides* are monocarbides and have the general formula MC, where "M" stands for metallic elements such as *titanium, tantalum, columbium,* or *tungsten*. These carbides are very stable and are believed to be formed just below the temperature where solidification begins. They dissolve with difficulty in the solid phase during solution heat treatment and restrict grain growth.

In $M_{23}C_6$ *carbides,* the "M" is usually *chromium,* but this element can be replaced by iron and to a smaller extent by tungsten, molybdenum, or cobalt, depending on the alloy. $M_{23}C_6$ carbides form during lower-temperature heat treatments and service in the temperature range 760 to 980°C. They can form either from the degeneration of MC carbides or from soluble carbon in the alloy matrix, and usually precipitate in the grain boundaries. $M_{23}C_6$ carbides have a complex cubic structure.

M_6C *carbides* form at temperatures in the range 815 to 980°C. They are similar to the $M_{23}C_6$ carbides and have a tendency to form when the *molybdenum* and *tungsten* contents of the base alloy are high. They are a similar carbide to the $M_{23}C_6$ and also have a complex cubic structure. When the nickel-base alloy contains more than about 6 to 8% Mo or W (e.g., in M252 and René 41), M_6C will form along with $M_{23}C_6$ in the grain boundaries.

Topologically close-packed (TCP) phases In nickel-based superalloys in which the composition has not been properly controlled, TCP phases can form either during heat treatment or during service. The most important of these are σ, μ, and laves or χ. These phases, which usually form as thin plates parallel to the $\{111\}_\gamma$, can lead to low rupture strengths and loss in rupture ductility. Figure 11-19 shows Widmanstätten σ which has been developed in alloy U-700 after holding 2500 h at 871°C.

Figure 11-19 U-700 given the standard commercial heat treatment and then held at 871°C for 2500 h. Structure shows acicular sigma in Widmanstätten pattern, which has undesirable effects on high-temperature tensile properties. (Kalling's reagent 2; X500.) (*After Metals Handbook, 8th ed., vol. 7, American Society for Metals, Metals Park, OH, 1972, p. 170.*)

Another undesirable effect of TCP phases is that they change the chemical balance of the nickel-base alloys by removing refractory elements such as Cr, Mo, and W so that solid-solution strengthening is reduced as well as the γ/γ' mismatch. TCP phases are now generally avoided in superalloys by using an alloy design technique known as *phase composition* (Phacomp), which was developed and used by Sims (Ref. 1, p. 274) and others. This technique essentially computes the average electron hole number N_v for the γ matrix. For example, if σ is assumed to form directly from γ, an N_v number between 2.45 to 2.50 is supposed to indicate that the alloy will not be susceptible to the formation of σ or other TCP phases. Phacomp has greatly facilitated the alloy design of commercial superalloys since the tendency of all alloying elements to form TCP phases is incorporated into the calculations.

Microstructure of Inconel X-750 Inconel X-750 alloy is a precipitation-hardenable nickel-base superalloy which is used for its corrosion and oxidation resistance and high-temperature strength (up to about 706°C). Although much of its strength is lost above 700°C, enough is retained to make the alloy useful up to 980°C.

Figure 11-20 Microstructures of Inconel X-750 nickel-base superalloy after different heat treatments. (*a*) Inconel X-750, solution annealed by holding at 1150°C and air cooling. Structure consists of gamma solid solution. (*b*) Alloy is solution annealed, air-cooled, aged at 843°C for 24 h and at 704°C for 24 h. Structure consists of a high-density fine precipitate of gamma prime; the grain boundary precipitate has been stabilized also. (This is a standard heat treatment for this alloy.) (*c*) Alloy is solution annealed, air-cooled and aged at 816°C for 24 h. Structure shows fine and uniformly dispersed gamma prime precipitate; large discontinuous particles are grain boundary carbide ($M_{23}C_6$). (*d*) Alloy is solution annealed, air-cooled, and overaged by holding 24 h at 927°C. Gamma prime is coarse, overaged, and not uniformly dispersed. The large particle at the lower left is MC carbide. (Glyceregia etched; replica electron micrographs; X15,000.) (*After Metals Handbook, 8th ed., vol. 7, American Society for Metals, Metals Park, OH, 1972, p. 166.*)

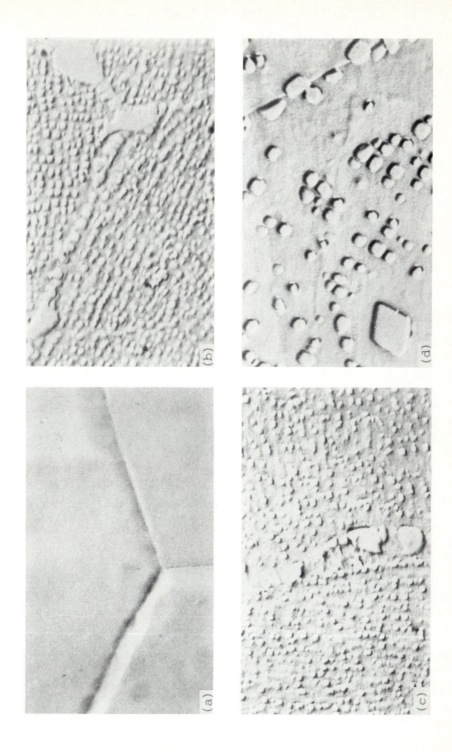

(b)

(d)

(a)

(c)

479

The microstructure of Inconel X-750 is shown in Fig. 11-20 for four heat treatments. When the alloy is solution-heat-treated and air-cooled, the γ' and carbides are retained in solid solution, as shown in Fig. 11-20a. If, after solutionizing and air cooling, this alloy is given a *double-aging treatment*, first at 843°C for 24 h and then at 704°C for 24 h, which is the standard procedure, a relatively fine, dense precipitate of γ' particles is produced and the grain boundary carbide ($M_{23}C_6$) is stabilized (Fig. 11-20b). However, if this alloy is solutionized, air-cooled, and just given a *single-aging treatment* at 816°C for 24 h, the γ' precipitate is fine and evenly distributed but large, discontinuous carbide precipitates form at the grain boundaries (Fig. 11-20c). Thus, without the second aging treatment the carbide precipitation in the grain boundaries cannot be controlled.

Finally, if the Inconel X-750 alloy is solutionized, air-cooled, and *overaged* by aging at 927°C for 24 h, the γ particles are not uniformly dispersed and are coarse (Fig. 11-20d). Such a structure has poor mechanical properties at elevated temperatures and is undesirable.

High-Temperature Stress-Rupture Properties

In general, the nickel-base superalloys are used in the 760 to 980°C temperature range. Table 11-7 lists the high-temperature stress-rupture values of selected

Table 11-7 Rupture strengths of wrought and cast nickel-base superalloys at 650, 815, and 982°C†

| | Characteristic rupture strengths, ksi | | | | | |
| | 650°C | | 815°C | | 982°C | |
Alloy	100 h	1000 h	100 h	1000 h	100 h	1000 h
			Wrought			
Inconel X-750	80	68	26	16	3.5	
Udimet 500	· · ·	· · ·	44	32		
Udimet 700	· · ·	102	58	43	17	8
Waspaloy A	108	88	40	25	6.8	
Astroloy	· · ·	112	59	42	15	8
René 41	110	102	45	29	10	
			Cast			
B-1900	· · ·	· · ·	73	55	26	15
MAR-M200	· · ·	· · ·	76	60	27	19
IN-100	· · ·	· · ·	73	55	25	15
IN 738	· · ·	· · ·	76	52	26	14
MAR-M246	· · ·	· · ·	82	65	27	18

† After "ASM Databook," *Metal Progress*, vol. 114, no. 1, mid-June 1978.

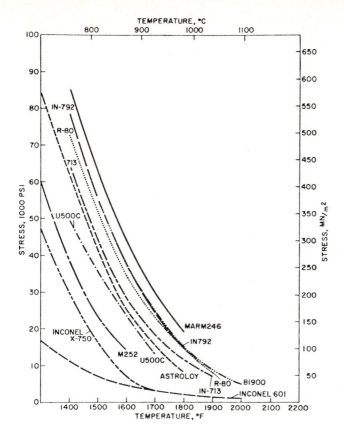

Figure 11-21 Stress-versus-temperature curves for rupture in 1000 h for selected nickel-base alloys. (*After C. T. Sims and W. C. Hagel, "The Superalloys," Wiley, New York, 1972, p. 592. Reproduced by permission of John Wiley & Sons, Inc.*)

nickel-base superalloys at 650, 815, and 968°C for 100 and 1000 h in test. Figure 11-21 shows the stress rupture in 1000 h versus temperature curves for selected nickel-base superalloys. It should be noted that the cast alloys maintain the highest strengths at the higher temperatures. For example, MAR-M246 casting alloy has a rupture strength of 18 ksi after 1000 h at 982°C, which is the highest stress value at this temperature for all the alloys shown.

Effect of Heat Treatment on Stress-Rupture Properties

Wrought alloys Heat treatment can affect the rupture properties of nickel-base superalloys, as has previously been discussed. Early heat treatments for wrought alloys such as Nimonic 80A and M-252 consisted principally of only a high-temperature solution heat treatment followed by a low-temperature age. This provided good tensile strengths and short-time rupture properties, but did not

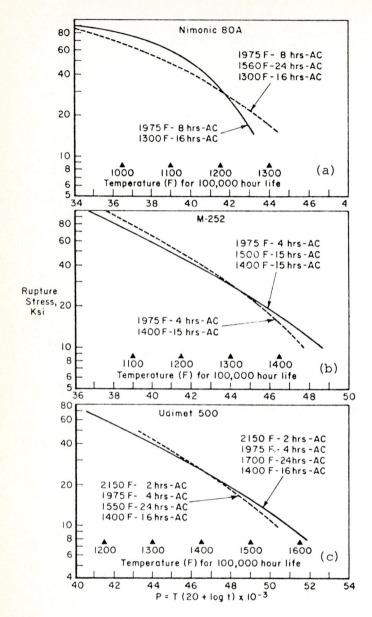

Figure 11-22 Effect of inserting intermediate aging on rupture properties of several wrought nickel-base superalloys. (*After C. T. Sims and W. C. Hagel, "The Superalloys," Wiley, New York, 1972, p. 70. Reproduced by permission of John Wiley & Sons, Inc.*)

sufficiently stabilize the structure to produce optimum long-time rupture properties. By adding another intermediate temperature age, the long-time rupture strengths were substantially increased (Fig. 11-22a). The extra age drives the MC reaction

$$MC + \gamma \rightarrow M_{23}C_6 + \gamma'$$

forward so that grain boundaries of coarse particles in $M_{23}C_6$ carbides are formed engloved in a layer of γ'. The intermediate aging treatment was also effective in extending the long-time rupture properties of alloys M-252 and Udimet 500, as shown in Fig. 11-22b and c.

Cast alloys When cast nickel-base superalloys were first developed, they were given simple heat treatments. They were cooled in their investments and then aged at approximately 760°C for about 12 h to fully develop the γ' phase. Today, cast superalloys used in industrial turbines and jet engines are given multistage heat treatments to homogenize their structure and to increase their strength and ductility.

To illustrate the effects of heat treatment on improving rupture properties, the following comparison of two heat treatments is given for René 77. Treatments A and B listed below were given René 77 alloy. The difference in structure and rupture properties are shown in Fig. 11-23.

Treatment A	Treatment B
1160°C, 2 h, furnace cool to 1085°C, air cool to room temperature	1160°C, 4 h, air cool
	1085°C, 4 h, air cool
	925°C, 24 h, air cool
760°C, 16 h, air cool	760°C, 16 h, air cool

In heat treatment B, by rapidly cooling from solution heat treatment at 1160°C, γ' particles nucleate but do not grow. However, by aging 4 h at 1085°C, γ' grows, producing a large number of medium-to-large homogeneously nucleated particles (Fig. 11-23). The lower-temperature ages at 925 and 760°C produce the fine "background" γ'. Treatment A, on the other hand, by slow cooling from 1160 to 1085°C, results in a coarse, unevenly distributed γ', which in turn results in lower rupture strengths.

Hot Corrosion[1]

Hot corrosion, which is sometimes called *sulfidation*, may be defined as an accelerated, often catastrophic, surface attack of superalloy hot-gas-path components. This corrosive attack is particularly severe in the temperature range 760 to

---

[1] A. M. Beltram and D. A. Shores, in Ref. 1, chap. 11, "Hot Corrosion."

Heat Treatment A	Heat Treatment B
2125F (1160C), 2hrs, furnace cool to 1975F (1085C), air cool to R.T. 1400F (760C), 16 hrs, air cool	2125 F (1160C), 4 hrs, air cool 1975F (1085C), 4 hrs, air cool 1700F (925C), 24 hrs, air cool 1400F (760C), 16 hrs, air cool

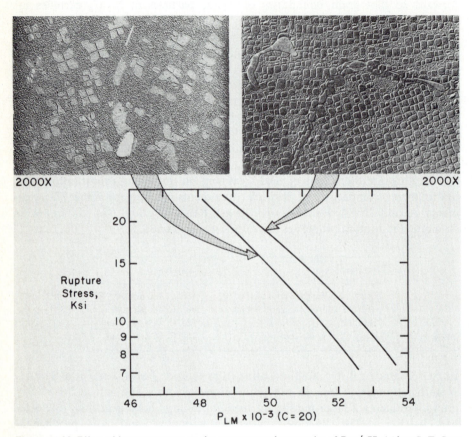

Figure 11-23 Effect of heat treatment on the structure and properties of René 77. (*After C. T. Sims and W. C. Hagel, "The Superalloys," Wiley, New York, 1972, p. 72. Reproduced by permission of John Wiley & Sons, Inc.*)

1000°C, and often affects aircraft turbine engines and industrial gas turbines. It is believed that the presence of condensed alkali metal salts, namely, Na_2SO_4, is a prerequisite for hot corrosion.

The sources of the alkali metal salts for hot corrosion could be

1. The direct ingestion of sea salt in a marine environment
2. The formation of Na_2SO_4 during the combustion of fuels containing both sodium and sulfur
3. The formation of Na_2SO_4 during combustion from sodium-contaminated dust in the air and sulfur from the fuel

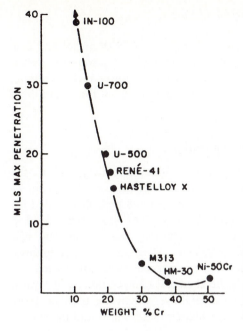

Figure 11-24 Idealization of the effect of chromium on the hot-corrosion resistance of nickel-base alloys in burner rig tests. (*After C. T. Sims and W. C. Hagel, "The Superalloys," Wiley, New York, 1972, p. 332. Reproduced by permission of John Wiley & Sons, Inc.*)

Considerable effort has been expended to develop nickel-base superalloys with improved hot-corrosion resistance, and alloying elements have been adjusted to improve it. The effect of chromium is dominant in establishing hot-corrosion resistance in nickel-base superalloys, as is shown in Fig. 11-24. For those alloys, Sims et al.[2] found that at least 15% Cr was needed to produce reasonably good resistance. The practical limitation for the addition of chromium to superalloys is the solubility limit of σ or other TCP phases.

Aluminum has been found to be detrimental for hot-corrosion resistance, while titanium or a high Ti-to-Al ratio is slightly beneficial. Thus, many contemporary nickel-base superalloys that have good hot-corrosion resistance have a high titanium-to-aluminum ratio, as shown below (IN 738 is an exception).

Alloy	% Cr	% Ti	% Al	Ti : Al ratio
IN 738	16	3.4	3.4	1
MAR-M432	15.5	4.3	2.8	1.5
Udimet 710	18	5	2.2	2.3
René 80	14	5	3	1.7

The increase in hot-corrosion resistance of the above alloys is attributed to the formation of protective surface oxides such as TiO_2 and Cr_2O_3.

---

[2] C. T. Sims, P. A. Bergman, and A. M. Beltran, ASME Preprint 69-GT-16, March 1969.

11-6 NICKEL-IRON-BASE SUPERALLOYS[1]

Nickel-base superalloys containing substantial amounts of both nickel and iron form a second important class of superalloys. In these alloys, lower-cost iron is substituted in part for nickel. However, because of their lower nickel content, they are not able to be utilized at as high temperatures as the nickel-base superalloys.

Chemical Compositions and Typical Applications

Knowledge of the stainless steels and the nickel-base superalloys led to the development of the nickel-iron-base superalloys. Most of them contain from 25 to 45% Ni and from 15 to 60% Fe. Chromium from 15 to 28 percent is added for oxidation resistance at elevated temperatures, while 1 to 6% Mo is also added to most of them for solid-solution strengthening. Titanium, aluminum, and columbium are added to combine with nickel for strengthening precipitates. Carbon, boron, zirconium, cobalt, and some other elements are added for various complex effects. Table 11-8 lists the chemical compositions and typical applications for selected nickel-iron-base superalloys, which are used in many gas turbine engines and in steam turbines for blades, disks, shafts and fasteners.

Microstructure

The austenite matrix Most nickel-iron-base superalloys are designed so that they have an austenitic FCC matrix. Since they contain less than 0.1% C and relatively large amounts of ferrite stabilizers such as chromium and molybdenum, the minimum level of nickel required to maintain an austenitic matrix is about 25 wt%. Additions of cobalt or other austenite stabilizers can slightly lower this nickel level. High-nickel contents are associated with higher useful temperatures and improved stability, but also with higher cost. High-iron contents lower the cost and improve malleability, but also considerably lower the oxidation resistance of these alloys.

Solid-solution strengtheners The solid-solution-strengthening elements added to nickel-iron superalloys are 10 to 25% Cr, 0 to 9% Mo, 0 to 5% Ti, 0 to 2% Al, and 0 to 7% Cb. Of these, molybdenum is the most useful. It expands the nickel-iron γ matrix and also enters carbides and γ'. (Molybdenum has a +12 percent difference in atomic diameter from nickel.)

Chromium is also a solid-solution strengthener of the γ matrix and enters the γ'. However, its chief function is to provide oxidation resistance. Columbium, titanium, and aluminum also provide some solid-solution strengthening of the austenite matrix, but this is not their primary function in nickel-iron

---

[1] D. R. Muzyka, in Ref. 1, chap. 4, p. 113.

Table 11-8 Chemical compositions and typical applications of nickel-iron-base superalloys[†]

Alloy	% Ni	% Fe	% Cr	% Mo	% Al	% Ti	% Mn	% Si	% C	% Other	Typical applications
Inconel 706	41.5	40.0	16.0	0.5	0.2	1.75	0.2	0.2	0.03	2.9 Cb, 0.5 Co	Gas turbine components.
Inconel 718	53.0	18.5	18.6	3.1	0.4	0.9	0.2	0.3	0.04	5.0 Cb	Jet engines, rocket motors.
Incoloy 800	32.5	44.5	21	...	0.4	0.4	0.8	0.5	0.05	0.4 Cu	Furnace, heat exchanger parts.
Incoloy 801	32	46	20.5	...	...	1.1	0.8	0.5	0.05	0.2 Cu	Heat exchangers.
Incoloy 901	42.5	36.0	12.5	5.7	0.2	2.8	0.1	0.1	0.05	0.015 B	Gas turbine rotors, blades, bolts.
Pyromet 860	43.0	30.0	12.6	6.0	1.2	3.0	0.05	0.05	...	4.0 Co, 0.01 B	Turbine engine parts.
A-286	26.0	53.6	15.0	1.3	0.2	2.0	1.3	0.5	0.05	0.015 B	Gas turbine parts, blades bolts.
Discaloy	26.0	54.3	13.5	2.7	0.1	1.7	0.9	0.8	0.04	0.005 B	Gas turbine parts, bolts.
V-57	27.0	52.0	14.8	1.2	0.2	3.0	0.3	0.7	0.08	0.01 B, 0.5 V	Jet engine rotors.

† After "ASM Databook," *Metal Progress*, vol. 114, no. 1, mid-June, 1978.

base alloys. Small amounts of carbon and boron are also potent solid-solution strengtheners.

Precipitation strengtheners The most important precipitation strengtheners in nickel-iron-base alloys are titanium, aluminum, and columbium since they combine with nickel to form intermetallic phases. The most important precipitating phases that occur in nickel-iron-base alloys are given here.

Phase	Composition	Structure
γ'	$Ni_3(Al,Ti)$	Ordered FCC
γ''	Ni_3Cb	Ordered BCT
η	Ni_3Ti	HCP
δ	Ni_3Cb	Orthorhombic

An important difference in the structure of γ'- and γ''-strengthened nickel-iron-base superalloys from the nickel-base alloys is that the Ni-Fe alloys are all susceptible to the precipitation of one or more secondary phases such as η, δ, μ, or laves. These phases can be detrimental *or* beneficial to rupture properties, depending on their morphology and distribution.

Titanium is the major γ'-forming element in γ'-strengthened nickel-iron superalloys, while in contrast most nickel-base superalloys are strengthened principally by aluminum-rich γ'. Aluminum, however, does provide some oxidation resistance to nickel-iron alloys. Columbium is the principal γ''-forming element in γ''-strengthened nickel-iron-base superalloys.

Structure of Inconel 901 and 718 alloys

Inconel 901 Inconel 901 is an example of a nickel-iron-base superalloy which is strengthened by ordered FCC γ'. When this alloy is solution-heat-treated for 2 h at 1066°C, water-quenched and aged for 2 h at 802°C, and then air-cooled and aged for 24 h at 732°C, a fine precipitate of γ' is developed in the γ matrix (Fig. 11-25a and b). Extended service exposure at 650 to 760°C can cause some γ' ordered FCC to transform to η HCP phase (Ni_3Ti), which is the needlelike precipitate shown in Fig. 11-25c.

For γ'-strengthened nickel-iron alloys, the titanium content is higher than the aluminum since higher titanium levels create higher strengths while they minimize the tendency to form unwanted phases. The antiphase boundary energy also increases in the γ' when the titanium-to-aluminum ratio is high. For example, in alloy 901, there is 2.5% Ti with only 0.2% Al. It is believed that γ' is coherent with the matrix but that coherency strains are not a major source of strength. The antiphase boundary energy, which makes it difficult for dislocations to pass through the γ', is believed to be the main precipitation-strengthening mechanism.

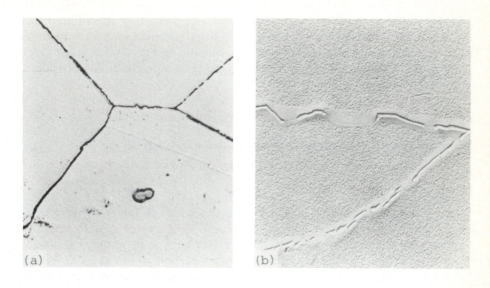

Figure 11-25 Microstructure of Inconel 901 after (*a*) and (*b*) precipitation strengthening and (*c*) after being creep-tested to rupture at 20,000 psi for 7380 h at 732°C. Note the eta phase (Ni$_3$Ti HCP) needles that have been developed during the creep test. (*a*) Alloy was solution heat-treated 2 h at 1066°C, water-quenched, aged 2 h at 802°C, air-cooled, aged 24 h at 732°C, air-cooled. Structure is a gamma matrix; grain boundary envelope; and large particles are MC carbide. (HCI, H$_2$O (1:1); X1000.) (*b*) Same heat treatment as *a* but at higher magnification. The grain-boundary constituents (MC) contributed to low ductility. The gamma matrix contains gamma-prime precipitate. (Replica electron micrograph; Electrolytic: H$_2$SO$_4$, H$_3$PO$_4$, HNO$_3$; X10,000.) (*c*) Inconel 901 creep-tested to rupture at 20,000 psi for 7380 h at 732°C. Needlelike constituent is eta phase (Ni$_3$Ti) while the remainder of microstructure consists of gamma prime in a gamma matrix. (Replica electron micrograph; Glyceregia; X15,000.) (*After Metals Handbook, 8th ed., vol. 7, American Society for Metals, Metals Park, OH, 1972, p. 161.*)

Inconel 718[1] Inconel 718 is an example of a nickel-iron-base superalloy that is strengthened by columbium-rich γ' (Ni_3Cb,FCC) precipitates. Some aluminum and titanium atoms may substitute for the columbium. This type of precipitate is in contrast to that found in other nickel-iron-base superalloys in which the γ' precipitate is $Ni_3(Ti,Al)$.

According to Barker et al.,[1] FCC γ' is the main phase which is initially present in the matrix of alloy 718 heat-treated in the standard precipitation-strengthened condition.[2] The γ' particles were found to be 75 to 300 Å in size and were both spherical and disklike in morphology, as shown in Fig. 11-26. When the samples of alloy 718 were exposed for long periods of time at elevated temperatures, the γ' phase transformed into a BCT phase of uncertain composition designated Ni_xCb. Upon even longer exposure times, part of the Ni_xCb phase transformed into orthorhombic Ni_3Cb, which is lamellar (needlelike). After prolonged exposure in the 650 to 700°C range, three distinct structural shapes were identified: spheroids, small plates, and large plates (Fig. 11-27). X-ray diffraction analysis identified the spherical precipitates as FCC γ' (Fig. 11-27*a*), the BCT Ni_xCb as the small plates, and orthorhombic Ni_3Cb as the large plates (Fig. 11-27*b*).

High-Temperature Stress-Rupture Properties

In general, the nickel-iron-base superalloys cannot be used at as high temperatures as the nickel-base alloys. Nickel-iron-base alloys that are strengthened by ordered FCC γ' (such as A-286 and V-57, which contain about 25 to 26 wt% Ni) can be used to about 650°C, while alloys which have higher nickel contents (such as 860 and 901, with 42 to 43 wt%) can be used to about 815°C. Inconel 706 and 718, which are strengthened by a columbium-containing γ', can be used to about 650°C. Table 11-9 lists the rupture strength of selected nickel-iron-base superalloys at 650, 735, and 815°C for times of 100 and 1000 h.

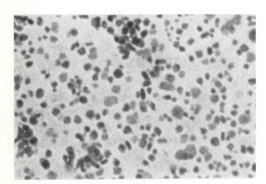

Figure 11-26 Electron micrograph of extracted gamma prime particles from Inconel 718 alloy in the fully precipitation strengthened condition. (X130,000.) (*After J. F. Barker, E. W. Ross, and J. F. Radavich, "Long-Time Stability of Inconel 718," J.Metals, Jan. 1970, p. 40.*)

---

[1] Ref. 10.
[2] Solution heat treatment: 982°C at 2 h, then air cool. Aging treatment: 720°C at 8 h, then furnace cool at 50°C per hour to 620°C, hold 8 h, and air cool.

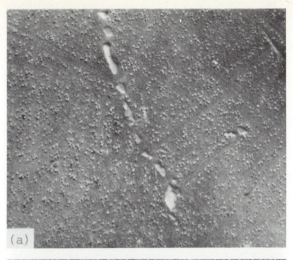

(a)

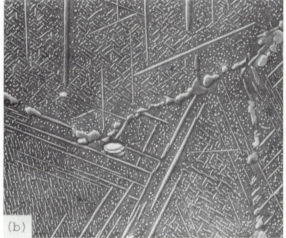

(b)

Figure 11-27 Electron micrographs of Inconel 718 sample exposed at 705°C at 37 ksi for 6,048 h. (*a*) Immersion etched in 20% HCl-methanol. (Selective gamma prime etch.) (*b*) Electrolytically etched at 2 V in a chromic-phosphoric sulfuric solution. (*After J. F. Barker, E. W. Ross, and J. F. Radavich, "Long-Time Stability of Inconel 718," J. Metals, Jan. 1970, p. 40.*)

Table 11-9 Characteristic rupture strengths of some iron-nickel-base superalloys†

| | Rupture strengths, ksi | | | | | |
| | 650°C | | 735°C | | 815°C | |
Superalloy	100 h	1000 h	100 h	1000 h	100 h	1000 h
Inconel 706	101	84				
Inconel 718	102	86				
Incoloy 800	32	23	13	9.3	9.2	6.0
Incoloy 801	40		20		10	
Incoloy 901	80	64	49	31	19	11
Pyromet 860	95	81	60	45	33	17
A-286	61	46	35	21	13	8.0
Discaloy	52	41	30	20	15	
V-57	85	70	50	29		

† After "ASM Databook," *Metal Progress*, vol. 114, no. 1, mid-June, 1978.

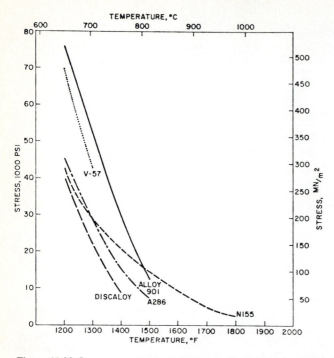

Figure 11-28 Stress-versus-temperature curves for rupture in 1000 h for selected nickel-iron base superalloys. (*After C. T. Sims and W. C. Hagel, "Superalloys," Wiley, New York, 1972, p. 595. Reproduced by permission of John Wiley & Sons, Inc.*)

Figure 11-28 shows the stress-versus-temperature curves for rupture after 1000 h for selected nickel-iron-base superalloys. It should be noted that these alloys rupture at considerably lower stress than the nickel-base superalloys. The lower nickel content of the nickel-iron alloys is mainly responsible for their decreased high-temperature strengths.

11-7 COBALT-BASE SUPERALLOYS

Elemental cobalt is a transition metal of the fourth period immediately preceding nickel and having the atomic number 27. Cobalt has many physical properties similar to nickel such as atomic size, melting point, and density (Table 11-10).

Table 11-10 Comparison of some physical properties of cobalt and nickel

	Atomic diameter, Å	Melting point, °C	Density, g/cm³	Crystal structure
Cobalt	2.497	1495	8.85	HCP $\overset{417°C}{\rightleftharpoons}$ FCC
Nickel	2.491	1453	8.90	FCC

At room temperature cobalt has a HCP crystal structure, but at 417°C it undergoes an allotropic transformation and changes to a FCC structure. By alloying cobalt with chromium, nickel, tungsten, carbon, and other elements, complex cobalt-base superalloys have been progressively developed since they first began to be used in gas turbine engines about 1943.

Chemical Compositions and Typical Applications

Table 11-11 lists the chemical compositions and typical applications for some selected cobalt-base superalloys. As a class, cobalt-base superalloys are much less chemically complex than the nickel-base alloys. Cast cobalt-base superalloys contain about 50 to 60% Co, 20 to 30% Cr, 5 to 10% W, and 0.1 to 1% C. The balance of these alloys is made up of nickel, tantalum, iron, columbium, and other elements. Wrought cobalt-base superalloys contain about 40% Co and increased nickel (about 20 percent) for workability, as well as many other alloy additions.

Cobalt-base superalloys are used for some industrial turbine parts because they are less subject to hot corrosion than the nickel-base alloys even though their oxidation resistance is not as good. Also, cobalt-base superalloys have very flat stress-rupture, time-temperature properties, which make them valuable for long-lived static parts that are at relatively low stresses and high temperatures. For this reason, they are predominant in nozzle-guide vane-partition applications for industrial turbines and for some aircraft engines.

Microstructure

The structure of cobalt-base superalloys is simpler than those of the nickel and nickel-iron types. It is common in cobalt-base superalloys to find a microstructure of only a FCC γ matrix and carbides of various types. Strengthening in cobalt-base superalloys is obtained primarily through a combination of solid-solution strengthening and carbide precipitation.

Austenitic matrix The austenitic matrix of most cobalt superalloys consists of about 50% Co and about 25% Cr, with the balance mostly nickel and refractory elements such as tungsten, tantalum, iron, or molybdenum. The austenite of cobalt-base superalloys has a FCC structure.

However, since unalloyed cobalt transforms from the FCC to HCP structure at 417°C, there is a tendency for FCC stabilized cobalt alloys to form stacking faults of HCP structure in the FCC matrix. Thus, in cobalt superalloys, which are normally FCC, smaller layers or volumes of HCP structure are created by stacking faults. It can be assumed that, for cobalt alloys, the existence of stacking faults indicates there is a tendency for the FCC structure to transform to the HCP one.

The tendency to form stacking faults, in turn, can itself be related to the stacking-fault energy,[1] and is affected by alloying elements in cobalt-base

---

[1] P. S. Kotval, *Trans. AIME* 236(1968):519.

Table 11-11 Chemical compositions and typical applications of cobalt-base superalloys†

Alloy	% Co	% Cr	% W	% Ni	% C	% Fe	% Ta	% Other	Typical applications
								Cast	
X-40‡	54.0	25.5	7.5	10.5	0.50	1.5	···	0.5 Si; 0.5 Mn	Gas turbine parts; nozzle vanes.
FSX-414	52.0	29.0	7.5	10.0	0.25	1.0	···	0.01 B	Gas turbine vanes.
MAR-M302	58.0	21.5	10.0	···	0.85	···	9.0	0.005 B, 0.20 Zr	Jet engine blades, vanes.
MAR-M322	61.0	21.5	9.0	···	1.00	···	4.5	0.75 Ti, 2.25 Zr	Jet engine blades, vanes.
MAR-M509	55.0	23.5	7.0	10.0	0.6	···	3.5	0.2 Ti; 0.50 Zr	Jet engine blades, vanes.
AR-213	66.0	19.0	4.7	···	0.18	···	6.5	3.5 Al, 0.15 Zr, 0.1 Y	Sheets, tubing, resistant to hot corrosion.
								Wrought	
S-816‡	42.0	20.0	4.0	20.0	0.38	4.0	···	4.0 Mo, 4.0 Cb, 1.2 Mn, 0.4 Si	Gas turbine blades, bolts, springs.
HS 188	39.2	22.0	14.0	22.0	0.10	1.5	···	0.75 Mn, 0.40 Si, 0.08 La	High oxidation resistance.

† After "ASM Databook," *Metal Progress*, vol. 114, no. 1, mid-June, 1978.
‡ Early developed alloy.

Figure 11-29 Stacking faults in MAR-M509 alloy. Dark areas are $M_{23}C_6$ carbides at fault intersections. (X25,000.) (*After J. M. Drapier, in C. T. Sims, "A Contemporary View of Cobalt-Base Alloys," J. Metals, Dec. 1969, p. 36.*)

superalloys. Elements such as nickel, iron, zirconium, and tantalum, which increase the stacking-fault energy, stabilize the FCC structure. Elements such as chromium, molybdenum, and tungsten, which lower the stacking-fault energy, stabilize the HCP structure and hence increase the tendency for cobalt alloys to form stacking faults. Figure 11-29 shows stacking faults formed in the cast cobalt-base superalloy MAR-M509.

Carbides Carbides are the most important "second phase" in cobalt-base superalloys. A fine dispersion of carbides contributes significantly to the strength of these alloys, in which there are no γ' intermetallic precipitates such as Ni_3Al and Ni_3Ti that are so effective in strengthening the nickel-base alloys. The carbon content of the cobalt-base alloys is therefore relatively high (i.e., 0.1 to 1 percent).

In general, there are three main types of carbides found in cobalt-base superalloys:

$M_{23}C_6$ carbides, where "M" is mostly chromium but can be substituted for by tungsten and molybdenum.

MC carbides, where "M" stands for the reactive metals tantalum, titanium, zirconium, and columbium.

M_6C carbides, where "M" stands for tungsten or molybdenum; these carbides form when the tungsten or molybdenum content exceeds about 5 atomic percent.

Carbides strengthen cobalt-base superalloys in several ways. First, they (principally $M_{23}C_6$) precipitate at grain boundaries in both cast and wrought

Figure 11-30 Stacking faults and $M_{23}C_6$ carbides in MAR-M302 rupture-tested at 870°C. Note excessive precipitation of secondary $M_{23}C_6$ in the stacking fault traces. (Electron replication micrograph; X7000.) (*After C. T. Sims, "A Contemporary View of Cobalt-Base Alloys," J. Metals, Dec. 1969, p. 36.*)

alloys and thereby decrease grain-boundary sliding and thus prolong rupture life. Second, some of these carbide particles precipitate in stacking faults, as shown in the microstructure of MAR-M509 (Fig. 11-29). Dislocation movement is strongly impeded by such barriers, and hence the alloy is strengthened. Extensive precipitation of $M_{23}C_6$ carbides on stacking faults during service is shown in Fig. 11-30 for MAR-M302 rupture tested at 870°C. However, these precipitates can also lead to a significant decrease in ductility, which is a problem with most cobalt-base superalloys.

Effect of heat treatment on microstructure The effect of heat treatment on the microstructure of cobalt-base superalloy MAR-M509 is shown in Fig. 11-31. The as-cast structure consists of MC carbides and $M_{23}C_6$ colonies (Fig. 11-31a). After solution heat treatment at 1275°C for 4 h, the grain boundaries are cleaned up (Fig. 11-31b). The principal residual carbide is M_6C, but some of the larger $M_{23}C_6$ particles are still not dissolved.

Aging 24 h at 925°C causes precipitation of $M_{23}C_6$ particles in large amounts (Fig. 11-31c). Some of this appears as a finely dispersed semicoherent precipitate, whereas other $M_{23}C_6$ particles precipitate as Widmanstätten plates on the {111} planes of the matrix. MC is also precipitated but is not visible in these optical microstructures. Also, the undissolved $M_{23}C_6$ is agglomerated.

After 732 h at 1100°C, which is a typical gas-turbine-partition operating temperature, blocky agglomerated $M_{23}C_6$ is developed (Fig. 11-31d). The fine background precipitate of $M_{23}C_6$ is coarsened and more evenly distributed. The rupture ductility of this alloy at the service temperature after service is good (10 to 20 percent). However, lower-temperature tensile ductility of this structure is

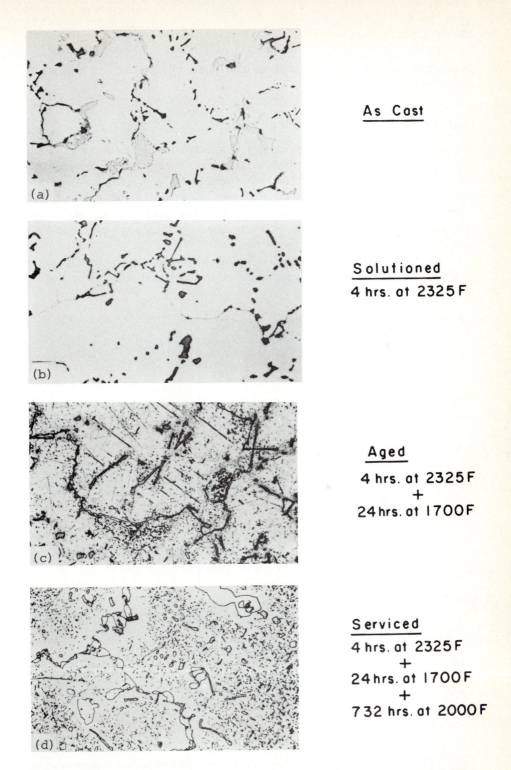

As Cast

Solutioned
4 hrs. at 2325 F

Aged
4 hrs. at 2325 F
+
24 hrs. at 1700 F

Serviced
4 hrs. at 2325 F
+
24 hrs. at 1700 F
+
732 hrs. at 2000 F

Figure 11-31 Effect of heat treatment on the microstructure of cobalt-base superalloy MAR-M509. (*After C. T. Sims and W. C. Hagel, "The Superalloys," Wiley, New York, 1972, p. 164. Reproduced by permission of John Wiley & Sons, Inc.*)

497

poor, which is a disadvantage of most cobalt-base alloys after long-time high-temperature exposure.

High-Temperature Stress-Rupture Strengths

The characteristic stress-rupture strengths of selected cobalt-base superalloys are listed in Table 11-12, while the stress-versus-temperature curves of rupture after 1000 h for some selected alloys are given in Fig. 11-32. A comparison of the stress-rupture properties of several cobalt-base alloys with those of nickel-base superalloys is given in Fig. 11-33. These data show that, at lower and intermediate temperatures, cobalt alloys do not possess the tensile properties required for root sections of rotating turbine blades.

The lower strength of the cobalt alloys at intermediate temperatures is due to a lack of a coherent γ'-type precipitate, which all nickel alloys have. Cobalt

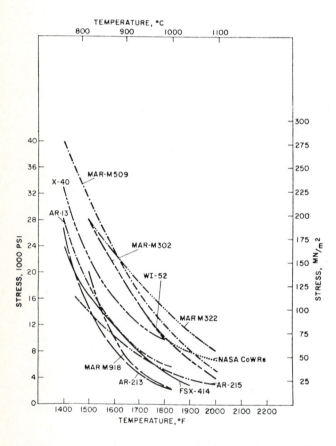

Figure 11-32 Stress-versus-temperature curves for rupture in 1000 h for selected cobalt-base alloys. (*After C. T. Sims and W. C. Hagel, "The Superalloys," Wiley, New York, 1972, p. 594. Reproduced by permission of John Wiley & Sons, Inc.*)

Table 11-12 Characteristic rupture strengths of selected cobalt-base superalloys†

Alloys	Rupture strengths, 1000 psi					
	815°C		985°C		1095°C	
	100 h	1000 h	100 h	1000 h	100 h	1000 h
	Cast					
X-40	26	20	11	8	4.0	
FSX-414	22	17	8	5	3.1	
MAR-M302	40	30	16	11	6.0	4.0
MAR-M322	40	28	20	15	10	
MAR-M509	39	33	17	13	8.0	5.5
AR-213	20	13	5	3.5	2.8	
	Wrought					
S-816	25	18				
HS-188	22	16	6	3.6	2.2	

† After "ASM Databook," *Metal Progress*, vol. 114, no. 1, mid-June, 1978.

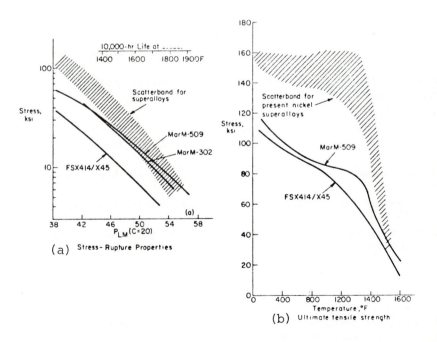

Figure 11-33 Mechanical properties of representative cobalt alloys compared with contemporary nickel-base superalloys. (*After C. T. Sims and W. C. Hagel, "The Superalloys," Wiley, New York, 1972, p. 169. Reproduced by permission of John Wiley & Sons, Inc.*)

alloys, by having only carbide precipitates and solid-solution strengthening as strengthening mechanisms, cannot compete at high temperatures where the γ' mechanism operates in nickel-base alloys. Hence, cobalt alloys are used for low-stress high-temperature long-life parts such as vanes in industrial turbines. For example, MAR-M509 possesses superior rupture properties when the time-temperature parameter (P_{LM}) is greater than 50 (Fig. 11-33a).

PROBLEMS

1. What engineering advantages does the metal nickel have? What is its chief disadvantage?

2. What are the engineering property advantages of Monel 400 alloy? What is its microstructure in the annealed condition?

3. What are the age-hardening precipitates in Monel K-500?

4. What is the microstructure of Inconel 600 in the solution treated and water-quenched condition? If this alloy is water-quenched to 870°C and held for 4 h, what change in the microstructure occurs?

5. How is the excellent high-temperature oxidation resistance of Inconel 601 explained in terms of its metallurgical structure?

6. What are the superalloys? What are the three main classes of superalloys?

7. What are the main engineering applications of the superalloys?

8. What are the major phases present in the nickel-base superalloys?

9. What are the principal solid-solution strengthening elements added to the nickel-base super-alloys?

10. What are the two major factors that affect the degree to which these elements are effective in solid-solution strengthening nickel-base superalloys?

11. How do additions of molybdenum and tungsten reduce high-temperature creep in nickel-base superalloys?

12. How does cobalt increase the high-temperature stability of nickel-base superalloys?

13. What is the structure of γ' in the nickel-base superalloys?

14. What is the chemical composition of γ' in nickel-base superalloys?

15. Describe the mismatch and coherency between γ and γ' in nickel-base superalloys.

16. What is believed to be the main strengthening mechanism resulting from γ' in nickel-base superalloys?

17. How can the increase in strength of some nickel-base superalloys with an increase in temperature up to about 800°C be explained?

18. What is the role of carbides in increasing the creep resistance of nickel-base superalloys at high temperatures?

19. Describe the chemical composition and stability of the MC, $M_{23}C_6$, and M_6C carbides in nickel-base superalloys.

20. What are TCP phases? Why are they undesirable in nickel-base superalloys?

21. What is the purpose of two-stage aging Inconel-X750 alloy? What type of structure is produced?

22. How can multistage heat treatments improve the stress-rupture properties of René 77 alloy? What is a desirable microstructure for this alloy?

23. What is hot corrosion of superalloys? What is believed to be its cause? What is one way in which hot-corrosion resistance can be increased?

24. What are nickel-iron-base superalloys used for engineering applications? What are their nickel and iron composition ranges?

25. What is the main disadvantage of high-iron-containing nickel-base superalloys?

26. What are the precipitating phases that can be present in nickel-iron-base superalloys? How do the precipitating phases present in them mainly differ from those in nickel-base superalloys?

27. What precipitating phases can be present in Inconel 718 after long-time exposure at elevated temperatures?

28. What are the high-temperature stress-rupture limitations of the nickei-iron-base superalloys?

29. Compare the following physical properties of cobalt to nickel: atomic diameter, melting point, density, and crystal structure.

30. What allotropic transformation occurs in cobalt at 417°C?

31. Why are cobalt-base superalloys used for some parts of industrial turbines?

32. How does the precipitation strengthening of the cobalt-base superalloys differ from that of the nickel-base superalloys?

33. Why are stacking faults commonly found in cobalt-base superalloys?

34. Why are carbides the most important "second phase" in cobalt-base superalloys?

35. How do carbides strengthen cobalt-base superalloys?

36. Why do cobalt-base superalloys have lower strengths at high temperatures than the nickel-base superalloys?

REFERENCES

1. C. T. Sims and W. C. Hagel (eds.): "The Superalloys," John Wiley & Sons, New York, 1972.
2. W. Betteridge and J. Heslop (eds.): "The Nimonic Alloys," 2d ed., Edward Arnold, London, 1974.
3. J. L. Everhard: "Engineering Properties of Nickel and Nickel Alloys," Plenum Press, New York, 1971.
4. "Handbook of Nickel Alloys," Huntington Alloys, Inc., 1970; "Monel Alloys," 1978; "Inconel Alloy 600," 1978; "Inconel Alloy 601," 1978; "Inconel Alloy 718," 1978; "Inconel Alloy X-750," 1977; and "Incoloy Alloys 800, 800H, 801," 1978.
5. C. Hammond and J. Nutting: "The Physical Metallurgy of Superalloys," *Met. Sci.* 11(1977):474.
6. C. T. Sims: "A Contemporary View of Nickel-Base Superalloys," *J. Metals,* Oct. 1966, p. 119.
7. C. T. Sims: "A Contemporary View of Cobalt-Base Alloys," *J. Metals,* Dec. 1969, p. 27.
8. J. A. Burger: "Heat Treating Nickel-Base Superalloys," *Metal Progress,* July 1967, p. 61.
9. J. P. Stroup and L. A. Pugliese: "How Low-Carbon Contents Affect Superalloys," *Metal Progress,* Feb. 1968, p. 97.
10. J. F. Barker, E. W. Ross, and J. F. Radavich: "Long-Time Stability of Inconel 718," *J. Metals,* Jan. 1970, p. 31.
11. W. L. Mankins, J. C. Hosier, and T. H. Bassford: "Microstructure and Phase Stability of Inconel Alloy 617," *Met. Trans.* 5(1974):2579.
12. P. S. Kotval: "Carbide Precipitation on Imperfections in Superalloy Matrices," *Trans. AIME* 242(1968):1651.
13. L. R. Woodyatt, C. T. Sims, and H. J. Beattie: "Prediction of Sigma-Type Phase Occurrence from Compositions in Austenitic Superalloys," *Trans. AIME* 236(1968):519.
14. J. R. Mihalisin, C. G. Bieber, and R. T. Grant: "Sigma—Its Occurrence, Effect, and Control in Nickel-Base Superalloys," *Trans. AIME* 242(1968):2399.
15. R. F. Decker: "Strengthening Mechanisms in Nickel-Base Superalloys." In *Steel Strengthening Mechanisms,* Climax Molybdenum Co., Greenwich, CO, 1969.
16. J. H. Moll, G. N. Maniar, and D. R. Muzyka: "The Microstructure of 706, a new Fe-Ni-Base Superalloy," *Met. Trans.* 2(1971):2143.
17. A. A. Tavassoli and G. Colombe: "Effect of Minor Element Variation on the Properties of Alloy 800," *Met. Trans.* 8A(1977):1577.

18. V. Biss and D. L. Sponseller: "The Effect of Molybdenum on γ' Coarsening and on Elevated-Temperature Hardness of Nickel Base Superalloys," *Met. Trans.* 4(1973):1953.
19. R. G. Dunn, D. L. Sponseller, and J. M. Dahl "Ductility Improvements in Superalloys." In *Toward Improved Ductility and Toughness*, Climax Molybdenum Development Co. (Japan) Ltd., 1971.
20. C. T. Sims, "Superalloys for Advanced Energy Systems," *J. Metals*, Dec. 1976, p. 7.
21. "Source Book on Materials for Elevated-Temperature Applications," American Society for Metals, Metals Park, OH, 1979.

INDEX

PERIODIC TABLE

According to latest reports including Commission on

BASED

Mass of proton $(m_p) = 1.007595 \pm 0.000002$
Mass of neutron $(m_n) = 1.008982$
Mass of deuteron $(m_d) = 2.014190 \pm 0.000004$
Mass of electron $(m_e) = (5.48760 \pm 0.00004) \times 10^{-4} = (9.1084 \pm 0.0004) \times 10^{-28}$ g
Electronic charge $(e) = (4.8029 \pm 0.0001) \times 10^{-10}$ esu
Avogadro's number $(N) = (6.0248 \pm 0.0003) \times 10^{23}$ mole^{-1}
Velocity of light *in vacuo* $(c) = (2.997923 \pm 0.000008) \times 10^{10}$ cm/sec
Faraday $(F = Ne/c) = (9652.2 \pm 0.2)$ emu/(g equivalent)
Planck's constant $(h) = (6.6253 \pm 0.0003) \times 10^{-27}$ erg-sec
Boltzmann constant $(k) = (1.38041 \pm 0.00007) \times 10^{-16}$ erg/deg
Gas constant per mole $(R_o = Nk) = (8.3167 \pm 0.0003) \times 10^7$ erg/mole deg
1 electron volt $= (1.60207 \pm 0.00007) \times 10^{-12}$ erg
$= 11605.8°K\ (E = kT)$
$= 1.78256 \times 10^{-33}$ g $(E = mc^2)$
$= 2.41813 \times 10^{14}$ sec^{-1} $(E = h\nu)$

I A

1	1.00797
H	
Hydrogen	
0.09 g/l	
−252.77	−259.15
1s	

	I A		II A
3	6.939	4	9.012
Li		**Be**	
Lithium		Beryllium	
0.53		1.86	
1336	186	1500	1284
(He) 2s		(He) 2s²	
11	22.990	12	24.31
Na		**Mg**	
Sodium		Magnesium	
0.97		1.74	
882.9	97.9	1107	650
(Ne) 3s		(Ne) 3s²	

III B — IV B — V B — VI B — VII B — VIII

19	39.102	20	40.08	21	44.96	22	47.90	23	50.94	24	52.00	25	54.94	26	55.85	27	58
K		**Ca**		**Sc**		**Ti**		**V**		**Cr**		**Mn**		**Fe**		**Co**	
Potassium		Calcium		Scandium		Titanium		Vanadium		Chromium		Manganese		Iron		Cobalt	
0.86		1.55		3.1		4.5		5.96		7.1		7.2		7.86		8.9	
757.5	63.5	1482	851	2730	1397	3130	1812	3530	1730	2482	1903	2087	1244	2800	1535	2900	1
(Ar) 4s		(Ar) 4s²		(Ar) 3d 4s²		(Ar) 3d² 4s²		(Ar) 3d³ 4s²		(Ar) 3d⁵ 4s		(Ar) 3d⁵ 4s²		(Ar) 3d⁶ 4s²		(Ar) 3d⁷ 4s
37	85.47	38	87.62	39	88.905	40	91.22	41	92.91	42	95.94	43	(98)	44	101.1	45	102.9
Rb		**Sr**		**Y**		**Zr**		**Nb**		**Mo**		**Tc**		**Ru**		**Rh**	
Rubidium		Strontium		Yttrium		Zirconium		Niobium		Molybdenum		Technetium		Ruthenium		Rhodium	
153		2.6		4.34		6.4		8.4		10.2		11.487		12.43		12.5	
679	39.0	1384	771	3230	1475	3580	1852	3300	1950	4804	2610		2200	4111	2506	3960	1
(Kr) 5s		(Kr) 5s²		(Kr) 4d 5s²		(Kr) 4d² 5s²		(Kr) 4d⁴ 5s		(Kr) 4d⁵ 5s		(Kr) 4d 5s²		(Kr) 4d⁷ 5s		(Kr) 4d⁸ 5s
55	132.905	56	137.34			72	178.49	73	180.95	74	183.85	75	186.2	76	190.2	77	19
Cs		**Ba**				**Hf**		**Ta**		**W**		**Re**		**Os**		**Ir**	
Cesium		Barium				Hafnium		Tantalum		Tungsten		Rhenium		Osmium		Iridium	
1.9		3.59				13.30		16.6		19.3		21.0		22.48		22.4	
690	28.4	1537	850			5230	2230	6000	2977	5630	3380	5630	3147	4400	2700	4350	2
(Xe) 6s		(Xe) 6s²				(Xe) 4f¹⁴5d²6s²		(Xe) 4f¹⁴5d³6s²		(Xe) 4f¹⁴5d⁴6s²		(Xe) 4f¹⁴5d⁵6s²		(Xe) 4f¹⁴5d⁶6s²		(Xe) 4f¹⁴5d⁷6
87	(223)	88	(226)			104		105		106		107		108			
Fr		**Ra**															
Francium		Radium															
		5															
		1140	960														
(Rn) 7s		(Rn) 7s²															

KEY

Atomic number	*Atomic weight
Atomic symbol	
Name of element	
density (g/ml)	
BP °C	MP °C
Ground state symbol (electronic configuration)	

EXPLANATION

*in ( )=mass number of the most stable isotope
Atomic symbols cross-hatched as Br and Hg=liquids at 25°C
Atomic symbols outlined as He, Ne, etc.=gases

←LANTHANIDE SERIES
←ACTINIDE SERIES

57	138.91	58	140.12	59	140.91	60	144.24	61	(147)	62	150.
La		**Ce**		**Pr**		**Nd**		**Pm**		**Sm**	
Lanthanum		Cerium		Praseodymium		Neodymium		Promethium		Samarium	
6.15		6.9		6.5		7.0				6.93	
2730	887	2527	785		932		840				13
(Xe) 5d 6s²		(Xe) 4f 5d 6s²		(Xe) 4f³ 6s²		(Xe) 4f⁴ 6s²		(Xe) 4f⁵ 6s²		(Xe) 4f⁶ 6s²	
89	(227)	90	232.04	91	(231)	92	238.03	93	(237)	94	(24
Ac		**Th**		**Pa**		**U**		**Np**		**Pu**	
Actinium		Thorium		Protactinium		Uranium		Neptunium		Plutonium	
		11.7				19.05		19.5		19	
		4230	1730			3500	1132	640			
(Rn) 6d 7s²		(Rn) 6d² 7s²		(Rn) 6d³ 7s²		(Rn) 5f³ 6d 7s²		(Rn) 5f⁴ 6d¹ 7s²		(Rn) 5f⁶ 6d¹ 7